Technologies for Sustainable Agriculture
in the Tropics

Technologies for Sustainable Agriculture in the Tropics

Proceedings of two international symposia sponsored by Division A-6 of the American Society of Agronomy in San Antonio, TX, and Denver, CO, 1990 and 1991, respectively.

Editors
John Ragland and Rattan Lal

Organizing Committee
John Ragland (1990)
Uzo Mokwunye (1991)

Editorial Committee

Kenneth Cassman   Rattan Lal, *cochair*
Charles Francis   Uzo Mokwunye
Peter Hildebrand   Richard Morris
Jagmohan Joshi   John Ragland, *chair*
Anthony Juo

Editor-in-Chief ASA
Gary A. Peterson

Editor-in-Chief CSSA
P.S. Baenziger

Editor-in-Chief SSSA
Robert J. Luxmoore

Managing Editor
Jon M. Bartels

ASA Special Publication Number 56

American Society of Agronomy, Inc.
Crop Science Society of America, Inc.
Soil Science Society of America, Inc.
Madison, Wisconsin, USA

1993

Cover Design: Photograph provided by Dr. Rattan Lal.

American Society of Agronomy, Inc.
Crop Science Society of America, Inc.
Soil Science Society of America, Inc.
677 South Segoe Road, Madison, WI 53711 USA

Library of Congress Cataloging-in-Publication Data

Technologies for sustainable agriculture in the tropics: proceedings
    of an international symposia sponsored by Division A-6 of the
    American Society of Agronomy in San Antonio, TX, and Denver,
    CO, 1990 and 1991 respectively / organizing committee, John Ragland
    (1990), Augustine Uzo Mokwunye (1991); editorial committee,
    Kenneth Cassman . . . [et al.]; editor[s] in chief, Gary A. Peterson,
    P.S. Baenziger, Robert J. Luxmoore.
        p.   cm. —— (ASA special publication; no. 56)
    Includes bibliographical references.
    ISBN 0-89118-118-0
        1. Sustainable agriculture—Tropics—Congresses. 2. Sustainable
    agriculture—Developing countries—Congresses. I. Ragland, John.
    II. Mokwunye, A. Uzo. III. Peterson, Gary, 1940-   . IV.
    Baenziger, P. Stephen. V. Luxmoore, R. J. VI. American Society
    of Agronomy. Division A-6. VII. Series.
    S1.A453 no. 56
    [S481]
    630 s--dc20
    [338.1 '0913]   93-21161
                                                             CIP

Printed in the United States of America

CONTENTS

v

FOREWORD

The sustainability of agriculture, including the land, the natural resources and the people is of concern throughout the world. Nowhere is the sustainability issue of more immediate importance than in the Tropics, where population pressures, a fragile environment and the need for foreign currency combine to put incredible pressures on the natural resources. To protect these resources and maintain the capacity of these lands to produce food and fiber requires that we know why these trends are occurring and how they can be prevented or reversed.

The American Society of Agronomy has a significant membership in tropical countries. The Society thus was most pleased to be able to support symposia on the topic of sustainable agriculture in the Tropics and to publish this book. The publication represents the assessment of many international experts in soils, crops and agronomy of the Tropics. This will be a valuable reference to the literature and an outstanding reference text for those who wish to develop and promote the technologies that will help sustain this critical resource—the Tropics.

D.R. KEENEY, *president,*
American Society of Agronomy

C.W. STUBER, *president,*
Crop Science Society of America

D.W. NELSON, *president*
Soil Science Society of America

PREFACE

Land and its associated natural resources provide our basic needs and is the source of most of the world's accumulated wealth. Agronomists and ecologists share a special responsibility for helping maintain the natural resource base because they understand why these resources are rapidly deteriorating and have the technical knowledge necessary to reverse negative trends.

In 1990, the American Society of Agronomy appointed a workgroup to find ways of better serving its growing international membership. Because sustainable agriculture was at that time an emerging and internationally recognized concept, the workgroup sought and gained approval to prepare a publication on the subject, not only for the membership of the Society, but also for the larger scientific community. The material for the publication *Technologies for Sustainable Agriculture in the Tropics* was accumulated from two symposia convened by the International Division (A-6) of the American Society of Agronomy in 1990 and 1991. The tropical region was chosen because its lands are more easily degraded than perhaps any other part of the inhabited world and population pressures there are the most intense.

In addition to the introductory and concluding chapters prepared by the editors, all multiauthor contributions are organized in seven parts. Part I, has two chapters, the first by H. Eswaran et al., examines the constraints, challenges, and choices facing the establishment of sustainable agriculture in developing countries, with the greatest emphasis on the quality of soil resources in the Tropics. In the second chapter, J. Ragland argues that agricultural sustainability is a task for all humanity, consumers and farmers alike, and that once the development process begins it must be managed to produce substantial wealth. Otherwise, the natural resources will continue to be degraded indefinitely to satisfy basic needs.

Part II also contains two chapters and is entitled "Technological Options." The first chapter, by P.A. Sanchez, presents convincing arguments and strong evidence that modern agronomic technologies can be linked with indigenous farming traditions to provide systems which are both high yielding and sustainable. In the second chapter D.P. Garrity assesses the state of knowledge in developing sustainable systems for the extensive sloping uplands of Southeast Asia, and the ways in which the pathway of land degradation in this ecosystem may be reversed.

The three chapters in Part III are concerned with agroforestry and nutrient cycling. These chapters provide valuable information on the use of woody and herbaceous species in building sustainable agriculture production systems but, at the same time, show that more research and interpretation of existing research will be required to resolve conflicting results form alley cropping and agroforestry experiments. The paper by D.P. Garrity in Part II should also be read in connection with the chapters in Part III.

Part IV contains one chapter on erosion management by J.W. Smyle and W.B. Magrath. It is a thorough review of Vetiver grass, which is one of the most useful plants available for the control of soil erosion in the tropics. Part V of the book also contains only one chapter, which presents the status of crop and crop–soil computer models and their integration into decision-support systems for sustainable agriculture. The argument is advanced and examples are given of how modern computer technology will play an increasingly major role in developing and gaining the adoption of sustainable agriculture systems. Such approaches are promising shortcuts to finding sustainable agriculture systems.

Part VI consists of three chapters which develop the socioeconomic aspects of sustainability, which must be fully integrated with the biophysical elements. This part is entitled "Macroscale Influences."

Part VII consists of six chapters and deals with the incredibly complicated and difficult problems of sub-Saharan Africa (SSA). The first chapter by A. Kesseba is a comprehensive review of the socioeconomic and political aspects of agricultural sustainability in SSA. It enumerates strategies for developing a viable and sustainable agricultural sector in SSA. The following three chapters by B.N. Okigbo, R. Lal and P. Vlek are comments on strategies outlined by A. Kesseba and critique pros and cons of available options. In the fifth chapter R. Lal summarizes technological options for agronomic management of soil and water resources of different agroecoregions of SSA. The last chapter of this section by V. Balasubramarian and Nguimgo K.A. Blaise presents two case studies on the importance of improved follows on sustaining agronomic productivity.

The concluding summary by the editors is an attempt to indicate knowledge gaps and prioritize research and development needs for achieving agricultural sustainability.

Taken as a whole, the chapters are an excellent report on the status of agronomic research and scholarship pertaining to sustainable agriculture-particularly of the Tropics. At the same time, the book makes no pretense of being a manual for producing sustainable agriculture systems throughout the Tropics. The resources and political will to achieve such systems will exist only when the public demands them. Hopefully, this book will hasten the coming of that day by revealing what is possible for one of the world's most problem-ridden regions—the Tropics.

JOHN RAGLAND, *chair*
University of Kentucky
Lexington, Kentucky

RATTAN LAL, *cochair*
The Ohio State University,
Columbus, Ohio

CONTRIBUTORS

A. N. Atta-Krah	Agroforester, International Institute of Tropical Agriculture, P.M.B. 5320, Ibadan, Nigeria
V. Balasubramanian	Senior Maize Agronomist, International Institute of Tropical Agriculture, IRA/NCRE Project, B.P. 2067, Yaounde, Cameroon
Nguimgo K. A. Blaise	Agronomist, Institut de Recherche Agronomique, B.P. 2067, Yaounde, Cameroon
W. G. Boggess	Professor, Department of Food and Resource Economics, University of Florida, P.O. Box 110240, Gainesville, FL 32611-0240
W. T. Bowen	Scientist, International Fertilizer Development Center, P. O. Box 2040, Muscle Shoals, AL 35662
Kenneth Cassman	Division Head, Agronomist, Division of Agronomy, Plant Physiology, and Agroecology, International Rice Research Institute, P. O. Box 933, Manila 1099, Philippines
C. B. Davey	Professor Emeritus, Department of Forestry, North Carolina State University, Raleigh, NC 27695-8008
H. Eswaran	National Leader, USDA-SCS, P. O. Box 2890, Washington, DC 20013
Erick C. M. Fernandes	Visiting Assistant Professor, North Carolina State University, Tropical Soils Research-Brazil, EMBRAPA-CPAA, Manaus, AM 69.001, Brazil
Charles A. Francis	Professor, Department of Agronomy, University of Nebraska, 279 Plant Science, Lincoln, NE 68583-0910
Dennis P. Garrity	Systems Agronomist and Coordinator, Southeast Asian Regional Research Programme, International Centre for Research in Agroforestry, P.O. Box 161, Bogor, Indonesia
Peter E. Hildebrand	Professor, Department of Food and Resource Economics, University of Florida, 2126 McCarty Hall, Gainesville, FL 32611-0240
James W. Jones	Professor, Agricultural Engineering Department, University of Florida, Gainesville, FL 32611
Jagmohan Joshi	Director, Soybean Research Institute, University of Maryland-Eastern Shore, Princess Anne, MD 21853
Anthony S. Juo	Professor, Department of Soil and Crop Sciences, Texas A&M University, College Station, TX 77843
B. T. Kang	Soil Scientist, International Institute of Tropical Agriculture, Oyo Road, P.M.B. 5320, Ibadan, Nigeria
Abbas M. Kesseba	Director, International Fund for Agricultural Development, Technical Advisory Division, Via Del Serafico, 107, Rome 00142, Italy

Clyde F. Kiker — Professor of Resource Economics, Food and Resource Economics Department, University of Florida, McCarty Hall, Gainesville, FL 32611-0240

Rattan Lal — Professor of Soil Science, Department of Agronomy, The Ohio State University, 2021 Coffey Road, Columbus, OH 43210

W. B. Magrath — Natural Resources Economist, The World Bank, 1818 H Street NW, Washington, DC 20034

Richard Morris — Professor, Department of Soil Science, Oregon State University, Corvallis, OR 97331

Uzo Mokwunye — Director, International Fertilizer Development Center, Agro-Economic Division, P. O. Box 2040, Muscle Shoals, AL 35660

S. K. Mughogho — Associate Professor, Crop Science Department, Bunda College of Agriculture, University of Malawi, P.O. Box 219, Lilongwe, Malawi

L. A. Nelson — Professor Emeritus, Department of Statistics, North Carolina State University, Box 8203, Raleigh, NC 27695-8203

Bede N. Okigbo — Director, United Nations University Programme on Natural Resources in Africa, c/o UNESCO/ROSTA, P.O. Box 30592, Nairobi, Kenya

John Ragland — Professor of Agronomy, Agronomy Department, University of Kentucky, Agriculture Science Building-North, Lexington, KY 40546

J. T. Ritchie — Professor, Homer Nowlin Chair, Department of Crop and Soil Sciences, Michigan State University, A570 Plant and Soil Sciences Building, East Lansing, MI 48824

P. A. Sanchez — Director General, International Center for Research in Agroforestry, P. O. Box 30677, Nairobi, Kenya

J. W. Smyle — Land Resources Management Specialist, The World Bank, 1818 H Street NW, Washington, DC 20034

L. D. Spivey, Jr. — Soil Scientist, USDA-SCS, P. O. Box 2890, Washington, DC 20013

Walter W. Stroup — Professor of Biometry, Department of Biometry, University of Nebraska, 103 Miller Hall, Lincoln, NE 68583-0712

S. M. Virmani — Agroclimatologist, International Crops Research Institute for the Semiarid Tropics, Patancheru P.O., Hyderabad, Andhra Pradesh, India

Paul L. G. Vlek — Professor, Institute of Agronomy in the Tropics, Georg-August University, Goettingen, Germany

Ray R. Weil — Professor of Soil Science, Agronomy Department, University of Maryland, H.J. Patterson Hall, College Park, MD 20742

Conversion Factors for SI and non-SI Units

Conversion Factors for SI and non-SI Units

To convert Column 1 into Column 2, multiply by	Column 1 SI Unit	Column 2 non-SI Unit	To convert Column 2 into Column 1, multiply by
Length			
0.621	kilometer, km (10^3 m)	mile, mi	1.609
1.094	meter, m	yard, yd	0.914
3.28	meter, m	foot, ft	0.304
1.0	micrometer, μm (10^{-6} m)	micron, μ	1.0
3.94×10^{-2}	millimeter, mm (10^{-3} m)	inch, in	25.4
10	nanometer, nm (10^{-9} m)	Angstrom, Å	0.1
Area			
2.47	hectare, ha	acre	0.405
247	square kilometer, km^2 (10^3 m)^2	acre	4.05×10^{-3}
0.386	square kilometer, km^2 (10^3 m)^2	square mile, mi^2	2.590
2.47×10^{-4}	square meter, m^2	acre	4.05×10^3
10.76	square meter, m^2	square foot, ft^2	9.29×10^{-2}
1.55×10^{-3}	square millimeter, mm^2 (10^{-3} m)^2	square inch, in^2	645
Volume			
9.73×10^{-3}	cubic meter, m^3	acre-inch	102.8
35.3	cubic meter, m^3	cubic foot, ft^3	2.83×10^{-2}
6.10×10^4	cubic meter, m^3	cubic inch, in^3	1.64×10^{-5}
2.84×10^{-2}	liter, L (10^{-3} m^3)	bushel, bu	35.24
1.057	liter, L (10^{-3} m^3)	quart (liquid), qt	0.946
3.53×10^{-2}	liter, L (10^{-3} m^3)	cubic foot, ft^3	28.3
0.265	liter, L (10^{-3} m^3)	gallon	3.78
33.78	liter, L (10^{-3} m^3)	ounce (fluid), oz	2.96×10^{-2}
2.11	liter, L (10^{-3} m^3)	pint (fluid), pt	0.473

Mass

To convert Column 1 into Column 2, multiply by	Column 1 SI Unit	Column 2 non-SI Unit	To convert Column 2 into Column 1, multiply by
2.20×10^{-3}	gram, g (10^{-3} kg)	pound, lb	454
3.52×10^{-2}	gram, g (10^{-3} kg)	ounce (avdp), oz	28.4
2.205	kilogram, kg	pound, lb	0.454
0.01	kilogram, kg	quintal (metric), q	100
1.10×10^{-3}	kilogram, kg	ton (2000 lb), ton	907
1.102	megagram, Mg (tonne)	ton (U.S.), ton	0.907
1.102	tonne, t	ton (U.S.), ton	0.907

Yield and Rate

To convert Column 1 into Column 2, multiply by	Column 1 SI Unit	Column 2 non-SI Unit	To convert Column 2 into Column 1, multiply by
0.893	kilogram per hectare, kg ha^{-1}	pound per acre, lb acre^{-1}	1.12
7.77×10^{-2}	kilogram per cubic meter, kg m^{-3}	pound per bushel, lb bu^{-1}	12.87
1.49×10^{-2}	kilogram per hectare, kg ha^{-1}	bushel per acre, 60 lb	67.19
1.59×10^{-2}	kilogram per hectare, kg ha^{-1}	bushel per acre, 56 lb	62.71
1.86×10^{-2}	kilogram per hectare, kg ha^{-1}	bushel per acre, 48 lb	53.75
0.107	liter per hectare, L ha^{-1}	gallon per acre	9.35
893	tonnes per hectare, t ha^{-1}	pound per acre, lb acre^{-1}	1.12×10^{-3}
893	megagram per hectare, Mg ha^{-1}	pound per acre, lb acre^{-1}	1.12×10^{-3}
0.446	megagram per hectare, Mg ha^{-1}	ton (2000 lb) per acre, ton acre^{-1}	2.24
2.24	meter per second, m s^{-1}	mile per hour	0.447

Specific Surface

To convert Column 1 into Column 2, multiply by	Column 1 SI Unit	Column 2 non-SI Unit	To convert Column 2 into Column 1, multiply by
10	square meter per kilogram, m^2 kg^{-1}	square centimeter per gram, cm^2 g^{-1}	0.1
1000	square meter per kilogram, m^2 kg^{-1}	square millimeter per gram, mm^2 g^{-1}	0.001

Pressure

To convert Column 1 into Column 2, multiply by	Column 1 SI Unit	Column 2 non-SI Unit	To convert Column 2 into Column 1, multiply by
9.90	megapascal, MPa (10^6 Pa)	atmosphere	0.101
10	megapascal, MPa (10^6 Pa)	bar	0.1
1.00	megagram per cubic meter, Mg m^{-3}	gram per cubic centimeter, g cm^{-3}	1.00
2.09×10^{-2}	pascal, Pa	pound per square foot, lb ft^{-2}	47.9
1.45×10^{-4}	pascal, Pa	pound per square inch, lb in^{-2}	6.90×10^3

(continued on next page)

Conversion Factors for SI and non-SI Units

To convert Column 1 into Column 2, multiply by	Column 1 SI Unit	Column 2 non-SI Unit	To convert Column 2 into Column 1, multiply by
Temperature			
$1.00 \, (\text{K} - 273)$	Kelvin, K	Celsius, °C	$1.00 \, (°\text{C} + 273)$
$(9/5 \, °\text{C}) + 32$	Celsius, °C	Fahrenheit, °F	$5/9 \, (°\text{F} - 32)$
Energy, Work, Quantity of Heat			
9.52×10^{-4}	joule, J	British thermal unit, Btu	1.05×10^{3}
0.239	joule, J	calorie, cal	4.19
10^{7}	joule, J	erg	10^{-7}
0.735	joule, J	foot-pound	1.36
2.387×10^{-5}	joule per square meter, J m^{-2}	calorie per square centimeter (langley)	4.19×10^{4}
10^{5}	newton, N	dyne	10^{-5}
1.43×10^{-3}	watt per square meter, W m^{-2}	calorie per square centimeter minute (irradiance), cal cm^{-2} min^{-1}	698
Transpiration and Photosynthesis			
3.60×10^{-2}	milligram per square meter second, mg m^{-2} s^{-1}	gram per square decimeter hour, g dm^{-2} h^{-1}	27.8
5.56×10^{-3}	milligram (H$_2$O) per square meter second, mg m^{-2} s^{-1}	micromole (H$_2$O) per square centimeter second, μmol cm^{-2} s^{-1}	180
10^{-4}	milligram per square meter second, mg m^{-2} s^{-1}	milligram per square centimeter second, mg cm^{-2} s^{-1}	10^{4}
35.97	milligram per square meter second, mg m^{-2} s^{-1}	milligram per square decimeter hour, mg dm^{-2} h^{-1}	2.78×10^{-2}
Plane Angle			
57.3	radian, rad	degrees (angle), °	1.75×10^{-2}

Electrical Conductivity, Electricity, and Magnetism

To convert Column 2 into Column 1, multiply by	Column 1 SI Unit	Column 2 non-SI Unit	To convert Column 1 into Column 2, multiply by
10	siemen per meter, $S\ m^{-1}$	millimho per centimeter, $mmho\ cm^{-1}$	0.1
10^4	tesla, T	gauss, G	10^{-4}

Water Measurement

To convert Column 2 into Column 1, multiply by	Column 1 SI Unit	Column 2 non-SI Unit	To convert Column 1 into Column 2, multiply by
9.73×10^{-3}	cubic meter, m^3	acre-inches, acre-in	102.8
9.81×10^{-3}	cubic meter per hour, $m^3\ h^{-1}$	cubic feet per second, $ft^3\ s^{-1}$	101.9
4.40	cubic meter per hour, $m^3\ h^{-1}$	U.S. gallons per minute, $gal\ min^{-1}$	0.227
8.11	hectare-meters, ha-m	acre-feet, acre-ft	0.123
97.28	hectare-meters, ha-m	acre-inches, acre-in	1.03×10^{-2}
8.1×10^{-2}	hectare-centimeters, ha-cm	acre-feet, acre-ft	12.33

Concentrations

To convert Column 2 into Column 1, multiply by	Column 1 SI Unit	Column 2 non-SI Unit	To convert Column 1 into Column 2, multiply by
1	centimole per kilogram, $cmol\ kg^{-1}$ (ion exchange capacity)	milliequivalents per 100 grams, meq $100\ g^{-1}$	1
0.1	gram per kilogram, $g\ kg^{-1}$	percent, %	10
1	milligram per kilogram, $mg\ kg^{-1}$	parts per million, ppm	1

Radioactivity

To convert Column 2 into Column 1, multiply by	Column 1 SI Unit	Column 2 non-SI Unit	To convert Column 1 into Column 2, multiply by
2.7×10^{-11}	becquerel, Bq	curie, Ci	3.7×10^{10}
2.7×10^{-2}	becquerel per kilogram, $Bq\ kg^{-1}$	picocurie per gram, $pCi\ g^{-1}$	37
100	gray, Gy (absorbed dose)	rad, rd	0.01
100	sievert, Sv (equivalent dose)	rem (roentgen equivalent man)	0.01

Plant Nutrient Conversion

To convert Column 2 into Column 1, multiply by	Column 1 Elemental	Column 2 Oxide	To convert Column 1 into Column 2, multiply by
2.29	P	P_2O_5	0.437
1.20	K	K_2O	0.830
1.39	Ca	CaO	0.715
1.66	Mg	MgO	0.602

1   Agricultural Sustainability in the Tropics

Rattan Lal

Department of Agronomy
The Ohio State University
Columbus, Ohio

John Ragland

Agronomy Department
University of Kentucky
Lexington, Kentucky

Agriculture in the Tropics is at a crossroads. While there is an urgency to enhance per capita production and establish a positive trend in productivity, there also is a need to preserve ecologically friendly aspects of the traditional systems, characterized by low productivity and subsistence farming. Transforming resource-based subsistence agriculture into science-based commercial farming must be brought about without endangering the productive capacity and life support processes of the resource base, and without polluting the environment. The gradual evolution from subsistence to commercial farming would be an ecologically favorable strategy. However, a gradual evolution is not an option because of rapidly increasing demographic pressure on highly weathered soils of inherently low fertility and because there is a great need for enhancing production immediately and substantially. The rapid transformation which is needed is difficult because of small landholders and resource-poor farmers who lack the availability of essential inputs at affordable prices, poor infrastructure, and extremely difficult logistics. Bringing about a sustainable quantum leap in per capita productivity that will be at a rate higher than that of the population growth is a challenging task for planners, policy makers, scientists, and farmers of the Tropics.

ISSUES OF AGRICULTURAL SUSTAINABILITY IN THE TROPICS VERSUS TEMPERATE REGIONS

The principal objective of agricultural sustainability is to meet human needs by improving standards of living, alleviating drudgery and human suffering, and providing a respectable way of life for a majority of the population. Consequently, sustainable agriculture is an important issue for both the modern agriculture of the temperate regions, as well as for the subsistence agriculture of the Tropics. However, basic issues of agricultural sustainability are different in each region.

1. Land resources. The shortage of prime agricultural land is not a major concern in North America, Western Europe and Australia. In contrast, per capita prime agricultural land is a major production constraint in the Tropics.

2. Per capita caloric intake and food availability. Per capita food availability remains a major issue in several regions of the Tropics but not in temperate regions. Food deficiency is a perpetual crisis in several regions of tropical Africa and Asia.

3. Off-farm employment. Agriculture is a principal source of employment in most countries of the Tropics. Other industries have not yet been developed to take pressure off the land resources, as in the temperate regions.

4. Land degradation. Although the problem of soil degradation is by no means limited to the Tropics, it is neither as widespread nor as severe in the temperate regions as in the Tropics.

5. Off-farm inputs. Science-based agricultural techniques for improving production and decreasing risks of land degradation are known for the major soils and principal ecoregions of the Tropics. However, the necessary inputs are not available. In contrast, excessive use of agriculture inputs is not uncommon in some temperate regions.

SOIL DEGRADATION AND SUBSISTENCE AGRICULTURE

Low-input subsistence agriculture and soil degradation often go hand-in-hand. Low yields, soil fertility depletion, and severe soil degradation form a vicious cycle which is difficult to break. The schematics in Fig. 1–1 depict the interdependence of soil degradation and low input or resource-based agriculture. The degraded resource base of depleted soils or soils of inherently low fertility coupled with high demographic pressure lead to intensive land use, enhanced depletion of the nutrient reserves, low biomass production and yields, and accelerated degradation of soil and water resources. Furthermore, poverty and a lack of resources lead to no-input or low-input subsistence farming that accelerates the mining of fertility and degradation of soil and water resources.

This vicious cycle can be broken only by enhancing the nutrient capital of the soil through off-farm inputs. Nutrients brought in from outside the ecosystem can be from organic or synthetic sources, however, a judicious balance of organic and inorganic inputs for fertility maintenance is an important strategy.

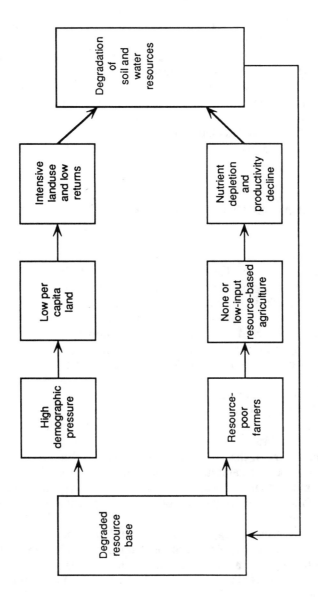

Fig. 1-1 Soil degradation and resource-based subsistence agriculture.

OFF-FARM INPUTS VERSUS MANAGEMENT SKILLS

Substituting management for inputs is a possible option only for soils of medium to high fertility. Exploiting nutrient reserves by nutrient cycling from subsoil layers, is viable only if nutrient reserves are available in the subsoil. Even the biological fixation of N by legumes must have an adequate amount of P and other nutrients in order to be of agronomic significance.

Some strategies for achieving agricultural sustainability are shown in Fig. 1–2. The first and foremost requirement is to enhance the nutrient capital of the soil. The result is not only to enhance the production efficiency, but also to increase the energy flux. Subsistence, low input, and resource-based agriculture are characterized by high entropy leading to a decline in structural attributes and depletion in fertility of the soil. This can be reversed by increasing the nutrient reserves and using other off-farm inputs.

TECHNICAL, SOCIOECONOMIC AND CULTURAL ASPECTS OF AGRICULTURAL SUSTAINABILITY

Achieving agricultural sustainability requires more than simple use of off-farm inputs. Improved technology must address the issues and constraints faced by the farming community, e.g., labor shortage during peak periods, marketing and infrastructure, institutional support, and basic requirements of an urban lifestyle (Fig. 1–3). Development of improved technology, if it is to be readily accepted, must be done in full cooperation with the farming community. It is extremely important that researchers and farmers work together in problem diagnosis, in developing a hypothesis, and in formulating a research program to solve the problem. Research programs must be undertaken with a sustainability perspective. Technology adoption is difficult in a top–down approach.

Translating scientific data into problem-solving technology requires other inputs. These include:

1. Knowledge of the soil, water, and climatic resources, their potential and constraints, and judicious management options for sustained use.
2. Economics of the production system is an important prerequisite. Technology can only be adopted if it brings about tangible economic returns.
3. Improved technology also must fit within the social, cultural and political framework of the farming community. Social and political acceptability are important considerations. Gender issues are important considerations in regions where women play an important role in farm operations.

The outline presented in Fig. 1–3 indicates the interdependency of all components and subsystems. The challenge of the scientific community in developing sustainable options lies in identifying the crucial components and finding a synthesis that is acceptable to the farming community.

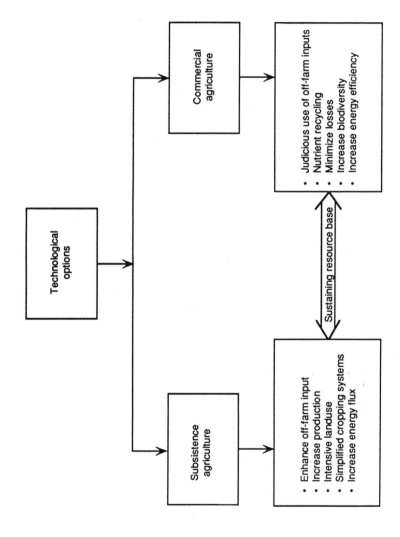

Fig. 1–2 Strategies for achieving agricultural sustainability.

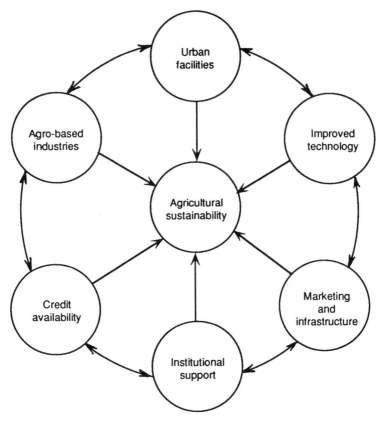

Fig. 1-3. Technical and socioeconomic requirements of agricultural sustainability.

This book synthesizes the available information on technological options for achieving agricultural sustainability in the Tropics. It also provides case studies with specific references to tropical Africa.

2

Sustainable Agriculture in Developing Countries: Constraints, Challenges, and Choices

H. Eswaran

USDA-SCS
Washington, District of Columbia

S. M. Virmani

International Crops Research Institute for the Semiarid Tropics
Hyderabad, India

L. D. Spivey, Jr.

USDA-SCS
Washington, District of Columbia

The finitude of the biophysical resource base is increasingly apparent, particularly in the last two decades. The burgeoning population coupled with rampant human-induced degradation of the resource base has forced a re-evaluation of the human population-carrying capacity of the earth. Concerns are highlighted in a report by the World Commission on Environment and Development (WCED, 1987). This report provided the political awareness and incentives for a new approach.

CONCEPTS AND DEFINITIONS

Sustainable agriculture means a number of things to different people. For some, it is more of the same, while at the other extreme, the focus is on ecological integrity at the expense of any other concern. In industrialized nations, it includes basic economic concerns including imposition of purely aesthetic values such as pleasant scenery and noise abatement. Dahlberg (1991) posed a question—will these industrial societies seek to maintain current lifestyles, privileges, and consumption patterns—ironically using the label, sustainability—or will they seek to restructure into genuinely sustainable, equitable patterns? The same concerns apply to a lesser extent in developing countries. In the developing countries, the farmer's immediate concerns are crop yield improvement, crop diversity, and income enhancement which are the farmer's individual values. On the other hand, even in these countries, the visionary segment of the population or the institutional values may focus on things such as efficient cropping systems, pest control methods, potable water supplies, support for agrobased industries and related infrastructure. Consequently, it must be emphasized at the outset that it is necessary to define the concept of sustainable agriculture in the context of the society, and seek to attain a balance among the different segments of the society including individuals, institutions, markets, and governments.

Probably, the most elementary but important element of the concept is that its origin is in the value system of social, political, economic, religious, and other institutions. In time, as the idea of sustainable agriculture expands and strengthens, there will be divergent trends and sharp conflicts with some values, needs, and goals of people as individuals.

In the last few years there have been several meetings and publications dealing with the broad perspectives of sustainable agriculture (Edwards et al., 1990; Singh et al., 1990). The basic thrust of sustainable agriculture is the improvement of the quality of life in the context of an environmentally sound approach so that the resource base is maintained or enhanced for future generations. Depending on the institutions or countries involved, there are frequently additional points of emphasis and these are portrayed in the definitions given below.

From the FAO:

> Sustainable agriculture should involve the successful management of resources for agriculture to satisfy changing human needs while maintaining or enhancing the quality of the environment and conserving natural resources (FAO, 1989a).

From Agriculture Canada:

> Sustainable agriculture systems are those that are economically viable, and meet society's needs for safe and nutritions food, while conserving or enhancing Canada's natural resources and the quality of the environment for future generations (Baier, 1990).

From the Technical Advisory Committee:

> Sustainable agriculture should involve the successful management of resources to satisfy changing human needs while maintaining or enhancing the quality of the environment and conserving natural resources (TAC, 1988).

From the USAID:

> Sustainable agriculture is a management system for renewable natural resources that provides food, income, and livelihood for present and future generations and that maintaining or improving the economic productivity of these resources (USAID, 1990, unpublished data).

From Lynam and Herdt:

> A sustainable system is one with a non-negative trend in measured output (Lynam & Herdt, 1988)

The sentiments in these definitions are included in the three basic components of sustainability of Douglas (1984):

1. Sustainability as long-term food sufficiency, which requires agricultural systems that are more ecologically based and that do not destroy their natural resources.

2. Sustainability as stewardship, that is, agricultural systems that are based on a conscious ethic regarding humanity's relationship to future generations and to other species.

3. Sustainability as community, that is, agricultural systems that are equitable.

From these definitions, important institutional values (Eswaran & Virmani, 1990) are discriminatory use of land resources, conservation and enhancement of environmental quality, economic viability, increased and stabilized productivity, enhancement of quality of life, intergenerational equity, and buffered against risks.

The perception of sustainable agriculture is partly a function of the scale of operation, determined by the sphere of influence of the individual or group. The scale ranges from a farmer's field, a farm, a watershed, or a catchment, an ecosystem, a country and, finally, the continent, or the earth as a whole. Though the factors that affect sustainability at each of these levels and the kinds of interventions needed to make each more sustainable is different at each of these levels, the global picture must be appreciated to implement local actions.

To be effective, sustainable agriculture must include a concept of stewardship. While institutional values provide the framework for governments to respond, there are individuals and groups whose dedication and actions are needed to make sustainable agriculture work. Legal decree and economic subsidies without a strong sense of stewardship will result in conflicts, if not failure, in implementing sustainable agriculture projects.

Consequently the major ethic of sustainability at any level is stewardship and there can be no effort and incentive to implement a sustainable form of agriculture if appropriate stewardship is lacking. Thus, as shown in Fig. 2-1, stewardship is the most important component. There are many forms of stewardship and it has to start from the government through appropriate policy decisions and programs. Just as governments have invested to seek self-sufficiency in food and fiber production, they also must invest public funds to attain sustainability. This is particularly crucial in developing countries and in the absence of this support, sustainable agriculture is not tenable. Once appropriate policy and programs are in place, stewardship then becomes a responsibility of other levels in society such as social organizations and down to the farm family.

The motivation towards this form of agriculture listed in Fig. 2-1 applies to all levels of society though at any particular level, one may be more

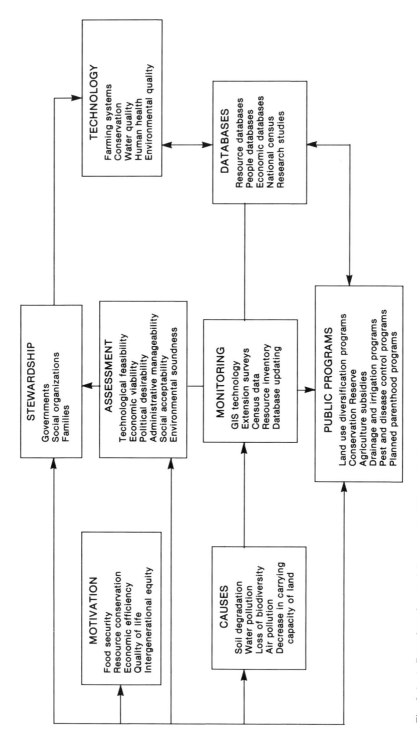

Fig. 2-1. A flow chart for attaining sustainability agriculture.

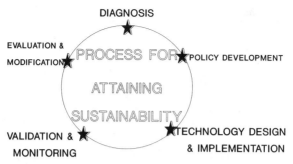

Fig. 2–2. Iterative process for attaining sustainability.

significant than the other. To attain some or all of these objectives in a sustainable manner, technologies are needed and each technology can be assessed (Virmani & Eswaran, 1989, 1990) to evaluate its appropriateness. In order to improve or introduce new technologies, a clear understanding of the causes of unsustainability is required and this is achieved through monitoring and development of relational data bases (Fig. 2–1). Based on such an assessment, public programs are launched.

This process for attaining sustainability is summarized in Fig. 2–2 and the sequence of iteration activities can be applied at any level, from the farm to a country or larger. Sustainable agriculture requires a slow stepwise approach commencing with a diagnosis stage where the system is critically evaluated for unsustainability. Figure 2–2 also can be used for designing research programs.

INDICATORS OF UNSUSTAINABILITY

Although there is a general consensus on the basic concepts of sustainability, measurement of sustainability is the immediate challenge. As indicated by Lynam and Herdt (1988), unless it is quantified, it will not be possible to evaluate a system, and consequently it wil not be possible to recommend remedial measures. Several kinds of indicators are needed, some of which are quantitative while others are qualitative. The latter, though not easily amenable to rigorous statistical analysis, are equally useful as they provide a first appraisal of the system.

The visible indicators of unsustainability are site-specific but, if available for defined ecosystems, become valuable tools particularly for extension workers. Jodha (1989) has developed a preliminary list of such indicators and this is excerpted in Table 2–1. Jodha (1989) working in the mountain areas of Asia, reports, "a persistent decline in crop yields in many areas. Mining activities have destroyed mountain flora, caused landslides, and choked water channels. In Tibet and Pakistan, highland pastures are overgrazed. High potential grazing areas are disappearing and cattle are being replaced by small ruminants.... The increasing scarcity of fuel and fodder is reflected in the longer distances and time involved in collection. Finally

Table 2-1 Negative changes as indicators of unsustainability.

Types of changes	Changes related to	
	Resource base	Resource use
Directly visible	Increased land slides and other forms of land degradation Abandonment of terraces Reduction in per capita availability of land Increased fragmentation of land holdings Changes in genetic composition of forest/pasture Reduction in water flows for irrigation and domestic use	Reduction in the intensity of crop rotation and other management practices Extension of cultivation to marginal lands Replacement of social sanctions for resource use by legal measures
Changes in response to resource degradation	Substitution of big animals with small animals: cattle with goats Change from deep rooted crops to shallow rooted crops Shifts in use of external inputs, from local manures to fertilizers	Unbalanced and high intensity of external input use Increased specialization in monocropping
Potentially negative changes due to development interventions	Introduction of new production systems without linkages to other diversified activities Promoting excessive dependence on outside resource, subsidies Ignoring traditional adaptive experiences Research and development focus on crops rather than resource	Ignoring ecosystem characteristics in designing programs and policies for area development Neglect of local wisdom and skills and excessive dependance on external expertise

there is an increasing dependence of people on government subsidies and inferior options." Jodha recommends that a reversal of the negative trends should be the primary focus of agricultural development strategies in mountain areas. This recommendation is applicable to most tropical ecosystems.

Quantified approaches are currently only available for a few components of systems. *There is an urgent need to develop methodologies that will provide reliable assessments of sustainability or unsustainability.* One of the greatest handicaps is the lack of data over a long time frame and methodologies to evaluate such data. Many agronomic experiments are conducted for 2 to 3 yr. The period of the research may be during a time when climatic conditions were favorable or a period of severe moisture stress or some situation in between. Agronomic research is seldom evaluated in this context of sustainability.

It is frequently necessary to evaluate the resource base itself for sustainability. This would permit the discriminatory use of the soil resources. Using simulated crop yields for different soils, Dumanski and Onofrei (1990) illustrated the different potential productivity of soils.

The previous examples only illustrate assessment of components of systems. There are other components such as soil loss, soil degradation, and biodiversity which are more difficult to quantify. The thrust of research in

Table 2-2. Constraints and related issues of sustainable agriculture in developing countries.

Soil Related
   Erosion and soil loss—watershed management
   Salinity and alkalinity
   Waterlogging
   Soil acidity
   Degraded lands
Pest and disease management
   Balance between pest population and acceptable crop loss
Forestry and agroforestry
   Changes in C cycle as a result of deforestation
   Biodiversity and gene pool
   Wildlife habitat
   Quality of life of forest-based communities
Rangeland
   Desertification
   Animal supporting capacities
Biodiversity/biological resource conservation
   Animal nutrition
   Human nutrition
Population supporting capacities of semi-arid tropics
   Food supply and security
Water resource demands
   Water quality
   Competition from agriculture, human,a nd industrial demands
Institutional
   Framework/operational
   Implications in legal, social, economic, and political
   Stewardship—decision makers, private sector, families
   Education and awareness
Technological
   Human and physical resources
   Training
   Appropriateness, acceptability
   Data bases, monitoring
Equity
   Worst year—period of survival
   Best year—amortization of benefits with time (response farming?)

---

the near future must be towards a quantification of sustainable land-use systems.

CONSTRAINTS

Developing countries have several constraints summarized in Table 2-2, not only in attaining sustainability but even to increase the production of their agriculture. Kellogg (1967) stated that he,

> ...fully expects that 'some day' the most productive agriculture of the world will be mostly in the tropics, especially in the humid parts...Whether 'some day' is 25, 50, 100, or some other number of years from now, depends on how

Table 2-3. Chemical constraints in Africa (FAO, 1986).

Chemical constraints	Area	
	millions ha	%
Low CEC	1 295.5	45.6
Al toxicity	635.0	22.3
P fixation	382.5	13.5
Low K supply	637.0	22.4
Salinity	75.6	2.7
Alkalinity	31.0	1.1

rapidly institutions for education, research, and the other public and private sectors of agriculture will develop.

Today, more than 20 yr after this statement was made, "some day" may have arrived at some plantations and a few other areas. For the most part of the tropics, however, "some day" remains a futuristic fantasy rather than a realistic scenario. Small-scale farming prevails in most of the region and, as Lal (1987) has pointed out,

the subsistence farmer who faces famine would consider a successful technology to be the one that produces some yield in the worst year rather than one that produces a high yield in the best.

The constraints in Africa are perhaps more urgent than the other continents. About 47% of Africa is dry and in these regions apart from some nomadic grazing, there is no potential for agriculture in the absence of massive capital investment. About 19% of the continent has a semiarid climate which is characterized by unreliable and low rainfall keeping agriculture at the mercy of the rain. Although moisture stress is not a major constraint in the remaining part of Africa, pests and diseases, and many soil-related factors (Table 2-3) keep agriculture at marginal levels. Steep slopes, sandy textures and other specific management problems also prevent a productive agriculture in large areas of Africa. Similar situations prevail in Asia and South America.

Table 2-4 lists the extent of degraded land in Asia and it is evident that it is of alarming proportions. Rehabilitation of degraded lands is an immediate necessity even prior to considerations of sustainable agriculture on such lands.

The introduction of high-yielding varieties of rice (*Oryza sativa*), wheat (*Triticum aestivum*), and to a lesser extent other grain crops, which signaled the dawn of the Green Revolution, also has introduced or resulted in a number of problems. A major impact has been the loss of genetic resources of land races of major crops. These natural varieties evolved through time and space and their losses are permanent (Singh, 1989). Monoculture, coupled with the use of pesticides, has other drawbacks as evidenced by the spread of rice brown plant hopper (*Landelphax striatellus*) in Asia. Even the multiple resistant IR36 was subject to pest damage under monocropping in Indonesia in 1986. The pesticides apparently eliminated the natural enemies of the plant hopper resulting in an imbalance of the biosystem.

Table 2-4. Estimated extent of degraded land (km^2) for selected countries in Asia-Pacific. Parenthesis indicates percentage of respective land area to total area (Dent, 1990, unpublished data).

Country	Total land area	Arable & cropped land area	Degraded land area
Bangladesh	13 391	9 164 (68)	989 (7)
China	932 641	96 976 (10)	280 000 (30)
India	297 319	168 990 (57)	148 100 (50)
Indonesia	181 157	21 220 (12)	43 000 (24)
Laos	23 080	901 (4)	8 100 (35)
Myanmar	65 754	10 000 (15)	210 (3)
Pakistan	77 088	20 760 (27)	15 500 (17)
Philippines	29 817	7 930 (27)	5 000 (17)
Western Samoa	283	122 (43)	32 (11)
Sri Lanka	6 474	1 887 (29)	700 (11)
Thailand	51 089	20 050 (39)	17 200 (34)
Tonga	72	48 (67)	3 (4)
Vietnam	32 536	6 470 (20)	15 900 (50)

Expanding population and the demand for agricultural land has denuded many countries of natural vegetation. The consequence has been erosion, loss of biodiversity, and a general decrease in the agricultural production as a whole. Increasing use of pesticides and herbicides is beginning to have consequences similar to those in many western countries. Pollution of the ecosystem is not the only problem. As shown earlier with the case of the rice plant hopper, an imbalance in the natural food chain is created.

Other constraints are listed in Table 2-2. Each country would have some or many of these constraints and a major problem is that there is little information on the kinds of constraints and their extent and magnitude in many countries.

Soil Resource-related Constraints

The soil qualities important for sustainable agriculture (Eswaran & Virmani, 1990) are many but can be reduced to 10: (i) Available water-holding capacity; (ii) nutrient retention capacity—cation exchange capacity; (iii) nutrient availability—pH and base saturation, P and K; (iv) nutrient fixation; (v) chemical constraints—acidity, sodicity, salinity; (vi) physical constraints—low hydraulic conductivity, permeability, high bulk density, crusting; (vii) effective soil volume—depth to root restricting layer, stoniness, structure; (viii) surface tilth; (ix) erodibility; and (x) water-logging.

In Table 2-5a,b,c,d, the soil qualities listed previously which are major determinants of sustainability are provided for four selected soil orders in soil taxonomy. These four soils together occupy about 60% of the total land surface in the Tropics. Similar tables may be constructed, and with greater degree of reliability, for the lower categories, such as subgroups, families, or even soil series. Eswaran (1977, p. 289–314) has developed similar tables of determinants for the subgroup category for a few orders of soils. Based on the determinants, "Sustainable Agriculture Success" in Table 2-5 attempts

Table 2–5a. Information on Oxisols.

Soils	Soil moisture regimes		
	Aridic	Ustic/Xeric	Udic
Oxisols			
Area			
World (millions sq. km)	0.01	4.2	7.7
Countries	USA	Brazil, Zaire Zambia, Kenya Uganda, Nigeria	Indonesia Malaysia Zaire
Population density			
Current		•†	••
Potential		••	••••
Population supporting capacity			
Low input		•	••
Medium input		•	•••
High input		•	•••••
Constraints-soil qualities			
Water holding capacity		•••••	•••••
Nutrient retention Capacity		•••••	•••••
Nutrient availability		•••••	•••••
Nutrient fixation (P)		••••	••••
Chemical constraints		•••	••••
Physical constraints		••••	••
Effective soil volume		•	•
Surface tilth		•	•
Erodibility		•••	•
Waterlogging		•	•
Management options			
Shifting cultivation		•••••	•••••
Annuals		•	•
Perennials		••	•••••
Agroforestry		•••••	•
Forestry		•••••	••••
Nutrient		•••••	•••••
Pests		•••	••••
Weeds		••••	•••••
Sustainable agriculture success			
Low input		•	•
Medium input		•	••
High input		••	••••

† Dot indicates relative importance to sustainability: • very low to ••••• very high.

to list the possibility of attaining sustainable agriculture under low-, medium-
and high-input technologies.

    Eswaran and Virmani (1990) have indicated the need to establish standards
for threshold values, with respect to specific uses, for each of the soil qualities. The USDA-SCS has established the soil loss tolerance "T value" as a
kind of soil quality standard. This standard is used to determine if a practice
or sets of practices are essential to meet resource management needs.

    Until and unless such soil quality standards are established, it will be
difficult to monitor soil degradation and consequently the effects of management. From the point of view of sustainable agriculture, soil quality standards provide a basis to evaluate changes in soil conditions due to
management, provide the tools for monitoring changes, provide the basis

Table 2-5b. Information on Ultisols.

Soils	Soil moisture regimes	
	Ustic/Xeric	Udic
Ultisols		
Area		
World (millions sq. km)	4.5	6.0
Countries	Kenya, Columbia, India Zaire, Ghana, Nicaragua	Malaysia, Thailand Cameroon, Panama Costa Rica
Population density		
Current	•••†	••••
Potential	••••	•••••
Population supporting capacity		
Low input	•	••
Medium input	••	•••
High input	••••	••••••
Constraints-soil qualities		
Water holding capacity	•••••	••••
Nutrient retention Capacity	•••	••
Nutrient availability	•••	••
Nutrient fixation (P)	••	•••
Chemical constraints	••	••••
Physical constraints	•••	••
Effective soil volume	••••	••••
Surface tilth	•••••	•••••
Erodibility	••••	•••
Waterlogging	•	••
Management options		
Shifting cultivation	••	•
Annuals	••	•
Perennials	•••	•••••
Agroforestry	•••••	•••
Forestry	•••••	•••••
Nutrient	•••••	•••••
Pests	•••	•••••
Weeds	••••	••••
Sustainable agriculture success		
Low input	•	•
Medium input	••	•••
High input	•••	•••••

† Dot indicates relative importance to sustainability: • very low to •••• very high.

for legislation for soil stewardship, provide a means of signaling potential problems in order to trigger research or development activities, and provide the criteria for evaluating sustainable agriculture.

There is an urgent need for the international scientific community, in collaboration with potential users, to develop guidelines for these standards. As the socioeconomic conditions vary, each country would probably need its own set of standards.

Detailed soil maps of countries would show the areas with major resource constraints and management options may be selected accordingly. An assessment of the success of sustainable agriculture purely from the point of view of soils, also is given in Table 2-5,a,b,c,d of the four soils, the Alfisols have fewer limitations and thus have a higher probability of attaining sus-

Table 2–5c. Information on Vertisols.

Soils	Soil moisture regimes		
	Aridic	Ustic/Xeric	Udic
Vertisols			
Area			
World (millions sq. km)	1.1	1.8	0.05
Countries	Sudan	Sudan, India Australia	Indonesia Uruguay
Population density			
Current	•†	••	•••
Potential	•	•••	•••••
Population supporting capacity			
Low input	•	••	•••
Medium input	•	•••	••••
High input	••	••••	•••••
Constraints-soil qualities			
Water holding capacity	•••••	•••••	•••••
Nutrient retention Capacity	•••••	•••••	•••••
Nutrient availability	•••••	•••••	•••••
Nutrient fixation (P)	•••	••••	•
Chemical constraints	•••••	•••	••
Physical constraints	•••••	•••••	•••••
Effective soil volume	•	•	•
Surface tilth	•••••	•••••	•••••
Erodibility	•••••	•••••	•••••
Waterlogging	•••••	•••••	•••••
Management options			
Shifting cultivation	•	•	•
Annuals	•	••	•••
Perennials	•	•••	••••
Agroforestry	•	•••	••••
Forestry	•	•••	••••
Nutrient	•	•••	••••
Pests	•	•••	••••
Weeds	•	•••	••••
Sustainable agriculture success			
Low input	•	••	•••
Medium input	•	•••	••••
High input	•	••••	•••••

† Dot indicates relative importance to sustainability: • very low to ••••• very high.

tainability even under low-input systems. The Oxisols and Ultisols on the other hand, due to their intrinsic properties behave and respond differently.

Detailed soil maps are needed to design appropriate farming systems, target soil conservation measures, recommend fertilizer policy and monitor nutrient and other needs of the farmer, and make efficient utilization of the extension services.

For a country to adopt a policy of sustainable agriculture and implement this policy, soil resource inventories at several scales are needed. At the national level, there must be an inventory of the Major Land Resource Areas (MLRA). The MLRA maps are used at national level (i) as a basis for making decisions about agricultural issues; (ii) as a framework for organizing and conducting resource conservation programs; (iii) for geographic

Table 2–5d. Information on Alfisols.

Soils	Ustic/Xeric	Udic
Alfisols		
Area		
World (millions sq. km)	9.5	7.6
Countries	< -------------- In most countries -------------->	
Population density		
Current	••••†	••••
Potential	•••••	•••••
Population supporting capacity		
Low input	•••	••••
Medium input	••••	••••
High input	•••••	•••••
Constraints-soil qualities		
Water holding capacity	•••••	•••••
Nutrient retention Capacity	•••••	•••••
Nutrient availability	•••••	•••••
Nutrient fixation (P)	•	•
Chemical constraints	•	•
Physical constraints	•	•
Effective soil volume	•••••	•••••
Surface tilth	•••••	•••••
Erodibility	••••	•••
Waterlogging	•	•
Management options		
Shifting cultivation	•••	••••
Annuals	•••••	••••
Perennials	•••	•••••
Agroforestry	•••••	••••
Forestry	••	•••
Nutrient	•	••
Pests	•••	••••
Weeds	•••	••••
Sustainable agriculture success		
Low input	••••	••••
Medium input	•••••	•••••
High input	•••••	•••••

† Dot indicates relative importance to sustainability: • very low to ••••• very high.

organization of research and conservation needs and the data from these activities; (iv) for coordinating technical guides among states and districts of a nation and among countries; (v) for organizing, displaying, and using data in physical resource inventories; and (vi) to aggregate natural resource data.

The MLRAs are most important for agricultural planning and have value for interstate, regional, and national planning and are most important tools for targeting sustainable farming systems technology. An assessment of the sustainability of agriculture for each of the MLRA units can be determined to guide national planning.

Soil resource inventories at intermediate scales between the MLRA map and farm level maps may be made if time, personnel, facilities and funds permit. In most less developed countries (LDCs) this is not the case and so emphasis must be on the farm-level maps. Farm-level maps are expensive to make and require highly trained personnel. There must be a national institution to coordinate the effort and develop and provide the standards for evaluation. In most LDCs, soil surveys are done on contract and by expatriates. Standards, methodology, and criteria all vary and few of these maps have long-term use. In addition to making maps, the extension service and decision makers must be trained to use the maps. The national institution must provide basic soil services to backstop potential users. These are some of the ingredients necessary for achieving sustainable agriculture.

There is practically no developing country in the world which has a program of systematic soil surveys at this scale and this is the major obstacle to developing the prerequisities for sustainable agriculture in these countries.

Constraints to Implementing Sustainable Agriculture

The efforts to improve the productivity of subsistence farming are just beginning to get the attention of the national and international agricultural research centers. Emerging technologies such as agroforestry, alley and multiple cropping, improved genetic material, N-fixing trees and crops, and biotechnology hold much promise. In the context of this paper, it also should be mentioned that it would be advantageous if the new farming systems developed are robust enough to be insensitive to the field-scale soil microvariability that tends to exist on tropical farms.

There are several constraints which prevent or hinder developing countries from being more receptive or even adopting sustainable agriculture. The overriding constraint is perhaps the absence of economic incentives. This transcends all levels from the government policy making level to that of the farmer. Reduction in soil loss and long-term environmental degradation are not tangible reasons for the small farmer whose immediate concerns are providing food for the family and assuring the quality of life in the immediate future. The farmer is aware and will contribute to the progress of his society. A good example is the success of social forestry in India. However, the farmer's contribution is extracurricular with the family concerns being the first priority.

Even if the farmer is willing and able, in most developing countries, extension services are poor or nonexistent. Creating an awareness among farmers is frequently an ad hoc exercise or absent. Lack of awareness, not only at the farm level but also at higher levels in the society is a second constraint in many developing countries.

The third major constraint relates to stewardship. No system becomes operational if it is not institutionalized. In many developing countries, particularly in Africa, research and development in agriculture including extension services are poor and suffer from lack of trained personnel, facilities, and motivation. In many African countries, donor-supported research is still the rule and in this environment, the country cannot build a research tradition and local expertise and consequently the benefits are project specific.

The fourth constraint relates to the information base. Implementing sustainable agriculture assumes that (i) reasons for unsustainability are known, (ii) there is sufficient information on the resource base to target sustainability related activities, and (iii) the resource base has been monitored to evaluate sustainability. In practically all developing countries, these three statements are false. In the past three decades or more, international donors supporting agricultural research and development focused on improving the genetic potential of the crop and related management practices to improve yields—a spin-off of the Green Revolution. There is practically no developing country with a systematic detailed soil resource inventory program. Agronomic research, including those by western expatriates, have been conducted and are still being conducted on soils for which they know nothing about. The so-called outreach research on farmers fields conducted by the Consultative Group on International Agricultural Research (CGIAR) institutions has many limitations, as few of the external variables that affect crop performance are monitored. In the absence of information on the resource base, it would be a waste of time and effort to institute sustainable agriculture.

A fifth constraint to implementing sustainable agriculture relates to research. Research methodology is not available and, until recently, not even the basic principles and concepts have been enunciated. Fundamental questions such as how long should an experiment be conducted, what are the treatments, what are the measurements, and how can the data be analyzed, have yet to be answered.

Challenges to Instituting Sustainable Agriculture in Less Developed Countries

There are many hurdles to overcome before a significant number of developing countries have sustainable agriculture programs in place. The responsibilities rest not only with the countries themselves but also with the international community, particularly donor countries. The basic question is how to incorporate a sustainability component in research and development programs.

Role of Donor Agencies

Donor agencies must make an unequivocal commitment towards a long-term support for sustainable agriculture. Many national leaders of developing countries. They view this as an "alka-seltzer" syndrome, where excitement and enthusiasm bubble up only to fizzle out within a short period. This image must be erased.

Over the last five or more years there has been a sufficient number of meetings, workshops, and symposia addressing the subject. It is now time to synthesize the subject and develop action plans for assistance. There is a tendency for many donor supported projects to give a "sustainability twist" to ongoing projects. In the long run this is a disservice and would lead to credibility problems.

From a research point of view, there is a need for a task force to evaluate not only the research needs but also to evaluate research methodologies. The projects must be multidisciplinary and the approach must have a broader context than conventional on-farm research. The area under consideration would be some kind of land unit, such as a catchment or watershed or a defined agro-ecological unit. The purpose of the research, apart from alleviating the low socioeconomic status of the farms will in addition enhance the quality of the resource base and minimize the environmental degradation processes. Consequently, monitoring the resources becomes an important component of any sustainable agriculture research project and methodologies for this must be developed. Funds must be committed so that the research is conducted over a sufficient time period, usually of the order of 10 or more years.

As sustainable agriculture cannot be achieved overnight, an important component of developmental assistance programs is institution building. Many developing countries still do not have detailed information on the resource base. Data bases must be developed, and techniques to monitor the resource must be instituted. This requires a cadre of highly trained professionals supported with adequate facilites.

An equally important contribution of developmental assistance programs is assistance in creating awareness. Nongovernmental organizations (NGO) are generally equipped to provide such services but require funds to cary out their activities. A casual evaluation shows that the impact of NGO activity is frequently greater and that they have more interaction with or influence on the rural community than their government counterparts. Sustainable agriculture would require NGO participation for greater impact.

CHOICES

Sustainable agriculture is not an option, it must be the goal of all developing countries. It must be viewed as a challenge and a motivating force to guide our research and development efforts in the foreseeable future.

The challenge of enhancing productivity while maintaining environmental soundness and attaining intergenerational equity is a new challenge for the low-input, resource-poor farmers of developing countries. Sustainable agriculture calls for an education of the farmers particularly on the long-term consequences of their traditional methods of agriculture. They must be assisted with innovative appropriate methodologies. They also must be provided with incentives to undertake this new thrust.

The application of the tenets of sustainable agriculture necessitates new thrusts in research, development, and implementation. The former demands innovative approaches and international collaboration in a network mode, while the latter requires institutional development, awareness at all levels, and socioeconomic motivation. A dedicated stewardship is the first step in the process of attaining sustainable agriculture. There must be a commitment at the highest levels of government and this must be coupled with an action program that addresses the needs of the farmer in the context of the environment.

In the absence of massive donor support, sustainable agriculture in developing countries becomes untenable in the immediate future. Donors must consider this as an investment to assure food security and social stability in the world. The impact of sustainability also can be used as a mechanism to evaluate project proposals and/or to evaluate ongoing projects. This may be one mechanism to introduce sustainability in all agricultural research and development projects.

In conclusion, the alternative to sustainable agriculture is degradation of the resource base, loss of biodiversity, environmental pollution, reduction of the population supporting capacity for humans or animals, and a general decrease in the quality of life for all living things on this planet. The choice is apparent, we have an obligation to move in this direction.

As the world will eventually have to feed 10 billion people, much of the as-yet underutilized land in the tropics will have to be brought under cultivation. Can there be sustainable agriculture on these lands? The four basic causes of land degradation—overgrazing on rangeland, overcultivation of cropland, waterlogging and salinization of irrigated land, and deforestation—all result from poor management of the land and can, therefore, at least in principle, be controlled. The record to date, however, is quite poor. Although effective technologies that prevent or reduce land degradation either exist or are being developed (Postel, 1989), their application is still constrained by institutional and societal barriers. The problem is aggravated by the fact that degradation control crosses all traditional boundaries and that lasting solutions must be rooted as much in social and economic reforms as in effective technologies (Postel, 1989). In the Tropics, as elsewhere, the current prospects for institutionalizing sustainable development strategies are not very encouraging. Only with massive developmental assistance can this situation be redeemed.

REFERENCES

Baier, W. 1990. Characterization of the environment for sustainable agriculture in the semi-arid tropics. Ind. Soc. Agron. 1:90–128.

Dahlberg, K.A. 1991. Sustainable agricultre—fad or harbinger. BioScience 41:337–340.

Douglas, G.A. 1984. Agricultural sustainability in a changing world order. Westview Press, Boulder, CO.

Dumanski, J., and C. Onofrei. 1990. Evaluating crop production risks related to weather and soils using crop growth models. p. 41–442. Proc. Int. Congr. Soil Sci., 14th. Vol. 5. Kyoto, Japan.

Edwards, E.A., R. Lal, P. Madden, R.H. Miller, and G. House (ed.). 1990. Sustainable agricultural systems. Soil and Water Conserv. Soc., Ankeny, IA.

Eswaran, H. 1977. An evaluation of soil limitations from soil names. Proc. Soil Resource Inventories. Agron. Mimeo no. 77-23. Cornell Univ. Ithaca, NY.

Eswaran, H., and S.M. Virmani. 1990. The soil component in sustainable agriculture. Ind. Soc. Agron. 2:64–86.

Food and Agriculture Organization. 1986. African agriculture: The next 25 years. Annex II. Main Rep. United Nations, Rome, Italy.

Food and Agriculture Organization. 1989. Sustainable agriculture production: For international agricultural research. Rep. of the Technical Advisory Com. Consultative Group on Int. Agric. Res. (CGIAR), Washington, DC.

Jodha, N.S. 1989. Mountain agriculture: Search for sustainability. Newsletter no. 11. Int. Center for Integrated Mountain Dev., Nepal.

Kellogg, C.E. 1967. Comment. p. 232–233. *In* H.M. Southworth and B.F. Johnston (ed.) Agricultural development and economic growth. Cornell Univ. Press, Ithaca, NY.

Lal, R. 1987. Managing the soils of Sub-Saharan Africa. Science (Washington, DC) 236:1069–1076.

Lynam, J.K., and R.W. Herdt. 1988. Sense and sustainability: Sustainability as an objective in international agricultural research. *In* Farmers and food systems. CIP-Rockefeller Conf. Lima, Peru.

Postel, S. 1989. Halting land degradation. p. 21–40. *In* P.D. Little et al. (ed.) Lands at risk in the Third World: Local level perspectives. Westview Press, Boulder, CO.

Singh, R.B. 1989. Conservation of fast eroding genetic resources. Rep. of FAO/RAPA. Bangkok, Thailand.

Singh, R.P., J.F. Parr, and B.A. Stewart. (ed.). 1990. Dryland agriculture: Strategies and sustainability. Vol. 13. Springer-Verlag, New York.

Technical Advisory Committee. 1988. Sustainable agricultural production: Implications for international agricultural research. Consult. Group on Int. Agric. Res., Washington, DC.

Virmani, S.M., and H. Eswaran. 1989. Concepts for sustainability of improved farming systems in the semi-arid regions of developing countries. p. 89–108. *In* J.B. Stewart (ed.) Int. Conf. Soil Quality in Semi-arid Agriculture, Saskatoon, Canada. Saskatchewan Inst. of Pedology, Saskatoon, Canada.

World Commission on Environment and Development. 1987. Our common future. Oxford Univ. Press, Oxford, England.

3 Sustainability and Wealth

John Ragland

Agronomy Department
University of Kentucky
Lexington, Kentucky

Population and environmental quality are moving in opposite directions. To avoid catastrophe, soil, water, air, and genetic resources must be managed, not only to produce more, but to produce on a sustained basis. Such production demands careful resource protection which often lowers annual profits; so who is to pay for maintaining the production base—producers or consumers? It is a task for all humanity and approaches must be found to persuade the consuming public to this point of view. Consumers cannot be expected to assume the responsibilities actually incumbent on them for natural resources unless they understand why the production base has been rapidly declining and what they must do to stop it. The scientific community should provide the explanations and take the leadership for creating a new world order as regards stewardship—a partnership of producers and consumers.

WHO BENEFITS FROM AGRICULTURE?

The most frequent answer is farmers—but this is not exactly true. Agricultural production in the developed world is dependent on large inputs of technology. Such production benefits consumers much more than farmers. This is the case because, once generally adopted, the product involved declines in price to exactly equal the benefit derived from the technology. This relationship is the principal reason for prices received by farmers having trended downward since 1910. Today they are only half as high as in 1910, when adjusted for inflation (Borlaug, 1986). Only those farmers who are early adopters profit. In contrast, consumers continue to benefit for the life of the technology. The U.S. consumer, who pays only 12% of his or her disposable income for food, is the prime example of this relationship. In contrast, a high proportion, often a majority, of the disposable incomes in developing economies must go for food. Until this changes and the developing countries become wealthy, the prospects for adequate resource protection are unlikely.

A major consequence of production which depends on high-input technology is to reduce the total number of farmers and to force those remaining to become large. The U.S. farm population has declined from approximately 15 million in 1960 to under 5 million today. The number of farms has gone from 3.4 million to approximately 2.1 million since 1965, but production has remained relatively constant (Borlaug, 1986).

The wealthy nations have tried to stop the decline in agricultural prices by giving subsidies to farmers. Part of these subsidies are to encourage conservation on a cost-sharing basis, but most are to supplement net farm income. The subsidies encourage overproduction of crops on marginal land, and inflate land prices and the costs of inputs. Such subsidies routinely disadvantage farmers in the underdeveloped world because of trade barriers and because of export subsidies which are initiated in an effort to solve the problem of overproduction caused by the production/conservation subsidies. If subsidies are to be given, they should go directly to the resource base and not be tied to crop production.

WHAT AGRICULTURE LEADERS SHOULD DO

Not understanding the constant economic squeeze on producers caused by a technology-driven free market economy, consumers blame farmers and agricultural scientists for poor natural resource protection. In turn, the farmers blame consumers for cheap food policies which encourage resource exploitation. Actually, the blame lies with a system that rewards production obtained at the lowest possible cost, regardless of how it is achieved.

To break this impasse, bold action will be required by farmers and the agricultural institutions which lead and serve them. The first step is to publicly acknowledge that the agricultural resource base has been declining for a long time, and that the decline is likely to get worse unless dramatic changes are made. The second step would be for farmers to signal a willingness to surrender some of their fiercely guarded independence in controlling how the land is managed in return for the general public paying a larger proportion of the cost of conservation and stewardship projects.

One can already hear farmers and agricultural scientists saying, "Why should we take the first steps to change things when it's the consumers who have grown wealthy from our hard work and ingenuity?" It should be said to them, "You must take the first steps because, unlike the general public, you understand the frightening decline in resources which has already occurred and know what must be done to stop it." An even more pragmatic reason for agriculture taking the initiative is because the farm and associated agriculture population is such a small minority, compared to the consuming public. Only consumers have the numbers and thus the political power to get funding for the massive investments which are needed.

As long as farm organizations view consumers and environmentalists as unwelcome intruders, the amount invested in making agriculture sustainable will be negligible in comparison to the need. Consumers first must be convinced that farmers sincerely wish to be good stewards before they will agree to the large expenditures needed. Being very open and forthcoming on this issue would help to prove the sincerity of the agriculture industry.

No apologies are expected from today's farmers and agriculturalists, because they have been extremely successful as engines of growth for other segments of the economy by providing abundant food at a low cost. This is what was asked of them and they did it. But they failed to protect and preserve the production base, for reasons previously described, and must change or it will be lost forever. In order to do this, agriculture must have access to the money and cooperation of the general public.

Agencies presently responsible for environmental sciences and programs should be merged with agriculture. Having agriculture and environmental protection in separate units, when they are opposite sides of the same coin, is no longer justifiable. Some universities have already abandoned the limitations of the name agriculture and become colleges of life sciences, or some other variation of an integrated approach. Similarly, new names and reorganization should occur at every level of society.

CONVINCING CONSUMERS

Too often agricultural products, offered at the lowest price, were produced by exploiting natural resources. Air and water pollution, erosion and siltation, and the destruction of genetic diversity are the consequences. Somehow these costs to the public for private gain must be stopped. Instituting a system of environmental shadow prices holds promise. The approach would be to calculate and display the shadow prices alongside market prices. The shadow price might work out to be one or two times greater than the market price, but it should nevertheless be posted. If such a dual pricing system were possible, think of the public pressure the shadow would generate for bringing the actual retail price and the environmental shadow price together. In some cases, the shadow price could be lowered by changing production and marketing methods. In other cases, the government would need to collect the difference between the two prices and invest it in restoring the environmental damage caused by producing the item.

Coupled with a strong educational effort, schemes like environmental shadow pricing could give the consuming public a heightened sense of their stewardship responsibilities.

Public support for environmental quality is growing, but needs acceleration. Primary and secondary schools, colleges, churches, extension and adult education programs, civic clubs and the mass media must do more. All are willing, but factual information must be prepared and packaged in a way that is more compelling.

UNEQUAL ENDOWMENTS

A key educational need is to make the public in wealthy nations understand that most undeveloped regions of the world are poor because their agricultural resources are poor. Too often inhabitants of chronically poor nations are dismissed as lazy or stupid. In fact, the agricultural potential of regions like Northeast Thailand and much of Africa were severely depleted by natural weathering forces long before humans evolved. Unlike Iowa, Illinois, or Central Europe, where it is difficult to destroy the land no matter how bad your farming methods, it is very difficult to avoid destroying the uplands of the Tropics, even with the best farming practices.

To a great extent, the developed nations occupy those regions of the world where the resource base is rich enough to withstand the degree of exploitation required to initiate a strong development process. Somehow wealthy nations must be convinced not only to reinvest in environmental quality at home, but also to help the less fortunate developing countries of the Tropics. Only carefully prepared arguments, skillfully presented, will be sufficient. For this reason, farmers must stop fighting the environmental movement and join it. Legitimate differences will continue to exist and have to be negotiated, but the focus of each group should be on halting resource depletion and rebuilding the damage. It is a task for all people.

INTEGRATING THE PIECES

Now that the sustainability concept is winning wide support, people want to know what to do about it. For some, it may be no more than recycling solid waste or reducing unnecessary trips in their automobiles. But agricultural scientists must establish goals, prepare guidelines and invent the tools required to build sustainable systems.

Sustainable agriculture cannot be achieved by technological manipulations alone. It involves finding a balance between production and preservation, natural and artificial, on-site and off-site, the present and the future, and fitting these competing needs into a system which will allow technicians to inform decision makers of the needs and the likely costs. Perhaps the process should consist of dozens of simultaneous equations being solved across time with feedback loops to properly adjust the final outcome. Someday computer programs will exist for "cranking out" the answers, but for now shortcuts must be taken. Perhaps a stepwise process is the best option for now.

Figure 3–1 is an old-fashioned, linear-reasoning approach which lists some of the physical and biological ingredients which must be sequenced into a building process for achieving sustainable agriculture (Ragland, 1991).

THE EDUCATIONAL AND PUBLIC AWARENESS STEP

Nationally and internationally, the public must come to accept its stewardship responsibilities for natural resources. Each country will have to build its own sustainable agriculture and resource management program, but the developed countries have a special responsibility. That comes from the

Fig. 3-1. A stepwise process for sustainable agriculture (from IBSRAM Proc. no. 12-2).

threat of global climate change caused by our overuse of fossil fuels. The awareness of responsibilities is growing, but must be accelerated at every opportunity through greater efforts with primary and secondary schools, colleges, churches, extension programs, civic organizations, environmental private volunteer organizations, and the mass media. Once firmly established, the public's concern for stewardship and the added responsibilities of being a developed nation will cause government planners and development workers to emphasize sustainable agriculture and natural resource management.

POLICY CHANGES

Policies must reward individual initiative and efficient production. The recent collapse of so many centrally planned economies in the early 1990s emphasizes this point. However, policies which determine prices in a free enterprise system need to distinguish between efficiency and exploitation. Too often the product offered at the lowest price was produced by practices which were not in the public interest. Air and water pollution, siltation of reservoirs, and the destruction of genetic diversity are examples. Somehow these public costs incurred for private gain must be stopped. In any event, there must be a continuing and systematic review of local, regional, national, and international policies in order to identify changes required for environmental compatibility (Kessaba, 1993). If the education and public awareness phase has been done properly, then the political support will exist to force legislators to make the changes identified.

AGROECOLOGICAL GROUPINGS

Agroecological zoning is a systematic method of gathering the parts of a region with similar soil and climatic conditions into common groupings. While agroecological zoning provides a basis for planning at the global, national, and provincial levels, it must be supplemented with site-specific information to make development projects effective at the village and farm levels.

SITE SPECIFIC INTERVENTIONS

A global network, using computer programs, has been developed by USAID for overcoming key constraints by prescribing appropriate technological interventions (Jones et al., 1993). The system integrates soils, climatic and agronomic response data using crop simulation models which predict production costs and returns at low, medium or high levels of inputs, as specified by the user. Compared with conducting all of the research in the field, this approach represents a quantum leap in reducing research cost and time. Actual field trials must be conducted, but only to validate predicted outcomes and to investigate unexpected problems. The performance of various combinations of crops and management practices can be simulated for a single growing season or for many years. By calculating the performance of a crop for many years, the sustainability of the system can be estimated. These simulations promise to result in sustainability modules which can be used to build sustainable agriculture at a fraction of the time and cost otherwise required. Although USAID initiated the work with the Decision Support System for Agrotechnology Transfer (DSSAT) project, collaboration from groups around the world has grown in recent years to become even larger than the USAID contribution. This growing support indicates a broad acceptance in the scientific and development communities.

EXPANDING DECISION SUPPORT SYSTEM FOR AGROTECHNOLOGY TRANSFER TO MINIMUM DECISION MAKING AREAS

The DSSAT is beginning to be used to expand to larger geographic regions the interventions prescribed for individual sites (see Jones et al., 1993). This is possible because of the geographic information system (GIS) which can link management modules constructed for a specific site to all parts of the world where matching soil, topography and climatic conditions exist. The information for expanding the DSSAT module on a regional basis is from agroecological zoning.

The impact of agriculture production systems upon various components of the natural resource base also can be predicted by combining DSSAT, GIS and expert system programs developed for managing natural resources.

The data from environmental and natural resource inventories will become increasingly valuable as investigators learn to expand their application through these new tools.

It may become possible to predict the cost of repairing the natural resource base and maintaining it by the use of this technology.

PROVIDING FOR ECOLOGICAL SERVICES

Although vital to food sufficiency and income generation, successful crop production and livestock enterprises alone do not ensure sustainable agriculture. Not only the cropland, but all land, water and biologic resources in a region must be properly managed. This means coping with deforestation and desertification, protecting biodiversity, and providing for ecological services, such as nutrient and water cycling. Ecologists, foresters, and others trained to understand the dynamics of whole systems are providing much of the impetus and knowledge for building sustainable agriculture at the watershed and larger geographic levels. This is possible because of the ability of these scientists to rationalize the larger system into which the work on component technologies and land-use schemes must fit.

SOCIOECONOMIC ADJUSTMENTS

Interventions to make agricultural and natural resource systems sustainable will fail unless they are consistent with the broader desires, needs, and capabilities of the people who must adopt and operate them. Therefore, sustainable agricultural programs must build on the indigenous capacity of local people to participate in research design and implementation. This will require close and regular exchanges between development workers and their client groups. Rapid rural appraisals, on-farm testing, and other methodologies have been developed recently for fitting biophysical interventions to the cultural and social realities of a village or region (Stroup et al., 1993).

Finally, perfectly constructed sustainability modules can succeed only if they are profitable to the farmers who are being urged to follow them. Therefore, markets must exist for what is produced, and the prices must yield a reasonable return on the labor and other inputs invested. The microeconomic analysis provided by the DSSAT models will have to be supplemented by macroeconomic analyses of the regional and national systems. Market and infrastructure changes may be required.

THE PROCESS SUMMARIZED

1. Build domestic and foreign support groups for sustainable agriculture through education and public awareness programs.

2. Determine the macroeconomic and public policy changes required for sustainable agriculture to succeed throughout the world and provide guidelines for doing such analyses.
3. Compile all relevant knowledge on soil, topography, climate and vegetative resources into an agroecological data base.
4. Expand the capacity of development organizations and educational institutions to utilize DSSAT software for deciding which interventions to test at any given site.
5. Match validated DSSAT sustainability modules to the universe of need by using the agroecological database and GIS mapping software. Estimate the cost and levels of development effort required to build sustainable agriculture throughout the village, province or country.
6. Review the intended interventions of compatibility with needed ecological services, such as nutrient and water cycling, biological pest control and the preservation of genetic diversity.
7. Adjust intended interventions for compatibility with farmer preferences and local customs. Determine if competitive local markets exist and, if not, try to establish them.
8. Validate the interventions prescribed by an on-farm testing program.

In conclusion the need must be emphasized for a process which will give planners and decision makers estimates of the policy changes and money required to produce sustainable agriculture. Until this is done at the national and international levels it is unlikely that the resources required at the local level will be provided. The important thing is to find an integrative approach which will be accepted and allow sustainable agriculture and natural resource management to move beyond the conceptual stage.

REFERENCES

Borlaug, N.E. 1986. Accelerating agricultural research and production in the Third World: A scientist's viewpoint. Agric. Human Values. 3(3):5–14.

Jones, J.W., W.T. Bowen, W.G. Boggess, and J.T. Ritchie. 1993. Decision support systems for sustainable agriculture. p. 123–138. In J. Ragland and R. Lal (ed.) Technologies for sustainable agriculture in the tropics. ASA Spec. Publ. 56. ASA, CSSA, and SSSA, Madison, WI.

Kesseba, A.M. 1993. Strategies for developing a viable and sustainable agricultural sector in sub-saharan Africa: Some issues and options. p. 211–243. In J. Ragland and R. Lal (ed.) Technologies for sustainable agriculture in the tropics. ASA Spec. Publ. 56. ASA, CSSA, and SSSA, Madison, WI.

Ragland, J. 1991. Stewardship and wealth. p. 481–490. In E. Pushparajah (ed.) Proc. Int. Workshop on Evaluation for Sustainable Land Management in the Developing World. Vol. 2. Int. Board for Soil Res. and Manage., Bangkok, Thailand.

Stroup, W.W., P.E. Hildebrand, C.A. Francis. 1993. Farmer participation for more effective research in sustainable agriculture. p. 153–186. In J. Ragland and R. Lal (ed.) Technologies for sustainable agriculture in the tropics. ASA Spec. Publ. 56. ASA, CSSA, and SSSA, Madison, WI.

4   Alternatives to Slash and Burn Agriculture[1]

P. A. Sanchez

International Center for Research in Agroforestry
Nairobi, Kenya

THE PROBLEM

Major adverse effects of global warming are expected to take place in the USA and other midlatitude countries within the next 30 to 60 yr. Increased concentration of greenhouse gases in the atmosphere (CO_2, nitrous oxides and CH_4) will raise mean annual temperatures in the continental USA by 2.2 to 5.6 °C (4–10 °F), flood some coastal areas due to a 1.22-m rise in sea level and increase the incidence of droughts. Little effect is expected in the Tropics, but the center of gravity of agricultural production may shift northwards towards Canada and the former Soviet Union.

The USEPA (USEPA, 1988) estimates indicate that 15 to 25% of the global warming is due to the clearing of tropical rainforests, which is occurring now at an annual rate of 7 million hectares. Current magnitude of these emissions (1 billion tons of C yr^{-1}) is equivalent to current emissions from industrial combustion in the USA. Research has shown that something can be done to decrease the rate of tropical deforestation, and thus attenuate the expected adverse effects.

Deforestation is driven by a complex set of demographic, biological, social and economic forces described in Fig. 4-1. Third World population growth continues at a high rate, while most of the fertile and accessible lands are intensively utilized. Government policies often exacerbate land scarcity by allowing gross inequities in land tenure. These factors result in an increasing landless rural population which essentially has three choices: stagnate where they are, migrate to the cities, or migrate to the rainforests that constitute the frontier of many Third World countries. Although urban migrations are spontaneous, national policies in key countries include the occupation of their tropical rainforests, notably Brazil, Peru and Indonesia via colonization programs.

---

[1] Reprinted from *Soils and the Greenhouse Effect*, edited by A.F. Bouwman. © 1990 John Wiley & Sons Ltd.

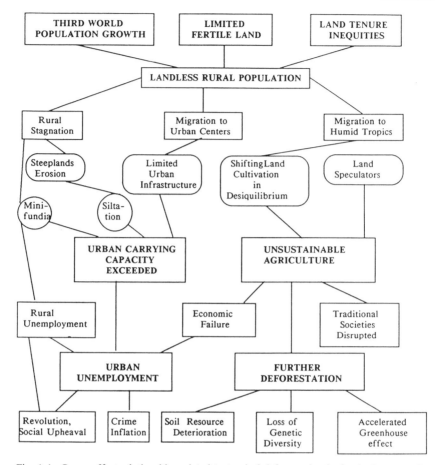

Fig. 4-1. Cause–effect relationships related to tropical deforestation in developing countries.

Densely populated rural environments such as the Andean valleys, northeast Brazil, and Java suffer from an ever-decreasing farm size and the overuse of steepland areas. This results in widespread soil erosion, siltation of reservoirs and other adverse off-site effects to urban centers. Migration to the cities in search of a better life results in bitter disappointments, and coupled with limited urban infrastructure, produces unmanageable cities with populations far exceeding their carrying capacity and infrastructure.

Migration to the humid Tropics seldom results in a bountiful cornucopia (with few notable exceptions). An equilibrium between the rainforest and shifting cultivation by traditional humid tropical societies is broken by the colonists, and in some countries by land speculators as well. The result is shifting cultivation in disequilibrium which quickly turns into various forms of unsustainable agriculture. Traditional societies are disrupted, economic failures abound and migration to urban centers follows.

The two endpoints are urban unemployment and further deforestation. The consequence of the former is abject urban poverty which leads to widespread crime, poor health and in many cases social upheaval. Deforestation depletes the ecosystems' limited nutrient capital, decimates plant and animal genetic diversity and accelerates global warming due to CO_2 and nitrous oxide emissions.

Can these trends be alleviated and eventually reversed? Our answer, based on long-term research is an emphatic *yes*. The key is an integrated approach consisting of (i) development and application of sustainable management technologies for tropical soils and (ii) appropriate government policies that will provide incentives to discourage the need for further deforestation.

People do not cut tropical rainforests because they like to, they clear land out of sheer necessity to grow more food. Deforestation, therefore, can be reduced by the widespread adoption of sustainable management practices that permit the use of cleared land on an indefinite basis. Sustainable management options for acid soils of the humid Tropics have been developed at Yurimaguas, Peru, and elsewhere to fit different landscape positions, soils and levels of socioeconomic infrastructure development (Fig. 4–2).

Most of the options are based on low-input systems which serve as transition technology for other options such as agroforestry systems, legume-based pastures, and continuous crop rotations. Other options developed in humid tropical Asia and Africa, include paddy rice (*Oryza sativa* L.), perennial crop production and forest plantations. In spite of the continuing need for additional research, there is no question that the technological basis for sustainable management options for acid soils of the humid Tropics is available now.

For every hectare put in these sustainable soil management technologies by farmers, 5 to 10 ha yr^{-1} of tropical rainforests will be saved from the shifting cultivator's ax, because of their higher productivity (Table 4–1). Such technologies are equally applicable to secondary forest fallows, where clearing does not contribute significantly to global warming because of the small tree biomass. Nevertheless the use of secondary forest fallows is of very high priority, because in many areas it is an excellent alternative to primary forest clearing. Many of the degraded or unproductive pastures or croplands resulting from extremely poor management practices also can be reclaimed using some, but not all, of these available technologies.

Such technologies, however, are useless without effective government policies that encourage, support and regulate them. Likewise, well-conceived policies will fail without sustainable technologies. Therefore the hope lies on a joint policy–technology approach—The Deforestation Reduction Initiative.

The present situation is analogous to when the world technical assistance community launched the Green Revolution in the late 1960s. At that time sustainable technologies for high-yielding rice and wheat (*Triticum aestivum* L.) production were sufficiently developed to be used. Key government officials were convinced of their importance by leading scientists, and instituted the necessary policies to make massive farmer adoption possible in India, Pakistan, Philippines and other countries. The Green Revolution became a

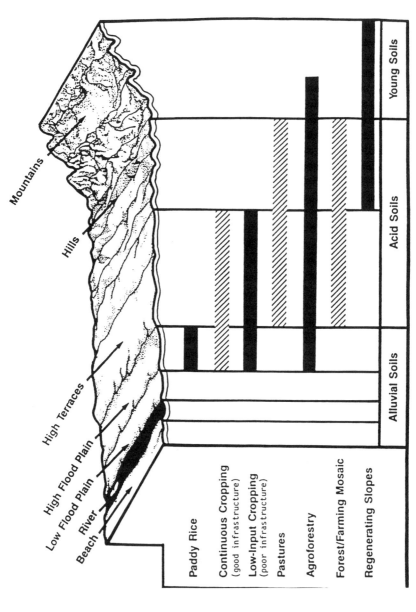

Fig. 4-2. Some soil management options for humid tropical landscapes dominated by Oxisols and Ultisols.

Table 4-1. Estimates of Yurimaguas for the various management options.

1 ha in sustainable management options	Hectares saved from deforestation annually
Flooded rice	11.0
Low-input cropping (transitional)	4.6
High-input cropping	8.8
Legume-based pastures	10.5
Agroforestry systems	not determined

worldwide success during the next 20 yr and the goal of arresting worldwide famine was definitely achieved.

The Deforestation Reduction Initiative is a similar challenge but with the major difference that it affects directly the livelihood of both Third World humid tropical countries and the developed countries as well. It also is likely to be more difficult to implement since the technology and policy issues are more complex than implementing high-input crop production technologies on the better lands of the Tropics.

THE GOAL

A worldwide, coordinated Deforestation Reduction Initiative is proposed to simultaneously accomplish the following goals during the next 25 yr:

1. Increase and stabilize crop, pasture and tree production in areas presently under shifting cultivation through the massive adoption of sustainable management technologies supported by appropriate government policies.
2. Decrease deforestation of primary tropical rainforest from the present annual rate of 7 million hectares to 1 million hectares.
3. Preserve the genetic diversity of the majority of the remaining areas of tropical rainforests by decreasing the demand for further land clearing.
4. Conserve the natural resource base of humid tropical regions through a reasonable, stable balance between sustainable agriculture and preserved rainforests.

APPROACH

Priority countries. There are 64 countries in the world with tropical rainforests and estimates of net C emissions due to land clearing in 1980 have been made by the Woods Hole Research Center (Table 4-2).

Twelve out of these 64 countries account for three-quarters of the net C emissions from clearing primary forests (Table 4-3).

Ten other countries account for much of the remaining deforestation— Burma, Cameroon, Venezuela, Madagascar, India, Nicaragua, Laos, Vietnam, Kampuchea and Bolivia.

Table 4-2. The breakdown by region, including emissions for primary forests, secondary forests and woodlands.†

Region	Forest cover	Net C emissions in 1980	
	billion ha	million tons	%
Tropical America	1.2	665	40
Tropical Asia	0.4	621	37
Tropical Africa	1.3	373	23

† Sources Houghton et al., 1987.

Table 4-3. Net C emissions from clearing primary forests.

Country	Net C emissions in 1980 from primary forests
	million tons
Brazil	207
Colombia	85
Indonesia	70
Malaysia	50
Côte d'Ivoire	47
Mexico	33
Thailand	33
Peru	31
Nigeria	29
Ecuador	28
Zaire	26
Philippines	21

As in the Green Revolution, dialogue may be initiated with top science policymakers in selected countries, to discuss the possibility of a cooperative program. In countries where the possibilities of collaboration are good, a dialogue between senior government officials should follow, with agreements signed and commitments for a 25-yr program.

Within each participating country, "hot spots" of deforestation can be identified where technology transfer and government policies would be focused. Examples of these hot spots are Rondonia, Acre and Southern Bahia in Brazil; Caquetá in Colombia; transmigration areas in Sumatra and Kalimantan in Indonesia; the Huallaga Valley, Pucallpa, and Pichis Palcazu in Peru, Chapare in Bolivia, Coca-Lago Agrio in Ecuador, and many others.

PROGRAM COMPONENTS

- An international workshop to launch the Tropical Deforestation Initiative
- Selection of participating countries and "hot spots" or target areas, with protocols signed
- Closer monitoring of deforestation rates in participating countries and target areas
- Dynamic, well-funded sustainable technology validation programs in each target area, including training of local research and extension specialists

- Assistance to participating countries in developing appropriate policies to support the Initiative
- Support continuing research on developing stable management options with reinforcement of one major research center in tropical America, Asia and Africa; emphasis will be on sustainability in the long run
- Support on-site research to quantify the nature of gas emissions from various sustainable management options, including clearing and common mismanagement options; major questions about nitrous oxides emissions exist, for some reason air above the Amazon has one of the highest nitrous oxides concentrations in the world; understanding the processes that regulate gaseous emissions at the ground level needs to be researched and linked with atmospheric measurements
- Support research to determine how the adoption of successful management technologies affect forest and soil preservation, conservation of genetic diversity and the hydrologic balance

REFERENCES

Houghton, R.A., R.D. Boone, J.R. Fruci, J.E. Hobbie, J.M. Melillo, C.A. Palm, B.J. Petersen, G.R. Shaver, and G.M. Woodwell. 1987. The flux of carbon from terrestrial ecosystems to the atmosphere in 1980 due to changes in land use: Geographic distribution of the global flux. Tellus 39B:122–139.

McElroy, M.B., and S.C. Wofsy. 1986. Tropical forests: Interactions with the atmosphere. p. 33–60. In G.T.Prance (ed.) Tropical rain forests and the world atmosphere. Westview Press, Boulder, CO.

Mooney, H.A., P.M. Vitousek, and P.A. Matson. 1987. Exchange of materials between terrestrial ecosystems and the atmosphere. Science (Washington, DC) 238:926–932.

Sanchez, P.A., and J.R. Benites. 1987. Low-input cropping for acid soils of the humid tropics. Science (Washington, DC) 238:1521–1527.

Sanchez, P.A., E.R. Stoner, and E. Pushparajah (ed.). 1987). Management of acid tropical soils for sustainable agriculture. IBSRAM Proc. no. 2, IBSRAM, Bangkok.

U.S. Environmental Protection Agency. 1988. Meeting summary. Stabilization Rep. Workshop on Agriculture and Climate Change. 29 February–1 March 1988. Office of Policy Analysis, USEPA, Washington, DC.

5

Sustainable Land-Use Systems for Sloping Uplands in Southeast Asia

DENNIS P. GARRITY

Southeast Asian Regional Research Programme
International Centre for Research in Agroforestry
Bogor, Indonesia

In Southeast Asia there are many serious problems of agricultural sustainability, associated with all of the major agroecosystems. But the sloping uplands are geographically the most extensive ecosystems, and the most threatened. The nonsustainability of land-use systems in the uplands is associated with increasing populations of subsistence farm families cultivating infertile sloping soils, accelerating land degradation and soil erosion, the impending loss of most tropical hardwood forest, and the failure of reforestation. Inequitable and insecure access to land resources exacerbates the tendency toward inappropriate land use. More productive and sustainable land-use systems must be developed under conditions of severe social and economic constraints. The crisis appears largely intractable unless systematically and innovatively addressed.

The problems of upland resource base deterioration extend across all national frontiers in the region, but they are only now beginning to be grappled with by the respective governments. Within each country the domain of the uplands is usually a complex division of responsibilities among the forestry department and the agriculture department. This greatly complicates technology generation, land tenure, and the delivery of infrastructure and services in these ecosystems.

The objectives of this chapter are: First, to characterize the humid sloping uplands of Southeast Asia as a distinct ecosystem, and to convey a sense of the problems of generating sustainable agricultural systems for them; second, to discuss the critical agricultural sustainability issues, and the technologies being evolved to meet those needs; third, to discuss the unique ways in which research must be organized, managed, and located, in order to address the problems comprehensively. Finally, I will propose a more comprehensive international effort to address sustainable land-use systems in the acid sloping upland ecosystems in Southeast Asia.

THE SETTLEMENT OF SOUTHEAST ASIAN ECOSYSTEMS

To begin, it is useful to put the contemporary problems of Southeast Asian farming systems in a brief historical and ecological perspective. The major populations of the Southeast Asian nations have based their livelihood predominantly on wet rice (*Oryza sativa* L.) culture. The great civilizations that evolved in the region tended to develop on the wide floodplains of the region's major river systems (Fig. 5-1). Wetland rice culture was a basis for comparatively sustainable farming systems. The unique properties of the flooded rice field, particularly the enhanced N fixation and greater P solubility in flooded soils, and minimal soil loss from bunded landscapes, has made continuous rice cultivation reasonably productive for centuries. These systems are amenable to a great degree of intensification.

Although they are of dominant economic importance, the lowlands represent only a small fraction of the total land area in the region. The up-

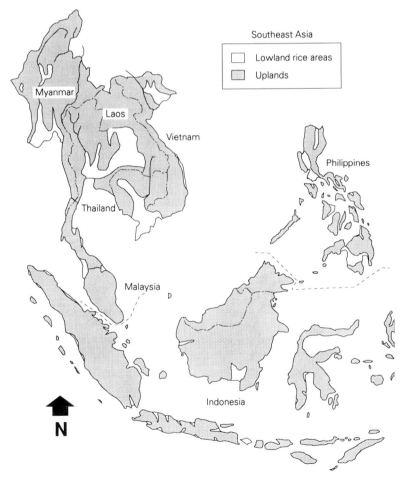

Fig. 5-1. The distribution of lowland rice and upland ecosystems in Southeast Asia.

land areas constitute from 60 to more than 90% of the total land area of the respective countries. Permanent upland settlement that occurred prior to recent decades was mainly restricted to the more fertile volcanic soils, as encountered in Java, and parts of the Philippines. Exceptionally favorable soil resources in these limited areas made highly productive, diverse upland farming systems possible. But the Southeast Asian uplands are predominantly sloping lands of strongly acid Ultisols and Oxisols. Of the 180 million hectares of acid uplands, about 118 million hectares are estimated to have pH values below 5.0 (IRRI, 1987).

The world's mountains have been given considerable attention in recent years (Rhoades, 1990). But sloping uplands or hilly lands are geographically dominant, and are arguably more ecologically fragile than the mountainous highland areas. The sloping upland ecosystems include the undulating and steep lands that range in elevation from near sea level to about 1000-m elevation. This zone, between the high mountains and the coastal plains remains poorly conceptualized and characterized, and lacks focus in research and development. The highlands of Southeast Asia, with their unique climatic characteristics and farming systems are relatively minor in area, and are not discussed here.

This chapter will argue that the sloping uplands demand a distinct ecosystem focus. As Rhoades (1990) stated, "without the slopeland perspective both our science and technological fixes will miss the mark...research that is removed from the ecosystem context often leads to shortsighted policies and programs."

Where crop cultivation was practiced on the acid, infertile sloping uplands, it was formerly limited to shifting cultivation systems as the only practical method of exploitation. However, during the last couple of decades, the lowlands and fertile uplands became densely settled in most countries, and the land frontier on favorable soils virtually closed. Millions of families that could not otherwise be absorbed in the national economies, began an accelerating migration to the sloping, acid uplands. As in past generations, they sought modest permanent landholdings to provide a livelihood, but encountered serious difficulties as increasing population pressure shortened or eliminated the fallows that were essential to accumulate nutrients for crop production.

In the Philippines, permanent cultivation or fallow rotation within the farm boundaries, is now observed throughout the hills of the entire archipelago. The number of people living in areas of 18% slope or greater, was recently estimated at 17.8 m, or almost 30% of the country's total population (Cruz & Zosa-Feranil, 1988). And the population growth rate in the uplands is significantly higher than that of the lowland population. In northern Thailand, the forest cover has predominantly disappeared within the past generation. The uplands have largely been transformed into farmland, and brush or grass fallows, by cultivation of upland rice and other subsistence food crops. In China, which is ecologically contiguous with northern Thailand, there is evidence of severe pressure on the sloping Ultisols and Oxisols, known locally as the red soils, throughout the southern one-third of the country. In Indonesia, settlement of the infertile uplands of the outer islands is proceeding rapidly, due both to spontaneous migration and a massive government-sponsored transmigration program.

Many of the prevalent cereal-based farming systems practiced by small-scale farmers are highly unsustainable. This is widely evident in negative trends in the condition of the land resource base, and in declining production flows. Due to the relative inaccessibility, fragility, and marginality of the sloping uplands, the sustainability of utilization and production flows are inseparable from the sustainability of the resource base itself (Jodha, 1990). Damage to the resource base is characteristically rapid. It is reversible only over a long period of time.

DEGRADATION OF THE LAND BASE

The extent of the massive rate at which the Southeast Asian land base is degrading may be deduced from data on the rate of soil loss in the region, compared to other areas of the world. The magnitude of the discharge of sediment from Southeast Asia's major river systems dwarfs that of other river basins around the world (Fig. 5-2). The river systems of mainland Southeast Asia discharge over 3.2 billion tons of sediment annually. The island nations of the region are nearly as prodigious as the mainland countries in production of river sediments, with 3.0 billion tons per year. Each of these areas produce more than an order of magnitude more river sediment than than the Mississippi River drainage system in North America, and twice the sediment of the Amazon River basin.

The factors responsible for the extraordinarily high soil losses in Southeast Asia are both man-induced, climatic, and geologic. The region's geographically steep young landscapes, occupied by extraordinarily dense human populations, experience abundant and intense annual rainfall (1500–3000 mm). Intensive rainfall interacts with relatively steep slopes and intense human land-use pressures. The net result is a regionwide pattern of soil loss of alarming extent. Most of the upland soils in Southeast Asia are physically fairly deep, and do not overlie shallow constricting layers. But they are chemically shallow due to high Al saturation in the subsoil. Roots are inhibited from downward penetration by Al toxicity. The rapid loss of topsoil further constricts crop rooting depth, confining the roots to a shallow zone of activity, and reducing the available nutrient and water reserves.

SUSTAINABLE FOOD CROP SYSTEMS ON SLOPES

The upland ecosystem is composed of a great diversity of land-use systems. These include large-scale ranching (open grasslands), cash cropping [eg, sugarcane (*Saccharam officinarum* L.)], perennial crop plantations [particularly coconuts (*Cocos nucifera* L.)], rubber [Hevea brasiliensis (Willd. ex Adr. Juss.) Mudl. Agr.], and oil palm (*Elaeis guineensis* Jacq.) and for-

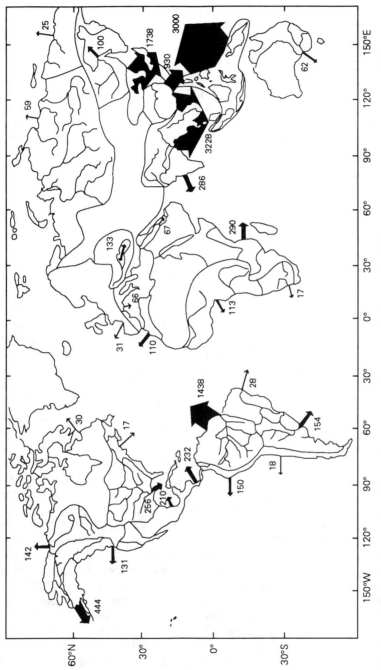

Fig. 5–2. Annual discharge of suspended sediment from various drainage basins of the world. Width of arrows corresponds to relative discharge, numbers refer to average input in millions of tons (Milliman & Meade, 1983).

estry (natural and plantation), in addition to small-scale subsistence farming. No attempt will be made to review the sustainability issues of all these systems. Rather, this chapter concentrates on the family farm sector in the sloping uplands, which relies on field crop production and agroforestry for its food supply and income. This sector dominates in terms of numbers of people involved, and land area affected, and it is unquestionably the most environmentally stressed.

A large and rapidly expanding portion of the upland landscape is being converted to permanent annual cropping. These farms are found in the more relatively accessible sloping areas, closest to the lowlands, and nearest to roads. They are predominantly planted to subsistence food crops, but are partly in perennial crop plantations. With increasing elevation, and more remote access, the land is predominantly grassland and brushland. The remaining forested areas are generally the secondary forest remnants of previous logging, or localized unlogged areas, found at the highest elevations and on the steepest slopes.

These three broad land-use types (permanently farmed, grassland, and forested land) tend to grade into one another. The permanently cultivated lands increase at the expense of the grasslands, as shifting cultivation on the grassland margins intensifies, and the grasslands advance at the expense of the forested land, as settlement and the relentless use of fire transform the forests, after they are logged and opened. The human and natural ecology of these three entities is distinct, and technology and policy instruments must be adapted to the realities of each. Garrity et al. (1993) review these issues in greater depth for the Philippines.

The food crop sector is characterized by large numbers of small-scale farmers whose primary imperative is the production of a family food supply. These systems are therefore predominantly based on the production of upland rice or maize (*Zea mays* L.), but in some areas there is more focus on root crops {viz. sweet potato [*Ipomoea batatas* Lamk.] and cassava (*Manihot esculenta)*}. Since most upland farm families have extremely limited access to capital, and/or have insecure tenure to the land they till, they are trapped with little capacity to bear the major risks associated with diversification into more ecologically sustainable perennial tree crop enterprises, or mixed farming with ruminant livestock.

There are many factors that limit the stability, productivity, and sustainability of upland farms, including climatic variation, biological stresses, insecure land tenure, and social and economic uncertainty. New technologies will be critical to sustaining agriculture in the uplands. But the technologies that are currently being extended are generally unproven in the diversity of environments and farmer circumstances encountered. Two issues need to be addressed in upland agriculture: What will it take to make small-scale farming permanently sustainable? What will it take to improve fallow-rotation systems where they are still practiced?

A fundamental factor is the nature and rapidity of soil degradation. Permanent small-scale upland farming systems are evolving in the sloping upland areas to replace shifting cultivation. Accelerating the trend toward permanent systems will fundamentally require: (i) simple, effective soil erosion control on open fields, through vegetative barriers and residue management; (ii) mineral nutrient importation to balance crop nutrient offtake, and stimulate greater biological N fixation; and (iii) enterprise diversification toward mixed farming systems that includes ruminant animals and perennials, in addition to subsistence food crops.

Some technologies to meet these needs are known (Garrity et al., 1993). Some even fulfill multiple requirements (e.g., trees in contour hedgerows may provide erosion control, fodder, and crop nutrients). However, research to evaluate and adapt them to the wide array of diverse ecological niches encountered by upland farmers is woefully inadequate.

CONTOUR HEDGEROW FARMING SYSTEMS

Since the mid-1970s there has been considerable interest in the concept of planting hedgerows of leguminous tree species along the contour on sloping fields, to provide a vegetative barrier to soil erosion while contributing green leaf manure to the cereal crops (rice or maize) grown in the alleys. This idea was not new in the region. Farmers in Nalaad, Cebu, have indigenously used a contour hedgerow system of *Leucaena leucocephala* in cultivating steep slopes since before 1923 (MacDicken, 1990). On Flores Island in eastern Indonesia, contour terracing with *Leucaena* was promoted by church and extension organizations in the 1970s, and was adopted on an area of over 20 000 ha (Parera, 1989, unpublished data).

Leucaena leucocephala is a commonly observed tree in the rural areas of Southeast Asia. It is widely grown in fencerows as a fodder source for cattle. Reports in the mid-1970s (Natl. Res. Council, 1977) indicated that the tree showed outstanding possibilities as a hedgerow intercrop that would supply large quantities of N and organic matter to a companion food crop. These observations stimulated applied research on hedgerow intercropping at several locations, with promising results. Guevara (1976) reported yield advantages of hedgerow intercroping of 23%. Vergara (1982, unpublished data) cited experiments in which yields were increased about 100%, with no advantage of inorganic N application beyond the N supplied by the green leaf manure. Alferez (1980, unpublished data) observed a 56% yield advantage when upland rice was grown in alleys between hedges of *Leuceana*.

On sloping land, hedgerows of *Leucaena* were observed by a number of researchers to provide an effective barrier to soil movement. Data from runoff plots on a steeply sloping site in Mindanao indicated that both runoff and soil loss were dramatically reduced (O'Sullivan, 1985). This study also observed a consistent yield advantage over a 4-yr period with corn fertilized by the *Leucaena* prunings obtained from adjacent hedgerows.

By the early 1980s, hedgerow intercropping was widely advocated as a technology to better sustain permanent cereal cropping with minimal or no fertilizer input, and as a soil erosion control measure for sloping lands. In the Philippines, a 10-step program was developed for farmer implementation of *Leucaena* hedgerows (Watson & Laquihan, 1987), known as Sloping Agricultural Land Technology (SALT). It was adopted by the Philippines Department of Agriculture (DA) as the basis for its extension effort in the sloping uplands. The SALT guidelines recommended that every third alleyway between the hedgerows of *Leucaena* be planted to perennial woody crops, such as coffee (*Coffea arabica* L.), with the majority of the alleys maintained in the continuous cropping of food crops. This concept offered the possibility of more diversified sources of farm income, and improved soil erosion control.

Some adoption of *Leucaena* hedgerows occured in high-intensity extension projects, but there was no strong evidence of spontaneous farmer interest. The lack of secure land tenure was implicated as one major constraint to the implementation of contour hedgerow systems or of other long–term land improvement among tenant farmers, or occupants of public lands. Among farmers with secure tenure, the large initial investment of labor, difficulty in obtaining planting materials, and the degree of technical training and backup required to install and manage the system, were observed to be serious constraints to adoption. In addition, the labor required to manage the hedges (pruning 3–10 times/year depending on the management system) was observed to absorb a large proportion of the household's available labor, and to compete with other income-generating tasks. This would normally limit the land area that can be farmed in this manner to less than about 0.5 ha.

After 1985 the extension effort on *Leucaena* hedgerows suffered a major setback when a psyllid leaf hopper (*Heteropsylla cubana*) naturally invaded the region and devastated hedgerows, killing or stunting the trees. *Gliricidia sepium* was the most common replacement species, but it must be propagated from cuttings in most areas, increasing the labor investment to establish the hedges.

Nongovernmental organizations made noteworthy contributions toward developing farmer participatory approaches to devising and implementing local solutions to the dominant constraints in crop cultivation on sloping land. A system of contour bunding was developed through efforts by World Neighbors (Granert & Sabueto, 1987). The bunds provided a base for the establishment of double contour hedgerows of leguminous trees or forage grasses, and were a barrier to surface runoff, which is carried off the field in contour ditches.

The experience with alley cropping discussed above was predominantly obtained on high-base status soils of recent volcanic or marine limestone origin. The contour hedgerow concept was more recently applied to strongly acid soils in the uplands of Mindanao by the International Rice Research Institute (IRRI) and the Philippine DA (Fujisaka & Garrity, 1988). After 3 yr of hedgerow intercropping, there was a distinct development of forward-sloping terraces. Regardless of species, the vegetative barriers' effects in reducing soil loss were striking, compared to soil loss in conventional open field cultivation (Table 5-1). All species induced rapid terrace formation. The rapid soil movement was largely attributed to the high frequency of animal-powered

Table 5-1. Soil loss as affected by contour hedgerow vegetation, Claveria, Misamis Oriental (Mindanao), Philippines, August 1988 to April 1990 (D.P. Garrity and A. Mercado, 1990, unpublished data).

Hedgerow species	Soil loss
	cm
Gliricidia + *Paspalum*	0.38
Napier grass	0.62
Gliricidia + *Napier*	1.38
Gliricidia alone	1.50
Open field (conventional practices)	4.20

tillage in the alleys (5–6 times annually). Tillage-induced downhill soil slumpage apparently contributed more to soil redistribution than water-induced erosion.

Modest yield benefits were observed in upland rice associated with hedgerows of *Cassia spectabilis* (Basri et al., 1990). Yield benefits were observed consistently for upland rice and corn associated with hedgerows of *Gliricidia sepium* trees or Napier grass (*Pennesitum purpureum*) (D.P. Garrity and A. Mercado, 1991, unpublished data). Crop yields were most seriously affected in the rows adjoining the hedges, with or without the application of external N and P fertilizers (Fig. 5-3). The roots of the two tree species were observed to spread laterally at a shallow depth (20–30 cm) in the alleyways, just beneath the plow zone. This indicated that a major potential existed for nutrient and water competition between the respective root systems of the trees and the crop.

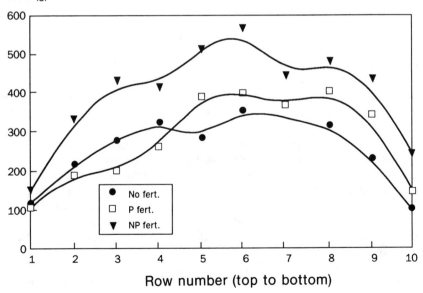

Fig. 5-3. Grain yield of upland rice on a row-by-row basis grown between hedges of *Cassia spectabilis* which supplied green manure for the crop (Basri et al., 1990).

    The sustainability of crop yields in tree hedgerow intercropping is a major concern on all soil types, but recent work raises particular questions about its viability on strong acidic soils. The high exchangeable Al in the subsoil inhibits the deep tree rooting patterns typically observed on higher base status soils. Phosphorus and other mineral elements are often more limiting than N in these soils. The strong subsoil acidity appears to promote intense root competition between the annual crop and the perennial hedge for mineral nutrients in the surface soil of the alleys, and prevents deep rooting by the perennial that would promote nutrient pumping from the deeper soil layers. The organic matter inputs of hedgerow prunings of *Gliricidia* and *Cassia* do not supply adequate quantities of P to meet the minimal requirements of the cereal crops (Mercado et al., 1989; Basri et al., 1990). Further, the prunings are composed of P that the tree may have captured predominantly from the crop root zone. Other recent alley cropping results on acid Ultisols in Peru (Fernandes, 1990) and Indonesia (Evenson, 1989) support the results obtained by IRRI in Mindanao. Under such circumstances, the intimate association of trees and crops in pruned hedgerow systems may not be advantageous.

    Grass strips also have received major attention as contour vegetative barriers for erosion control in Southeast Asia, and in other parts of the world (Lal, 1990). Considerable work has been done on the evaluation of grass strip technology in northern Thailand (Thai-Australian Highland Development Projects, 1989, unpublished data), in the central highlands of Java, Indonesia, and in the Philippines (Fujisaka & Garrity, 1988; Granert & Sabueto 1987). Predominant attention has been given to the use of vigorous forage grasses, since these provide high biomass production for ruminant fodder. Therefore, they are presumed to serve as a beneficial way to use the field area occupied by the hedgerows, which is lost to food crop production. They have the potential to markedly reduce erosion and rapidly develop natural terraces on slopes (Table 5–1). Therefore, the establishment of planted forage grasses has been extended as a practical alternative to tree legumes on contour bunds. Data from northern Thailand (M. Hoey, 1990, personal communication) indicated that erosion control may be a prevalent factor in systems sustainability, since grass strips were equally effective as *Leucaena* strips in increasing upland rice yields.

    Two major problems have emerged with the use of grass strips. Farmers tend to have difficulty keeping the tall, rapidly growing forage species trimmed frequently enough to prevent them from shading adjoining field crops. The biomass productivity of grass hedgerows exceeds the fodder requirements of most small-scale farm enterprises. The unnecessary foliage becomes a burden to cut frequently. High biomass production also tends to exacerbate resource competition with the adjoining food crops and reduces crop yields.

    The constraints observed with both trees and forage grasses have stimulated the concept of employing hedgerows that contain noncompetitive or relatively "inert" species (Garrity, 1989). This concept places primary emphasis on the rapid and effective development of terraces in order to improve the field hydrology and maximize soil and nutrient retention. An inert species is one with short stature and a low growth rate to minimize hedgerow-crop competition for resources, but provides an effective ground cover for filtering out soil particles. *Vetiver zizanoides* may exemplify this type of hedgerow species (Smyle et al., 1990). *Vetiver* is found throughout the Philippines under various names (*anias, ilib* or *anis de moras* in Luzon, *amoras* or *muda* in the Visayas). It tends to form a dense barrier and does not self propagate to become a weed in cultivated fields. However, it must be planted by vegetative slips, a somewhat laborious process.

An alternative approach that has received very little attention, is the installation of natural vegetative filter strips. These are narrow strips of field area laid out along the contours and left unplowed. They may be put in at the time that a piece of fallow land is brought into cultivation, or in the interval between crops in a continuous cropping system. The dominant species in natural vegetative filter strips are native weedy grasses—*Imperata cylindrica, Paspalum conjugatum, Chrysopogon aciculatus*, or others depending on the location, and the management regime the strips are subjected to. These natural grasses may be suppressed by grazing between crops, by periodic slashing, or by mulching them with crop residue. Natural vegetative filter strips have shown an outstanding capability to reduce soil loss, a characteristic in which they are potentially superior to the commonly recommended introduced species. They generally are less competitive as hedgerow components, and are ruggedly adapted to the locale.

There have been some isolated observations of the indigenous development of natural vegetative barriers by upland farmers in the Philippines (Ly, 1990; Fujisaka, personal communication, 1991). However, research has not been targeted to this option. There has, however, been extensive research on filter strips for sediment and chemical pollution control in the USA (Williams & Lavey, 1986).

Initial adoption of natural vegetative filter strips is comparatively simple—contour lines are laid out at the desired spacing with a simple A-frame or water level, and the field is plowed on the contour, avoiding the designated strips to be left in fallow vegetation. The advantages of this approach are the simplicity of installation, the absence of input costs or off-farm materials, and the low labor requirements for installation and maintenance. As terraces form, the farmer may diversify the terrace risers into other enterprises, including trees or perennial crops, as they fit management objectives. These characteristics suggest that the natural filter strip concept could be a comparatively practical basis for rapid, widescale dissemination of hedgerow technology. Simplicity is critical, given the constraints of very limited upland extension capability, and the necessity of reaching very large numbers of farmers. The one significant limitation of natural vegetative strips, which also is observed with other types of hedgerow vegetation, is that the development of natural terraces from the rapid redistribution of soil across the alleyways between strips tends to result in scouring of the topsoil in the upper alley. Serious degradation of the soil fertility on the upper half of the alley is often observed. Methods to alleviate soil degradation as the terraces develop are needed. A substantial effort in both strategic and farmer-participatory research on natural filter strips is warranted.

Institutional support will be crucial in fostering the expanded use of hedgerows as a soil conservation tool. There are promising signs that support mechanisms can be evolved. For example, the Philippine Crop Insurance Corporation, and local cooperatives, are implementing policies that enable the availability of crop insurance and credit for crop input purchase on sloping lands if the farmer has hedgerows installed.

Hedgerows also may be particularly suitable for the production of cash perennials. Examples of perennial crops that have been used in these systems include coffee, papaya (*Carica papaya* L.), citrus (*Citrus* spp.), mulberry (*Morus nigra* L.), and others. The suitability of the perennial species is limited by the degree of shading projected onto the associated food crops. Cash income from the hedgerow is a major advantage of this approach. The Robusta cultivars of coffee, for example, are architecturally well suited to hedgerow intercropping. Their stature is erect, providing minimal competition with the food crop. Erosion control is not adequately provided by the perennials, but is induced by a grass strip occupying the area between the widely spaced plants. Filipino farmers tend to view natural vegetative strips as the first stage in a process that leads to hedgerows of perennial crops.

Backyard production of cattle with cut-and-carry methods has become an important enterprise in many densely settled upland areas. Tree legumes, particularly *Leucaena* and *Gliricidia*, are widely used as high-protein forages, especially in the dry season. Backyard ruminant production will stimulate a more intensive husbandry of manure. One model of tree legume hedgerow development is to use the prunings as a source of animal feed, either for on-farm use or for off-farm sales (Kang et al., 1990). This will potentially increase the value of the hedgerow prunings, but it will even more rapidly deplete soil nutrient reserves. The manure can be returned, but borrowed nutrients will not be fully replaced. Nutrient augmentation through fertilizers may be made more necessary, but possibly more affordable, because of the added income generated by the animals.

The experience of the past 15 yr with alley cropping and contour hedgerows suggests that appropriate solutions are diverse, and must be uniquely tailored to variable soil and environmental conditions, farm sizes, labor availability, markets, and farmer objectives. Several basic types of hedgerow systems are recognized (Garrity, 1989), each adapted to a unique combination of ecological and market conditions. Table 5–2 illustrates some of the tradeoffs that enter in the choice of hedgerow enterprise among farmers. Such factors as relative effectiveness in soil conservation, potential hedgerow-crop competitiveness, and the value added by the hedgerow enterprise, are shown for a range of hedgerow choices.

Extension systems are strongly conditioned to employ a package approach in popularizing a new technology. This is not suitable for hedgerow technology, which needs an extension approach that recognizes the wide range of possible hedgerow species, and management systems, suitable for different farm circumstances (Fujisaka & Garrity, 1988). Unfortunately, there has

Table 5-2. Comparison of several major alternatives for contour hedgerow enterprises.

Hedgerow component	Soil loss control	Potential crop competition	Value added
Natural vegetative filter strips	High	Minimal-low	Very low
Planted grasses or legumes	Moderate-High	Low-high	Low-moderate
Multipurpose trees	Moderate	Moderate-high	Low-moderate
Perennial crops	Moderate	Moderate-high	Moderate-high

been very little attempt to clarify the technology–environment fit for different hedgerow systems, and inappropriate recommendations are very common. Elucidating the recommendation domains for the various hedgerow systems, in relation to the range of specific local physical and institutional settings, is a task needing urgent attention. Carson (1989) has recommended that a simplified land classification system be developed that is practical for research and extension workers to use.

PHOSPHORUS AS A CRITICAL CONSTRAINT

The acid upland soils of Southeast Asia are predominantly fine textured, with organic C contents often observed as high as 2 or 3% (Garrity et al., 1989). Phosphorus, rather than N, is commonly the most limiting nutrient. Typically, P deficiency must be overcome before there is any response to N (Garrote et al., 1986; Basri et al., 1990). Over 50% of the P taken up by the major food crops is removed in the grain, and a lower proportion of P is retained in the residue than for any of the other major crop nutrients (Sanchez & Benitez, 1987). This exacerbates the difficulty of maintaining adequate available P in a highly P-deficient agroecosystem through nutrient cycling. Phosphorus pumping from deeper soil layers also is limited by Al toxic subsoils and low subsoil P reserves. Since constant nutrient offtake is occurring, sustained crop yields and significant biological N fixation, will become dependent on the importation of P and lime.

Application of small or moderate amounts of P is a realistic strategy with a high marginal return for low-cash flow food crop farming on strongly acid upland soils (Garrity et al., 1989). Fertilizer P recommendations for upland rice range from 15 to 20 kg P/ha on P-deficient Ultisols in a number of Asian countries (Santoso et al., 1988). Local supplies of phosphate rock are an efficient source of both P and Ca (Briones & Vicente, 1985). The exploitation of phosphate rocks for upland farmers' use has been neglected, and should be expedited. This would require greater government and commercial recognition of the fundamental importance of these minerals to permanent agriculture in the uplands.

Nearly all Southeast Asian upland farmers participate in the national cash economy. Wherever they are capable of exporting agricultural commodities off-farm for sale, they are also physically capable of shipping them in. Fertilizers will not be used as long as nutrients can be obtained without charge in fallow rotations. But as shortened fallows fail to provide adequate nutrients, or as limited farm size makes it necessary to farm all available land permanently, there will be a progressive adoption of external fertilizer sources. On acid soils, the efficient use of on-farm biological nutrient sources will lower external fertilizer requirements, but seldom can eliminate them.

The serious limitations encountered in generating and conserving adequate quantities of nutrients within the cropping system, leads to debate on the appropriate balance between research attention on technology that will raise the marginal returns to fertilizer application and other inputs, vs. technology that attempts to better conserve and cycle nutrients on-farm. Odum (1989) examined this issue in his study of the state of Georgia, USA, as an ecosystem. Formerly a "wasted land" of shifting cultivation in a "soil-destroying agriculture" on acid Ultisols, the net primary productivity of the state more than doubled during the past 50 yr, increasing from 2.5 to 6.4 t/ha of dry matter, while the yield of food quadrupled. This was associated with a sevenfold increase in commercial fertilizer use. There was initial emphasis on increasing external nutrient inputs to the agroecosystem, followed by a more recent shift to input management efficiency that shall foster a reduced input agriculture.

Experience with hybrid maize and other cash crops by small-scale farmers in the Philippines lends support to this approach: Sloping upland farmers readily adopt fertilizers and lime with these crops because the perceived marginal returns to input use are promising, relative to their own cash investment opportunities. On subsistence food crops, the perceived marginal returns are not adequate to justify such levels of investment. Therefore, emphasis is needed on the development of nutrient responsive cultivars for food crops on unfavorable soils.

REDUCED TILLAGE

The practice of retaining surface residues through conservation tillage systems is wholly unexploited in Southeast Asian upland farming systems, yet its ecological value is nowhere as profound in reducing erosion than on the tropical sloping uplands. Clean cultivation is the universal practice of upland farmers, whether they use animal power or hand tillage on the slopes. Crop residues are either plowed under, burned, or removed as fodder. Many studies have shown significant benefits in maintaining a surface mulch. For example, Paningbatan (unpublished data, 1989) found that soil loss was reduced by over two-thirds by the presence of a vegetative barrier, but the maintenance of crop residues on the soil surface reduced soil loss by more than 95%.

Reduced tillage systems could be of enormous benefit in sustainable land management in the uplands, but as yet there is no practical approach to satisfactorily cope with weeds. The intense weed pressure on upland farms, and the tendency for weed populations to rapidly shift to resistent species, has severely constrained the development of herbicide-based solutions.

FARMING THE GRASSLANDS

The most common form of vegetation on vast areas of the region's uplands is grass, predominantly *Imperata cylindrica* (known as *cogon* in the Philippines and *alang-alang* in Indonesia and Malaysia) or *Themeda triandra*. These rhizomatous perennials are highly resistant to fire, in addition to being highly inflammable during dry periods. They readily invade abandoned swiddens, cleared forests, and forest openings. A small portion of the current grassland area may be a result of natural disturbances, but the overwhelming area was derived by repeated disturbance by fire (Bartlett, 1956).

Small-scale farming in grasslands is predominantly practiced with animal power. Settlers initially employ a migratory system, shifting their farm area as necessary to sustain crop yields. Greater density of population necessitates fallow-rotating the fields within permanent farm boundaries. As farm size decreases, permanent cropping evolves, in many cases with extremely low comparative returns, as is observed in the densely populated uplands of Cebu, Philippines (Vandermeer, 1963).

Can the productivity of intensive cropping be sustained on a strongly acid soil in *Imperata* grassland? McIntosh et al. (1981) showed that with balanced crop nutrition, an average annual crop yield of 12 to 24 t/ha of rice equivalent could be maintained on a strongly acid soil in an *I. cylindrica* (alang-alang)-infested area in southern Sumatra. The cropping system included as many as five crops per year, including upland rice, maize, cassava, peanut (*Arachis hypogara* L.) and rice bean [*Vigna umbellata* (Thunb.) Ohwi & H. Ohashi] in intercrop and sequential cropping patterns. This was an exceptional level of output for an upland environment, and disputed the concept that *Imperata* grasslands were necessarily waste areas. Productivity levels of this magnitude, however, are unlikely at higher latitudes due to less uniform annual rainfall.

FALLOW IMPROVEMENT SYSTEMS

Farming systems in the grasslands vary over a range of shifting and permanent cultivation systems. The technology appropriate in a shifting cultivation system will differ from that for permanent field cultivation, due to major differences in labor and land-use intensity. As Raintree and Warner (1986) have pointed out, shifting cultivators will maximize their returns to labor rather than to land, and resist inappropriate labor-intensive technologies. Pruned-hedgerow alley farming is a solution suitable to a more intensive stage of permanent cultivation. In shifting systems the management of fallow vegetation is a more relevant issue.

A fallow improvement crop must yield higher nutrient levels and accumulate more organic matter than the natural fallow it replaces. Fallowed fields are usually burned, or subjected to intensive grazing. Farmers acknowledge that these practices are ineffective in regenerating fertility, and this has been corroborated by sampling the nutrient status of grass fallow fields in the Philippines (S. Fujisaka, 1990, unpublished data).

Leguminous cover crops are candidates for enriched fallows, but empirical evidence of their practical utility is sparse. Participatory research with farmers in Leyte, Philippines, found that tropical kudzu (*Pueraria phaseoloides*) was successfully established by broadcasting seed in *Imperata* fallows, and it suppressed the *Imperata* in less than 1 yr. But the practice was only successful when fire can be excluded. Grass fires are exceedingly common in the dry season, and the individual farmer has great difficulty excluding fire from his farm. Communal grazing practices also will be a major constraint to the adoption of managed fallows in many areas.

Systems of fallow enhancement have been demonstrated with leguminous trees. MacDicken (1990) described a planned fallow that evolved indigenously on steep slopes in Cebu. Dense stands of naturally reseeded *L. leucocephala* form the fallow portion of the cycle. When the *Leucaena* is cut, the stems are placed on the contour and staked to create contour bunds. A fallow of 3 to 7 yr is followed by several years of cereal cropping. Stewart (1990) has studied the development of *Leucaena* fallows for use in charcoal production. He found the economic sustainability of the practice to be quite favorable as an added enterprise in the shifting cultivation systems of the Ati Tribal Reservation in Iloilo, Philippines. The substantial cash income generation would reduce the food crop area required per family. It also would enhance the utility of establishing tree fallows to regenerate soil fertility more rapidly.

Managed tree fallows may be practical for farming systems without animal power. However, they are not as suitable when draft power is available. Managed tree fallows have been given almost no research attention. Fire exclusion will be a dominant concern in their successful implementation.

SMALL-SCALE FARM DIVERSIFICATION

The most plausible model of sustainable independent small-holder farming in the uplands is one of diversification into mixed farming systems. Given the exceptionally high production and marketing risks in the uplands, and generally low marginal returns, the farm family must engage in a number of enterprises to provide nutritional and income security (Chambers, 1986). These enterprises seek to take maximum advantage of the complementarities among income-generating activities (e.g., leguminous trees for fodder, green leaf manure, and fuelwood; cattle for draft, cash income, and manure.)

The upland farm family must place primary emphasis on subsistence food crop production. But the land-use systems that result from pursuing these needs are the least sustainable alternatives. The issue is how to enable the farm enterprise to move profitably along a trajectory that will continually increase the area of perennial plants, and decrease the area devoted to annuals. The gradual expansion of home gardens, ruminant livestock production, and plantation and timber tree crops, will contribute to this end. Figure 5–4 presents a generalized model of the evolutionary development of a small-

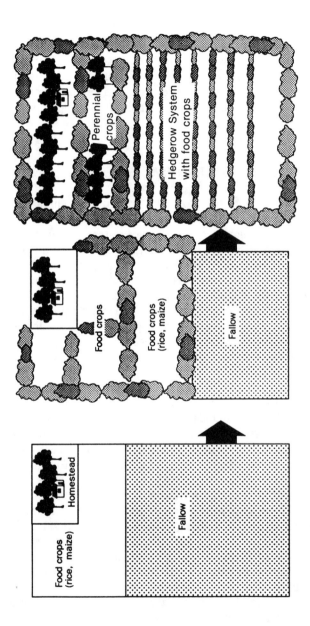

Fig. 5–4. Model of the evolutionary development of a small-scale upland farm on sloping land.

scale upland farm that illustrates this trajectory. The nature of the fallow vegetation, the food crop system, and the evolving enterprise mix will be highly specific to the particular agroecosystem. The model, therefore, emphasizes the tendency toward greater on-farm diversity, toward conservation farming systems to enable sustained food cropping, and toward an ever-enlarging area under commercial perennial vegetation.

At present there is very little predictive understanding of these evolutionary pathways. This has resulted in much confusion and wasted effort directed to solutions inappropriate to farmers' circumstances. Much more detailed agroecosystem-specific models are needed than the one given here (Fig. 5-4). They will provide research and development institutions a clearer predictive understanding of the directions that upland farming systems will follow in the future.

PERENNIAL CROPS: AN UNCERTAIN FUTURE

Proceeding north from the equator the climate generally becomes progressively harsher, with longer, and more severe dry seasons, cooler winters, and a greater threat of severe tropical storms. In the Equatorial regions there is an array of plantation crop alternatives. Above about 10° N lat, coconut is the only perennial grown on a large scale. Above about 15° there are few major cash perennials produced on sloping lands. This produces a south-to-north ecological gradient of declining perennial crop alternatives that progressively limits the available protective land-use options. Northward along this gradient also are ecological conditions that are associated with slower establishment of vegetation, lower primary productivity, and a greater tendency for upland soils to lack ground cover for substantial portions of the year.

Since perennial crops have generally proven to be an ecologically sustainable land use on tropical sloping uplands, while food crops have not, the prescription of inducing more perennial crop production by small-holders is easy to make. But this prescription is difficult to successfully implement. The perennial crop choices in any given environment are limited, prices are often highly volatile, and the long–term market outlook for most tropical commodities is generally weak.

A major exception has been in Peninsular Malaysia, where the small–holder rubber and oil palm sector has thrived for decades, mostly independent of government support (Vincent & Hadi, 1991). Approximately 3 million hectares are now planted there to these two perennial crops, while less than 0.1% of the land area is in shifting cultivation. Forty-six percent of the oil palm area was in small holdings in 1988, with a similar percentage for rubber.

What were the factors that induced a uniquely successful small-holder plantation sector in Malaysia? Several factors appear to have contributed. There was an excellent fit between the ecological requirements of these two crops and the climatic environments and marketing infrastructure developed due to the presence of a large industrial plantation sector. And there was long-term political stability, rapid industrial development, and a low rural population density.

Many of these advantages are lacking in the uplands of the other countries in the region. Coconut (*Cocos nucifera* L.) is a dominant plantation crop in the Philippines and Indonesia. They occupy much of the steepest nonarable land at lower elevations in these countries. The livelihood of millions of the poorest families, and the economic future of many parts of the uplands, is heavily dependent on the coconut industry. But there is a serious threat of economic collapse with the decline in international demand for coconut oil, due to preference for substitute vegetable oils with a lower saturated fat content.

The tropical perennials have received very little international research attention, but if agricultural sustainability is dependent on a shift toward perennials, a stronger case should be made for greater emphasis on them. For example, the demise of coconut farming may result in both an economic and ecological catastrophe, as more families are dispossessed, and more steep land is put in food crop farming. There must be a more determined effort to find substitute uses for coconut oil.

The need for more funding for product development research that will bolster the competitiveness of traditional and nontraditional tropical crops is evident. Current trends point toward greater substitution for tropical products through genetic engineering, and other technologies. This may tend to further undermine the markets for tropical crop commodities in the future. There also is potential for the development of novel tree crops that open up new opportunities for the tropical environment, or that have a comparative advantage in providing unique future products that the industrialized world will value. But greater private and public sector support for the development of perennial crop enterprises will be essential. This effort must be linked with the improvement of food crop production methods, in order to release land and labor to pursue other cash-generating activities.

ELEMENTS OF A STRATEGY

This paper has reviewed some of the technology issues that are central to the sustainability of small-scale farming in the sloping uplands. However, a comprehensive strategy for evolving sustainable land-use systems will involve three main elements: (i) tenure, (ii) delivery, and (iii) technology (Fig. 5-5). Tenure encompasses populations and their relationship to the land. Delivery involves the mechanisms which government institutions and the private sector employ to deliver policy and varied infrastructural support to facilitate and guide the process of change. Technology covers the technical solutions, and the institutional capabilities to develop them. Technical solutions are unlikely to be suitable unless they are researched and implemented within the context of the complex interconnections with these other dominant elements.

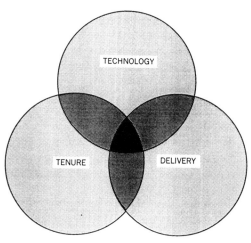

Fig. 5–5. Development of sustainable land-use systems in the sloping uplands will depend on integrated solutions in three key areas.

Future success in bringing sustainable land use to the uplands will be fundamentally dependent on major changes in the way public lands are managed. Most governments in the region have been incapable of managing the public domain entrusted to them. To harness the energies of the upland populations in creating sustainable land-use systems, and ensure success in reforestation and forest remnant conservation, governments must establish a new political relationship with the upland populations. They must proceed to recognize the boundaries of lands held by the indigenous occupants, and move to recognize their full ownership rights. The dominant issue is empowerment of the upland people: Giving them a fair stake in the land.

Any strategy to address the sustainable management of upland resources also must include a strong program to reduce the rate of population growth. The poorest households in the rural uplands have the highest birth and mortality rates. Governments must redirect health programs to ensure greater investments in village-level health work and paramedical personnel, with family planning support and education an integral part of this effort. This will necessitate strong and sustained international support. Demographic targets, and effective organization to meet them, must be highlighted as fundamental components of such support.

THE RESEARCH MODEL

Agricultural research basically follows two strategies (van Otten & Knobloch, 1990)—the commodity strategy, which selects specific commodities and attempts to expand their production; and the resource-base oriented strategy, which aims to find a production system that can best use the available resource base. The resource-base oriented strategy tends to aim at dealing with marginal land situations, because the more difficult the resource base, the greater the need for a resource-base approach in research (Plucknett et al., 1987).

Upland agricultural systems contrast starkly with the less heterogeneous lowland rice systems, that have historically received overwhelming attention. Interrelatedness and complexity are fundamental conditions of the land-use problems of the sloping uplands. Quick-fix, single-factor solutions have proven clearly inadequate. The government agriculture departments in the region have only recently begun to give significant attention to the task of understanding upland agricultural technologies. Reorientation of both the research and extension approach is underway, involving more decentralization to local levels. The farming systems research and development model is being adopted for upland technology generation, with strong emphasis on farmer participatory research.

Chambers (1986) argued that normal professionals have neglected the gaps and linkages between the central concerns of their different disciplines. This is best shown diagramatically in Fig. 5–6. The professions, and government departments, tend to preserve their specializations, focus narrowly, and overlook the linkages which are often important for resource-poor farmers. Agroforestry is an example of a more integrative approach to exploiting critical linkages, where both agronomists and foresters are concerned with the interactions between crops and trees. Much of the research success in the future will be related to exploiting these linkages.

It is at the interface between forestry and farming that many of the future research and development challenges will be encountered. For the upland farming populations to become effective partners with government in conserving, managing, and replanting forests, while meeting their basic subsistence food production needs, a deep understanding of the constraints, and workable solutions is needed. Only teamwork will generate it.

Farming-systems research evolved as a framework for a more comprehensive, multidisciplinary attack on the complex constraints in agroecosystems (Harrington et al., 1989). Ecosystem-based research should aim to target a

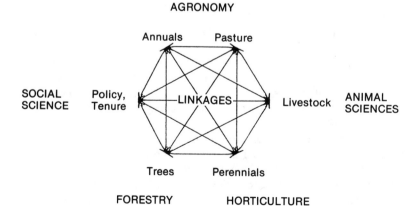

Fig. 5–6. Linkages among enterprises on small-scale upland farms that require interdisciplinary focus.

broader continuum that includes forest management and agriculture. Such work needs a methodology that provides foresters and agriculturists a common framework within which to interact.

Hart and Sands (1990) have proposed a strategy that may provide a starting point—sustainable land-use systems research (SLUSR). It focuses on the land-use system, targeting the land management unit, within the context of its biophysical and socioeconomic environments (Fig. 5–7). The research methodology of SLUSR is based on the farming systems methodology. It directs strong emphasis to the ecosystem as the starting point of problem analysis and design.

Research relevant to the sloping uplands will involve more emphasis on the parallel streams of on-farm research (Fig. 5–8): Conventional researcher-managed work will continue to be essential, but much more emphasis will be given to farmer-participatory research, including farmer-initiated experimentation. Research directed to the sloping uplands will be carried out largely at on-farm laboratories at representative sites, rather than at experiment stations. These sites will provide an explicit focal point for the disciplines to concentrate their respective skills. The Southeast Asian region needs a more complete network of key upland research sites where teams can focus their efforts in a critical mass. These key sites need sustained support, with a budgeting structure that keeps team members working together.

Although new technology will be a central focus of research, in the current climate of policy change two other areas need much greater field research emphasis: (i) land tenure issues, particularly the effects of different tenurial instruments on small farm viability and sustainable land management; and (ii) community organization and implementation of participatory management, emphasizing process-oriented research on effective grass-roots organization.

CONCLUSIONS

The complex issues of sustainable land use in the upland ecosystem are common to all the countries in Southeast Asia. The problems transcend national boundaries, and will require stronger international mechanisms to provide efficient research and development support to the respective nations. There are a number of institutions and networks that are concerned with components of upland resource management, and a regional assessment of the current work relevant to sustainable land-use systems has been made (Garrity & Sajise, 1991).

The key to eventual success in the uplands is in understanding the interrelatedness of the problems across sectors, and in developing the capacity to strengthen the efforts of each country to conceptualize, plan, and implement interventions appropriate to the unique realities of the ecosystem. This will require a new sloping upland ecosystem-based approach in international research. Such a focus does not fall wholly within the mandate of any of the International Agricultural Research Centers. In fact, a credible model

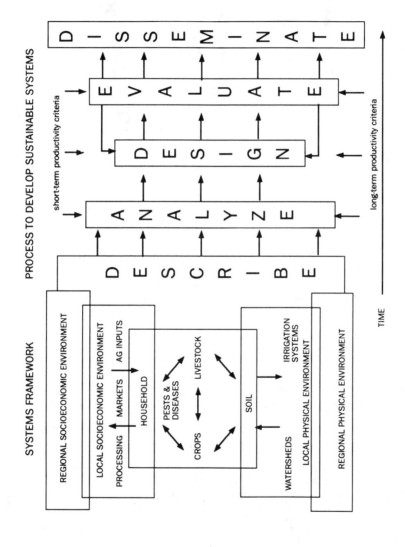

Fig. 5-7. A research and development process that could be used by a multidisciplinary team as a guide in the development of sustainable land-use systems (Hart & Sands, 1990).

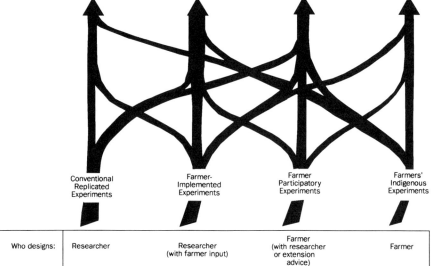

Who designs:	Researcher	Researcher (with farmer input)	Farmer (with researcher or extension advice)	Farmer
Who manages:	Researcher	Researcher (with farmer)	Farmer	Farmer
Who implements:	Researcher	Farmer	Farmer	Farmer
Examples:	Fertilizer trials Variety trials	Cropping pattern tests	Hedgerow farming systems	New crops or cultivars

Fig. 5–8. Four major streams of on-farm experimentation, with interactive information flow over time.

For this approach does not yet exist in the developing world, and cannot be recognized in the institutional settings of the developed countries. Such a model needs to be conceptualized and developed on an international basis, and adapted to the diverse ecological and institutional settings encountered in the Southeast Asian nations.

REFERENCES

Bartlett, H.H. 1956. Fire, primitive agriculture, and grazing in the tropics. p. 692–720. *In* W.L. Thomas (ed.) Man's role in changing the face of the earth. Univ. of Chicago Press.

Basri, I., A. Mercado, and D.P. Garrity. 1990. Upland rice cultivation using leguminous tree hedgerows on strongly acid soils. p. 140. *In* Agronomy abstracts. ASA, Madison, WI.

Briones, A.M., and P.R. Vicente. 1985. Fertilizer usage of indigenous phosphate deposits: I. Application of apatitic phosphate rock for corn and upland rice in a hydric dystrandept. Philipp. Agric. 68:1–17.

Carson, B. 1989. Soil conservation strategies for upland areas of indonesia. Occasional Pap. no. 9. Environ. Policy Inst., East–West Center, Honolulu, HI.

Chambers, R. 1986. Normal professionalism, new paradigms, and development. Disc. Pap. 227. Inst. Dev. Stud. Univ. Sussex, Brighton, England.

Cruz, M.C., and I. Zosa-Feranil. 1988. Policy implications of population pressure in Philippine Uplands. Pap. World Bank/CIDA Study on Forestry, Fisheries and Agric. Resource Management, Washington, DC.

Evensen, C.L.I. 1989. Alley cropping and green manuring for upland crop production in West Sumatra. Ph.D. Diss. Univ. Hawaii, Honolulu.

Fernandes, E.C.M. 1990. Alley cropping on acid soils. Ph.D. Diss. North Carolina State Univ., Raleigh.

Fujisaka, S. 1991. Has "Green Revolution" rice research paid attention to farmers' technologies? Social Sci. Div., IRRI, Los Baños, Laguna, Philippines.

Fujisaka, S., and D.P. Garrity. 1988. Developing sustainable food crop farming systems for the sloping acid uplands: A farmer-participatory approach. p. 182–193. *In* Proc. SUAN IV Regional Symp. Agroecosystem Res. Khon Kaen, Thailand. Khon Kaen Univ., Thailand.

Garrote, B.P., A. Mercado, and D.P. Garrity. 1986. Soil fertility management in acid upland environments. Philipp. J. Crop Sci. 11:113–123.

Garrity, D. 1989. Hedgerow systems for sustainable food crop production on sloping lands. Contour 2:18–20.

Garrity, D.P., and Percy E. Sajise. 1991. Sustainable land use systems in Southeast Asia: A regional assessment. *In* R.D. Hart and M.W. Sands (ed.) Sustainable land use systems research and development. Rodale Press, Emmaus, PA. (In press.)

Garrity, D.P., C.P. Mamaril, and G. Soepardo. 1989. Phosphrus requirements and management in upland rice-based cropping systems. *In* Phosphorus Requirements and Management in Asia and Oceania. IRRI, Los Baños, Laguna, Philippines.

Garrity, D.P., D.H. Kummer, and E.S. Geriang. 1993. Country profile: Philippines. *In* Sustainable agriculture and the environment in the Humid Tropics. Natl. Acad. Sci., Washington, DC.

Granert, W.G., and N. Sabueto. 1987. Farmers' involvement and use of simple methods: Agroforestry strategies for watershed protection. *In* N.T. Vergara and N. Briones (ed.) Agroforestry in the humid Tropics. East–West Center/SEARCA, Laguna, Philippines.

Guevara, A.B. 1976. Management of *Leucaena leucocephala* (Lam.) de Wet for maximum yield and nitrogen contribution to intercropped corn. Ph.D. diss. Univ. Hawaii, Honolulu.

Harrington, L.W., M.D. Read, D.P. Garrity, J. Woolley, and R. Tripp. 1989. Approaches to on-farm client-oriented research: Similarities, differences, and future directions. p. 37–53. *In* Developments in Procedures for Farming Systems Research. Proc. Int. Workshop, Puncak, Bogor, Indonesia. 13–17 March.

Hart, R.D., and M.W. Sands. 1991. Sustainable land use systems research. *In* R.D. Hart and M.W. Sands (ed.) Sustainable land use systems research and development. Rodale Press, Emmaus, PA.

International Rice Research Institute. 1987. Annual report for 1986. IRRI, Los Baños, Philippines.

Jodha, N.S. 1990. Mountain agriculture: The search for sustainability. J. Farming Systems Res.-Ext. 1:55–75.

Kang, B.T., L. Reynolds, and A.N. Atta-Krah. 1990. Alley farming. Adv. Agron. 43:315–359.

Lal, R. 1990. Soil erosion in the tropics. McGraw-Hill, New York.

Ly, T. 1990. FARMI newsletter. Farm and Resour. Manage. Inst., Visayas State College of Agric. Baybay, Leyte, Philippines.

MacDicken, K.G. 1990. Agroforestry management in the humid tropics. p. 99–149. *In* K.G. MacDicken and N.T. Vergara. Agroforestry: Classification and management. John Wiley & Sons, New York.

McIntosh, J.L., I.G. Ismail, S. Effendi, and M. Sudjadi. Cropping systems to preserve fertility of red-yellow podzolic soils in Indonesia. p. 409–429. *In* Int. Symp. on Distribution, Characterization, and Utilization of Problem Soils. Trop. Agric. Res. Center, Tsukuba, Japan.

Mercado, A.R., A.M. Tumacas, and D.P. Garrity. 1989. The establishment and performance of tree legume hedgerows in farmer's fields in a sloping acid upland environment. Pap. 5th Annual Scientific Meet. of the Federation of Crop Sci. Soc. of the Philippines, Iloilo City. 26–29 April.

Milliman, J.D., and R.H. Meade. 1983. World-wide delivery of river sediment to the oceans. J. Geol. 91:1–21.

National Research Council. 1977. Luecaena: Promising forage and tree crop for the Tropics. NAS, Washington, DC.

Odum, E.P. 1989. Input management of production systems. Science (Washington, DC) 243:177–182.

O'Sullivan, T.E. 1985. Farming systems and soil management: The Philippines/Australian development assistance program experience. p. 77–81. *In* E.T. Craswell et al. (ed.) Soil erosion management. ACIAR Proc. Ser. 6. ACIAR Canberra, Australia.

Plucknett, D.L., J.L. Dillon, and G.J. Ballaeys. 1987. Review of the concepts of farming systems research: The what, why and how. *In* Proc. Workshop on Farming Systems Res./Int. Agric. Res. Centers, Pantancheru, India: ICRISAT,

Raintree, J.B., and K. Warner. 1986. Agroforestry pathways for the intensification of shifting cultivation. Agrofor. Systems 4:39–54.

Rhoades, R.E. 1990. Thinking globally, acting locally: Science and technology for sustainable mountain agriculture. Int. Potato Center, Manila, Philippines.

Santoso, D., M. Sudjadi, and C.P. Mamaril. 1988. Integrated nutrient management for sustainable rice farming in upland rice environments. Pap. Int. Rice Res. Conf. 7-11 November. IRRI, Los Baños, Laguna, Philippines.

Sanchez, P.A., and J. Benitez. 1987. Low input cropping for acid soils of the humid tropics. Science, Washington (DC) 238:1521-1527.

Smyle, J., W. Magrath, and R.G. Grimshaw. 1990. Vetiver Grass—A Hedge against Erosion. p. 51. *In* Agronomy abstracts. ASA, Madison, WI.

Stewart, T.P. 1990. Economic analysis of land-use options to encourage forest conservation on the Ati Tribal Reservation in Nagpana, Iloilo Province, Philippines. M.Sc. thesis. North Carolina State Univ., Raleigh.

Vandermeer, C. 1963. Corn cultivation on Cebu: An example of an advanced stage of migratory farming. J. Trop. Geogr. 17:172-177.

van Otten, M., and Ch. Knobloch. 1990. Composite watershed management: A land and water use system for sustaining agriculture on Alfisols in the semiarid tropics. J. Farming Syst. Res. Ext. 1—37-54.

Vincent, J.R., and Y. Hadi. 1991. NRC country study: Peninsular Malaysia. *In* Proc. Natl. Res. Council Study of Agric. Sustainability and the Environment in the Humid Tropics. NRC, Washington, DC.

Watson, H.R., and W.A. Laquihon. 1987. Sloping agricultural land technology: An agroforestry model for soil conservation. In N.T. Vergara and N Briones (ed.) Agroforestry in the humid Tropics. East–West Center/SEARCA, College, Laguna, Philippines.

Williams, R.D., and E.D. Lavey. 1986. Selected buffer strip references. Water Quality and Watershed Res. Lab., Durant, OK.

6

Alley Farming as a Potential Agricultural Production System for the Humid and sub-Humid Tropics

A. N. Atta-Krah and B. T. Kang

International Institute of Tropical Agriculture
Ibadan, Nigeria

It is widely recognized that the biggest challenge facing agricultural research in the Tropics is the development of farming systems capable of ensuring increased and sustained productivity with minimum degradation of the soil resource base. Reversing the trend of declining per capita food produciton in sub-Saharan Africa, therefore, does not depend solely on the development of improved and high-yielding crop varieties. Development of sustainable production systems is necessary to foster and maintain advantages derived from such improved varieties. The issue of sustainability has received considerable attention in recent years in agricultural research and development (CGIAR, 1990).

The nature of much of the upland soils in the Tropics is such that overexposure and overcultivation can easily lead to their degradation. According to the Soil Management Support Services (1986), agricultural land in most humid and subhumid tropical regions is dominated by low activity clay (LAC) soils whose inherent characteristics and limitations make large upland areas dominated by these soils less suitable for conventional mechanized and high-chemical input farming methods. These soils have inherently low fertility, and are highly erodible when left unprotected. Nutrient loss through run-off and erosion can, however, be kept to a minimum by using appropriate production systems that minimize disturbance and ensure a protective cover of the surface soil through live or dead mulches (Lal, 1986). Management of vegetation within the cropping system also could contribute to maintenance of soil fertility and enhance the stability of the cropping system (Juo & Lal, 1977; Lal et al., 1978).

In the Tropics, trees have long been recognized as essential both for the stability of the environment and for maintenance of soil fertility for crop production. Trees have been recognized as major elements in soil fertility regeneration and conservation as reflected by their prominence in traditional farming systems (Okigbo, 1983). Shifting cultivation, the dominant farming system in the Tropics, relies mainly on fallow regrowth of trees and other vegetation for fertility regeneration (Nye & Greenland, 1960). This system remains viable only when alternating periods of cropping and fallowing maintain a balance between nutrient loss (during cropping) and nutrient gain (during fallowing).

The efficiency of shifting cultivation depends on two related factors: (i) the duration of the fallow period, and (ii) the nature and density of the fallow vegetation. Rapid increases in human population and the associated increases in demand for farmland and wood products have overstretched this traditional system. Long fallow periods, which in the past lasted 10 to 25 yr, have been shortened drastically or disappeared in areas. This has resulted in increasing degradation of farmland, increasing infestation of problem weeds, and declining yields and production of food crops. Fertilizer use has not been a viable option in much of the Tropics, because of its high cost and unavailability to most smallholder farmers, especially those in sub-Saharan Africa. Even when such inorganic fertilizers are available, continued use of high rates of N fertilizer may lead to soil acidity problems.

There is therefore a need for new directions in tropical agricultural research. Systems are needed that incorporate the biological stability and nutrient balance characteristic of the traditional shifting cultivation system, while allowing intensification of production over a long-term period of time.

The subsequent sections of this paper will describe the alley farming system and provide an overview of research results on alley farming as a potential farming system for the humid and subhumid Tropics.

AGROFORESTRY AND ALLEY FARMING

Agroforestry is an integration of a tree component into an agricultural production and land–use system (Steppler, 1982). Such systems have been widely acclaimed as a solution to tree depletion, soil degradation, and declining yields under shifting cultivation. As defined by King and Chandler (1978), agroforestry is a sustainable land management system which increases the overall yield of the land. It combines, simultaneously or sequentially, the production of trees and the production of crops and/or animals on the same unit of land, and it applies management practices that are compatible with the cultural practices of the local population. Lundgren and Nair (1985) add that in agroforestry there are both ecological and economic interactions between the different components, while Nair and Fernandes (1985) make a distinction between production-oriented and protection-oriented agroforestry.

One agroforestry system that has received a good deal of research attention and has shown great promise for sustainability is alley cropping (Kang and Reynolds, 1989; Kang et al., 1981, 1985, 1990). In alley cropping, trees and shrubs (preferably fast-growing N_2-fixing leguminous species) are established in hedgerows spaced 4 to 6 m apart on arable crop land with food crops cultivated in the alleys between the hedgerows (Wilson & Kang, 1981).

Hedgerows are oriented to minimize shading within the alleys (i.e., in an east–west direction) and also to minimize erosion (i.e., across slopes rather than along slopes). The trees are periodically pruned and managed as hedgerows during the cropping phase to prevent shading of the companion crops. The prunings of foliage and young stems are incorporated into the soil as green manure or used as mulch. Some portion of the tree foliage can be harvested and fed to livestock, particularly small ruminants. The integration and linkage of livestock production into the system is referred to as "alley farming" (ILCA, 1983; Sumberg & Okali, 1983).

POTENTIAL OF ALLEY FARMING

Evaluation of the alley cropping/alley farming concept has been a major research effort at the International Institute of Tropical Agriculture (IITA) and at the International Livestock Centre for Africa (ILCA) Humid Zone Programme, both in Nigeria, and also at the International Council for Research in Agroforestry (ICRAF), in Nairobi, Kenya.

Tree and shrub species that have been most commonly studied in the system include *Acacia auriculiformis, Cajanus cajan, Calliandra calothyrsus, Erythrina peoppigiana, Flemingia macrophylla, Inga edulis, Gliricidia sepium, Leucaena leucocephala* and *Senna siamea* (syn. *Cassia siamea*). Some indigenous tree species such as *Alchornea cordifolia*, and *Dactyldenia barteri* (syn. *Acioa barteri*) also have been studied in alley cropping trials. Results of investigations carried out on Alfisols and associated soils in southwestern Nigeria have demonstrated that for the low-altitude humid and subhumid Tropics, *L. leucocephala* and *G. sepium* are the best-performing hedgerow species for alley farming (Kang et al., 1981, 1984; Reynolds & Atta-Krah, 1989). *Acacia auriculiformis* has been observed to perform well on a highly eroded and low fertility Alfisol (Terre de bare degrade) in southern Benin Republic (M. Versteeg, 1990, personal communication).

Results of observations at the IITA high rainfall station at Onne have shown that *D. barteri* and *F. macrophylla* (Kang et al., 1989) have good potential for alley cropping on a Typic Paleudult. Evensen and Yost (1990) reported that *Paraserianthes falcataria* (syn. *Albizia falcataria*) is an effective hedgerow species for alley cropping on a Tropeptic haplorthox in South Sumatra.

For the highlands in Rwanda, it was reported that *S. spectabilis* (syn. *C. spectabilis*) is performing well in an alley cropping system (IITA, 1986, p. 34–35). Kass et al. (1988) showed that *E. poeppigiana* is performing well for alley cropping with *Phaseolus* beans on a Typic humitropept under high rainfall conditions.

Effect on Soil Properties

In alley farming, the periodic addition of hedgerow prunings to the soil helps to maintain soil fertility for crop production. Kang et al. (1984) showed that with continuous addition of *L. leucocephala* prunings, higher soil or-

Table 6-1. Effect of 6 yr of alley cropping maize and cowpea with *L. leucocephala* and N application on some chemical properties of surface soil (0–15 cm) of a Psammentic Ustorthent (adapted from Kang et al., 1985).

| Treatment | *Leucaena* prunings | pH–H_2O | Organic C | \multicolumn{3}{c}{Exchangeable cations} |||
				K	Ca	Mg
kg N ha^{-1}			g kg^{-1}	—	cmol kg^{-1}	—
0	Removed	6.0	6.5	0.19	2.90	0.35
0	Retained	6.0	10.7	0.28	3.45	0.50
80	Retained	5.8	11.9	0.26	2.80	0.45
LSD (0.05)		0.2	1.4	0.05	0.55	0.11

ganic matter and nutrient status are maintained than on plots receiving no prunings (Table 6-1). Similar results also were reported by Atta-Krah (1990), who showed that the organic matter content in soils under alley cropping and alley grazing (in which the alley cropping plot is left fallow for 2 yr and used as grazing paddock by small ruminats) was higher than in control plots without trees. The higher soil organic matter status under alley farming also enhances soil biotic activities (Kang et al., 1990).

Mulching is known to have favorable effects on physical properties of soil (Lal, 1974; Kang et al., 1985). In a study to determine the long-term effects of alley cropping on the physical properties of surface soil of an Oxic Paleustalf in southwestern Nigeria, Hulugalle and Kang (1990) observed that plots alley cropped with *L. leucocephala, G. sepium, A. cordifolia* and *D. barteri* had lower soil bulk density, lower diurnal temperature fluctuation, and higher water infiltration than in the control (no tree) treatment. Among the hedgerow species, soil compaction was lower with *L. leucocephala* and *D. barteri*. Alley cropping on sloping land, with hedgerows running across slopes, also is effective in controlling water run-off and soil erosion (Kang & Ghuman, 1988; Paningbatan, 1990).

Effect on Crop Production

The alley cropping system has been tested, with encouraging results, using a variety of crops, including cereal crops [maize (*Zea mays* L.), upland rice (*Oryza sativa* L.)], grain legumes [*Vigna unguicalata* (L.)], soybean [*Glycine max* (L.) Merr.]), root and tuber crops [(*Manihot esculanta* Crantz), yam (*Dioscores rotundata* Poir)), and plantain (*Musa paradiscra*) and vegetable crops, under both monocropping and intercropping systems (Kang et al., 1984, 1989; Atta-Krah et al., 1986; Chen et al., 1989; Leihner et al., 1988). In an alley farming trial on an eroded Alfisol (Oxic paleustalf) at Ibadan, southwestern Nigeria, maize yields under alley cropping with various hedgerow species were significantly higher, with or without applied N, than in the control plots (no trees) (Fig. 6-1). This trial also showed that, in addition to N benefit, the generally improved soil conditions resulting from alley cropping also had a positive effect on maize yield.

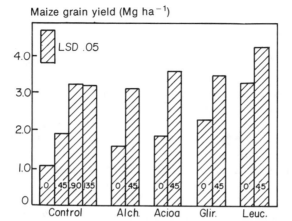

Fig. 6-1. Grain yield of maize on eroded Alfisol (Oxic Paleustalf) at Ibadan, southwestern Nigeria, as affected by alley cropping with woody species (*A. barteri*, *A. cordifolia*, *G. sepium*
and *L. leucocephala*) and N rates (kg N/ha) (Kang and Wilson, 1987).

Results of another long-term alley farming trial also conducted at Ibadan
have shown that with addition of *L. leucocephala* prunings, even without
N application, maize yields can be maintained at the reasonable level of ap-

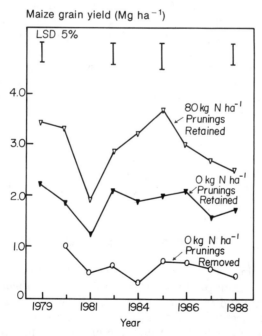

Fig. 6-2. Grain yield of first season maize in maize-cowpea sequential cropping on a Psammentic Ustorthent in alley cropping with *L. leucocephala* at Ibadan, southwestern Nigeria,
as affected by N application and prunings of hedgerows (N rates, 0 and 80 kg N ha^{-1}) (B.T.
Kang, 1987, unpublished data).

proximately 2 t ha^{-1} (Fig. 6–2). Higher yields were obtained when the prunings were supplemented with fertilizer N. The technique thus provides more flexibility in developing low- chemical input production systems. Removal of the prunings from the plot resulted in a reduction of the benefits accruing to the arable crop. Investigations by Atta-Krah and Sumberg (1987) with *G. sepium* hedgerows showed that even partial removal of the prunings lowers crop yields, though application of some inorganic N can compensate for the loss.

Choice of suitable hedgerow species is important for the success of the system. Using *I. edulis* on an Ultisol in Yurimaguas, Peru, Fernandes et al. (1990) reported significant reduction in crop yield, particularly in areas adjacent to hedgerows, due to root competition from the hedgerows.

The sustainability of the alley farming system can be enhanced through the integration of short fallows (which may be grazed by small ruminants) into the alley cropping cycle. Such fallows show positive effects on survival, growth, vigor, and productivity of hedgerows, and on their soil fertility maintenance and regeneration ability. Results of a long-term alley farming with a rotation system of alley farming and short (2-yr) grazing fallow, showed a consistent positive effect on soil organic matter and crop yields in previously grazed alley cropping plots as compared to the control plot (Atta-Krah, 1990). Maize grown in alley cropping plots under continuous cultivation also consistently out-yielded the control plot (no trees). In the first season of alley cropping following the 2-yr grazed fallow, these plots yielded about 35 and 55% more maize than plots under continuous alley cropping and the control plot respectively (Table 6–2). The yield superiority of the previously grazed alley plots over the continuously alley cropped plots was retained, but at decreasing levels, in 3 yr of postfallow cultivation. By the 3rd yr the crop yield advantage had dropped from the 55% level to 22% (Atta-Krah, 1990). Cowpea grain yield was lowered due to excess addition of N from *L. leucocephala* prunings (Table 6–2), this promoted higher plant biomass yield and resulted in lower seed yield.

Short fallow periods within alley farming also allow the satisfaction of the farmers' requirements for stakes, wood, and fodder. They also are effective in combating noxious and problem weeds such as *Imperata cylindrica* and *Chromolaena odoratum* (Aken 'Ova & Atta-Krah, 1986; Yamoah et al., 1986).

A Viable Alternative to Shifting Cultivation

Alley farming is thus a system with the potential for providing improved sustainability in crop productivity in the Tropics. Through alley farming, soil productivity and crop production could be maintained at higher levels with no requirement for large doses of inorganic fertilizer or long fallow periods. Short fallows of 1- to 2-yr duration in alley farming can enhance soil fertility regeneration (Atta-Krah, 1990) and allow production of wood for fuel and stakes, as well as providing fodder for livestock. The benefit to livestock from feeding portions of hedgerow prunings to small ruminants

Table 6-2. Yield of maize and cowpea (t ha^{-1}) on Psammentic Ustorthent under continuous cropping, continuous alley cropping, and alley cropping after grazed fallow, 1983 to 1986 (Atta-Krah, 1990).

Season/crop	Continuous cropping	Continuous alley cropping	Alley cropping after grazing	LSD (0.05)
		t ha^{-1}		
Period I (1983–1984)†				
Maize				
1983 main season	2.6	2.5		N.S
1984 main season	2.6	3.6		0.8
–minor season	1.2	1.4		N.S
Cowpea				
1984 minor season	0.6	0.4		0.2
Period II (1985–1986)				
Maize				
1985 main season	2.5	2.8	3.9	0.4
–minor season	0.9	1.7	2.1	0.4
1986 main season	1.3	2.0	2.6	0.5
–minor season	0.7	1.3	1.5	0.6
Cowpea				
1985 minor season	0.5	0.3	0.3	0.1

† During Period I (1983–1984), the alley grazing/cropping rotation treatment was under grazed fallow. Hence there was no cropping of these plots during this period. During Period II (1985–1986), the grazed fallow plots came into cropping.

has been shown to be positive (Sumberg, 1985; Atta-Krah et al., 1986; Reynolds & Adeoye, 1986; ILCA, 1987).

The socioeconomic and institutional aspects of transferring alley farming to farmers have received some research attention (Hoekstra, 1982; Francis, 1989; Francis & Atta-Krah, 1989), as have its economic potentials and implications (Verinumbe et al., 1984; Ngambeki, 1985; Sumberg et al., 1987). However, more research attention is needed in these areas to facilitate the transfer of the technology to farmers. Raintree (1983) mentioned the need for the development of strategies that will enhance adoptability of agroforestry innovations. Atta-Krah and Francis (1987) stress that a community-based, developmental on-farm research approach should be adopted for the assessment of the relevance, workability, and social acceptability of the system. This process requires the combined inputs of research and extension techniques, as well as a community focus. Atta-Krah and Francis (1987) reported the use in southwestern Nigeria. More development-oriented research activities are required to link the research process with farmer adoption and eventual extension.

CONCLUSIONS

Research conducted on the alley cropping/alley farming system in the humid and subhumid tropics has shown its potential as an improved production system for smallholder farmers in the humid and semihumid Tropics.

There is a recognized need to promote more research on the system in national research institutions, and also to increase the on-farm experimentation component. This will ensure that the technology is fully tested by client farmers and that any necessary adjustments in its management are implemented to make the system more practicable and acceptable to farmers.

To further explore the potential of alley farming, the Alley Farming Network for Tropical Africa (AFNETA) was established in 1988. Alley Farming Network for Tropical Africa's major objective is to promote and coordinate alley farming research in national institutions in the region, with a research mandate covering both on-station and on-farm investigations in the various agroecological zones. Alley Farming Network for Tropical Africa's research programs cover four main issues: multipurpose tree (MPT) screening and evaluation, alley farming management, livestock integration, and socioeconomic assessment. It is envisaged that the research activities of the network will help to identify the real potential, the adaptability and adoptability and the limitations of the alley farming concept in the various agroecological regions of tropical Africa.

REFERENCES

Aken'Ova, M.E., and A.N. Atta-Krah. 1986. Control of spear grass (*Imperata cylindrica* (L.) Beauv.) in an alley cropping fallow. Nitrogen Fixing Tree Res. Rep. 4:27–28.

Atta-Krah, A.N. 1990. Alley farming with leucaena: Effect of short grazed fallows on soil fertility and crop yields. Exp. Agric. 26:1–10.

Atta-Krah, A.N., and P. Francis. 1987. The role of on-farm trials in the evaluation of composite technologies: The case of alley farming in southern Nigeria. Agric. Systems 23:133–152.

Atta-Krah, A.N., and J.E. Sumberg. 1987. Studies with *Gliricidia sepium* for crop/livestock production systems in West Africa. Agrofor. Systems 6:97–118.

Atta-Krah, A.N., J.E. Sumberg, and L. Reynolds. 1986. Leguminous fodder trees in the farming system: An overview of research at the Humid Zone Programme of ILCA in southwest Nigeria. p. 307–329. *In* I. Haque et al. (ed.) Potential of forage legumes in farming systems of sub-Saharan Africa. Conf. Proc., ILCA, Addis Ababa.

Chen, Y.S., B.T. Kang, and F.E. Caveness. 198. Alley cropping vegetable crops with leucaena in southern Nigeria. HortSci. 24:839–840.

Consultative Group on International Agricultural Research. 1990. Food today and tomorrow. CGIAR, Washington, DC.

Evensen, C.I., and R.S. Yost. 1990. The growth and lime replacement value of three woody green manures produced by alley cropping in West Sumatra. p. 106–112. *In* E. Moore (ed.) Agroforestry land use systems. Nitrogen Fixing Tree Assoc., Wiamanalo, HI.

Fernandes, E.C.M., C.B. Davey, and P.A. Sanchez. 1990. Alley cropping on an Ultisol: Mulch, fertilizer and hedgerow root pruning effects. p. 56. Agronomy abstracts. ASA, Madison, WI.

Francis, P.A. 1989. Land tenure systems and the adoption of alley farming. p. 182–195. *In* Alley farming in the humid and sub-humid Tropics. IDRC, Ottawa, Canada.

Francis, P.A., and A.N. Atta-Krah. 198. Sociological and ecological factors in technological adoption: Fodder trees in south-east Nigeria. Exp. Agric. 25:1–10.

Hoekstra, D.A. 1982. *Leucaena leucocephala* hedgerows intercropped with maize and beans: An *ex ante* analysis of a candidate agroforestry landuse system for the semi-arid areas in Machakos district, Kenya. Agrofor. Systems 1:335–345.

Hulugalle, N.R., and B.T. Kang. 1990. Effect of hedgerow species in alley cropping systems on surface soil physical properties of an Oxic Paleustalf in south-west Nigeria. J. Agric. Sci. (Cambridge) 114:301–307.

International Institute for Tropical Agriculture. 1986. Annual report and research highlights 1986. IITA, Ibadan, Nigeria.

International Livestock Centre for Africa. 1983. Annual Report, 1983. ILCA, Addis Ababa, Ethiopia.

International Livestock Centre for Africa. 1987. Final report to the Federal Livestock Department, Federal Military Government of Nigeria. FLD Grant September 1986–September 1987. ILCA Humid Zone Programme, Ibadan Nigeria.

Juo, A.S.R., and R. Lal. 1977. The effect of fallow and continuous cultivation on the chemical and physical properties of an Alfisol in western Nigeria. Plant Soil 47:567–584.

Kang, B.T., and B.S. Ghuman. 1988. Alley cropping as a sustainable crop production system. p. 172–184. In W.C. Moldenhanger et al. (ed.) Development of conservation farming on hillslopes. Soil and Water Conserv. Soc., Ankeny, IA.

Kang, B.T., H. Grimme, and T.L. Lawson. 1985. Alley cropping sequentially cropped maize and cowpea with leucaena on a sandy soil in southern Nigeria. Plant Soil 85:267–277.

Kang, B.T., and L. Reynolds. (ed.) 1989. Alley farming in the humid and subhumid tropics. IDRC, Ottawa, Canada.

Kang, B.T., L. Reynolds, and A.N. Atta-Krah. 1990. Alley farming. Adv. Agron. 43:315–359.

Kang, B.T., A.C.B.M. van der Kruijs, and D.C. Couper. 1989. Alley cropping for food crop production in humid and subhumid tropics. p. 16–26. In B.T. Kang and L. Reynolds (ed.) Alley farming in the humid and sub-humid Tropics.

Kang, B.T., and G.F. Wilson. 1987. The development of alley cropping as a promising agroforestry technology. p. 227–243. In H.A. Steppler and P.K.R. Nair (ed.) Agroforestry, a decade of development. ICRAF, Nairobi, Kenya.

Kang, B.T., G.F. Wilson, and T.L. Lawson. 1984. Alley Cropping: A stable alternative to shifting cultivation. IITA, Ibadan, Nigeria.

Kang, B.T., G.F. Wilson, and L. Sipkens. 1981. Alley cropping maize (*Zea mays*) and leucaena (*Leucaena leucocephala* Lam.) in southern Nigeria. Plant Soil 63:165–179.

Kass, D.C.L., J. Sanchez, and G. Sanchez. 1988. Results of five consecutive years of production of maize and beans under alley cropping on a Typic Humitropept fine halloysitic isohyperthermis in Turrialba, Costa Rica. p. 57. Agronomy abstracts, ASA, Madison, WI.

King, K.F.S., and M.T. Chandler. 1978. The wasted lands: The programme of work of the International Council for Research in Agroforestry (ICRAF). ICRAF, Nairobi, Kenya.

Lal, R. 1974. Role of mulching techniques in tropical soil and water management. Tech. Bull., IITA, Ibadan, Nigeria.

Lal, R. 1986. Deforestation and soil erosion. p. 299–316. In R. Lal et al. (ed.) Land clearing and development in the Tropics. A. Balkema, Rotterdam, the Netherlands.

Lal, R., G.F. Wilson, and B.N. Okigbo 1978. No-till farming after various grasses and leguminous cover crops in tropical Alfisol. I. Crop performance. Field Crops Res. 1:71–84.

Lundgren, B., and P.K.R. Nair. 1985. Agroforestry for soil conservation. p. 703–717. In S.A. El-Swaify et al. (ed.) Soil erosion and conservation. Soil Conserv. Soc. Am., Ankeny, IA.

Nair, P.K.R., and E. Fernandes. 1985. Agroforestry as an alternative to shifting cultivation. p. 169–182. In Improved production systems as an alternative to shifting cultivation. Soils Bull. 53, FAO, Rome, Italy.

Ngambeki, D.S. 1985. Economic evaluation of alley cropping leucaena with maize-maize and maize-cowpea in southern Nigeria. Agric. Systems 17:243–358.

Nye, P.H., and D.J. Greenland. 1960. The soil under shifting cultivation. Tech. Commun. no. 51. Commonwealth Agric. Bureau, Farnham Royal, England.

Okigbo, B.N. 1983. Plants and agroforestry in landuse systems of West Africa. p. 25–92. In P.A. Huxley (ed.) Plant research and agroforestry. ICRAF, Nairobi, Kenya.

Paningbatan, E.P., Jr. 1990. Alley cropping for managing soil erosion in sloping lands. p. 376–377. In Int. Congr. Soil Sci. 14, 7. ISS, Kyoto, Japan.

Raintree, J.B. (1983). Strategies for enhancing the adoptability of agroforestry innovations. Agrofor. Systems 1:173–188.

Reynolds, L., and S.A.O. Adeoye (1986). Planted leguminous browse and livestock production. p. 44–54. In B.T. Kang and L. Reynolds (ed.) Alley farming in the humid and subhumid Tropics. IDRC, Ottawa, Canada.

Reynolds, L., and A.N. Atta-Krah. 1989. Alley farming with livestock. p. 27–36. In B.T. Kang and L. Reynolds (ed.) Alley farming in the humid and sub-humid Tropics. IDRC, Ottawa, Canada.

Soil Management Support Services. 1986. Proc. Symp. on Low Activity Clay (LAC) Soils. Tech. Monogr. no. 14, Soil Manage. Support Serv., USDA, Washington, DC, USA.

Steppler, H.A. 1982. An identity and strategy for agroforestry. p. 1–5. In L.H. MacDonald (ed.) Agroforestry in the African Humid Tropics. The United Nations Univ., Tokyo, Japan.

Sumberg, J.E. 1985. Small ruminant field production in a farming systems context. p. 41–46. *In* J.E. Sumberg and K. Cassaday (ed.) Sheep and goats in Humid West Africa. ILCA, Addis Ababa, Ethiopia.

Sumberg, J.E., and C. Okali. 1983. Linking crop and animal production. A pilot development program for smallholders in south-west Nigeria. Rural Dev. Nigeria 1:1–10.

Sumberg, J.E., J. McIntire, C. Okali, and A.N. Atta-Krah. 1987. Economic analysis of alley farming with small ruminants. ILCA Bulletin (Addis Ababa, Ethiopia), 28:2–6.

Verinumbe, I., H.C. Knipscheer, and E.E. Enabor. 1984. The potential of leguminous tree crops in zero-tillage cropping in Nigeria: A linear programming model. Agrofor. Systems 2:129–138.

Wilson, G.F., and B.T. Kang. 1981. Developing stable and productive cropping systems for the humid tropics. p. 193–203. *In* B. Stonehouse (ed.) Biological husbandry: A scientific approach to organic farming. Butterworth, London, England.

Yamoah, C.F., A.A. Agboola, and K. Mulongoy. 1986. Decomposition, nitrogen release and weed control by prunings of selected alley cropping shrubs. Agrofor. Systems 4:239–246.

7

Alley Cropping on an Acid Soil in the Upper Amazon: Mulch, Fertilizer, and Hedgerow Root Pruning Effects

Erick C. M. Fernandes and C. B. Davey

Department of Soil Science
North Carolina State University
Raleigh, North Carolina

L. A. Nelson

Department of Statistics
North Carolina State University
Raleigh, North Carolina

Alley cropping has received considerable attention over the last decade as an agroforestry technology with potential to sustain crop productivity via enhanced soil protection, nutrient cycling, and/or reduced weed pressure (Kang et al., 1981, 1985, 1990). The system involves the growing of food crops in the alleys formed by hedgerows of fast-growing trees. Hedgerow species are generally N-fixing and offer the potential for enhancement of the soil N status in addition to recycling soil nutrients. The hedgerows are pruned periodically to provide green manure or mulch for the crops in the alleys and to minimize shading and root competition by the hedgerows.

Experimental results from several studies on high-base status soils (mainly Alfisols and Entisols) show that alley cropping can sustain crop yields, maintain soil nutrient status and prevent organic matter decline (Atta-Krah et al., 1986; Yamoah et al., 1986; Lal, 1989; Kang et al., 1990). Hedgerows planted along the contours on sloping land can physically minimize runoff and soil erosion. In a study to evaluate soil erosion under conventional plowing, no-till, and alley cropping on a 7% slope in Nigeria, Lal (1989) reported mean rates of soil loss over 2 yr of 8.75 t ha^{-1} under plowing, 0.95 under alley cropping (prunings incorporated by tillage) and 0.02 under no-till. Alley cropping with *Gliricidia sepium* on slopes of 45 and 70% in Colombia (annual rainfall 4000 mm) reduced soil losses of 23 to 38 t ha^{-1} yr^{-1} under maize (*Zea mays* L.) to 13 t ha^{-1} yr^{-1} (van Eijk-Bos & Moreno, 1986). The effect of litter on the soil surface has been found to be more important than that of the tree canopy cover in reducing soil erosion (Wiersum, 1984).

On some high-base status soils (Alfisols, Entisols, and Andisols) nutrients recycled in hedgerow prunings vary between 50 to 180 kg N, 2 to 10 kg P, 10 to 150 kg K, 10 to 60 kg Ca and 5 to 25 kg Mg ha^{-1} pruning^{-1} on a 60- to 90-d pruning cycle (Duguma et al., 1988; Kang et al., 1990). The nutrient cycling potential of hedgerows on acid soils (Ultisols and Oxisols) is lower than for fertile soils (Szott, 1987; Sanchez, 1987). The nutrient content of prunings on acid soils varies from 20 to 100 kg N, 2 to 5 kg P, 10 to 90 kg K, 2 to 25 kg Ca and 1 to 10 kg Mg ha^{-1} pruning^{-1} on a 90- to 120-d pruning cycle (Evensen, 1989; Fernandes, 1990; Ghuman & Lal, 1990; Szott et al., 1991). Nutrient uptake by the trees is likely to result in greater nutrient competition between trees and agronomic crops on soils with low subsoil nutrient status (e.g., Ultisols) than on base-rich subsoils (e.g., Alfisols).

Decreased crop yields close to hedgerows is a common observation of many studies on alley cropping (Szott, 1987; Singh et al., 1989a,b; Evensen, 1989; Huxley et al., 1989; Fernandes, 1990). It is hypothesized that the lowered yields are due to aboveground competition for light and belowground competition for nutrients and water. Hedgerow shoot prunings can alleviate shading of crops and it is assumed that such shoot prunings also reduce root competition due to the death of some of the hedgerow roots immediately following shoot pruning. No direct evidence was found in published studies on root competition between hedgerows and crops on acid soils in the Tropics.

Szott et al. (1991) outlined the soil chemical constraints to nutrient cycling and increased crop yields in alley cropping on acid soils. Low levels of nutrients, especially P, can reduce N_2 fixation by hedgerow species. A high level of Al saturation (80–90%) in the subsoil restricts hedgerow root growth to the same rooting zone as the agronomic crop in the alleys and may lead to increased root competition between hedgerows and crops. Palm (1988) reported crop yield responses to mulch similar to that obtained with inorganic N fertilizer in a tree–crop interface study where crop yields were measured in plots outside the competitive influence of hedgerow roots. It is hypothesized that on acid mineral soils, high levels of Al saturation in the subsoil restrict hedgerow root growth largely to the topsoil. As the hedgerows age, the increasing tree root biomass in the crop rooting zone (close to the hedgerow and within the alleys) results in greater root competition between the hedgerows and crops and offsets the beneficial effects of mulch additions and the nutrients recycled in the mulch.

The objectives of this study were to evaluate the effects of mulch, addition of a moderate level of fertilizer nutrients, and root pruning of hedgerows on: (i) crop yields with time and distance from hedgerows, (ii) weed biomass, (iii) hedgerow biomass and nutrient content of prunings, with time.

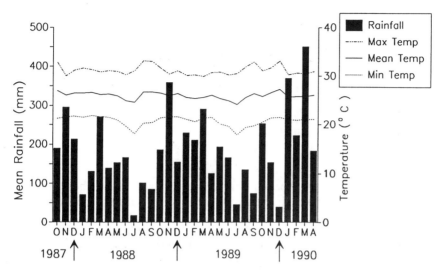

Fig. 7-1. Monthly precipitation, maximum temperature, mean temperature, and minimum temperature at Yurumaguas, Peru.

MATERIALS AND METHODS

Site Description

The study was conducted at the Yurimaguas Experimental Station located in the Amazon basin of Peru (76°05′ W, 05°45′ S, and 180 m above sea level). The area has a mean annual rainfall of 2200 mm and mean annual temperature of 26°C. Rainfall and temperature data for the period during which the study was conducted are presented in Fig. 7-1.

The dominant land-use practice on the upland terraces and slopes in the region is shifting cultivation and to a lesser extent pastures for beef and dairy cattle, resulting in a mosaic of cleared land, regenerating secondary forest, and scattered remnants of primary forest. The main crops on acid, upland soils are rice (*Oryza sativa* L.) followed by cassava (*Manihot esculenta* Crantz) and plantain (*Musa balbisiani*), often intercropped with pineapple [*Ananas comosus* (L.) Merrill.]. Cultivation of flooded rice and corn occurs on the more fertile alluvial sites.

An area of secondary forest on an upland terrace, cultivated and abandoned approximately 12 yr previously, was selected for the study. A portion of the site was on a north-facing slope of 21%. The soil was a fine-loamy, siliceous, isohyperthermic Typic Paleudult. Clay content in the top 15 cm averaged 16.5% with kaolinite as the predominant mineral in the clay fraction (Tyler, 1975). The vegetation was cleared manually in October 1987 and burned after in situ drying for 10 d. Partly burned logs and branches remain-

Table 7-1. Soil chemical properties after harvest of first rice crop.†

Depth	pH	Exchangeable				ECEC‡	Available P	Al saturation
		Acidity	Ca	Mg	K			
cm				cmol$_c$ kg^{-1}			mg kg^{-1}	%
0-15	4.3	4.15	0.92	0.38	0.17	5.62	8.9	74
15-30	4.2	6.19	0.20	0.13	0.12	6.64	3.2	93
30-45	4.2	6.84	0.17	0.09	0.08	7.18	2.4	95

† Available P by the modified Olsen extract.
‡ ECEC = effective cation exchange capacity.

ing after the burn were removed from the site by hand. The methods described for site preparation are identical to those used by shifting cultivators in the region. After burning, the topsoil (0–15 cm) had an effective cation exchange capacity of 5.62 cmol$_c$ kg^{-1}, acidity of pH 4.3, and Al saturation of 74%. Total C and N in the soil were 1.10 and 0.11%, respectively. Table 7–1 presents data on the soil chemical properties to a 45-cm depth.

Experimental Design and Statistical Analysis

The experimental design was a randomized complete block with four blocks and eight treatments. The site was blocked on the basis of clay content, Al saturation, and slope so as to maximize the variation of these variables between blocks and minimize the variation within blocks. Blocks 1 and 2 were located on a flat portion of the site whereas Blocks 3 and 4 were located on uniform north-facing slopes of 17 and 21%, respectively. The eight treatments comprised a 2^3 factorial, i.e., all combinations of the two levels (presence and absence) of each of the factors mulch, fertilizer, and hedgerow root pruning. The full factorial not only had precision advantages, but it also allowed the estimation of interaction effects. Except where stated otherwise, means were further examined only if the corresponding main or interaction effects were significant at $P < 0.05$.

An alley cropping plot layout involving double hedgerows of *Inga edulis* Mart. spaced 4 m apart was used. Previous studies at Yurimaguas (Szott, 1987; Palm, 1988) had shown that *Inga*, an acid tolerant, leguminous tree has good potential for alley cropping. The test crops grown in the alleys followed an annual rotation of upland rice (*O. sativa* L. africano desconocido)-rice-cowpea (*Vigna unguiculata* L. Vita 7) as proposed by Sanchez and Benites (1987).

The treatments which were imposed beginning with the second rice crop after burning of the fallow vegetation were as listed below.

Mulch. Inga prunings (leaves + branches) applied at a rate of 3.33 Mg ha^{-1} dry matter (DM) per crop. Prunings were applied approximately 4 wk after crop germination because prior experiments with *Inga* mulch (Palm, 1988) showed that the slowly decomposing mulch layer prevented emergence of germinating rice and cowpea crops. Plots not receiving mulch had the hedgerow prunings placed within the hedgerow so as to minimize export of nutrients from the site.

Fertilizer. 50 kg N ha^{-1} (urea), 25 kg P ha^{-1} (triple super phosphate), 20 kg K ha^{-1} (KCl), 35 kg Ca ha^{-1} and 16 kg Mg ha^{-1} (dolomite) broadcast evenly in the alley portion of the plot. Nitrogen was only applied to rice crops, not cowpea, and applications were split 30% prior to sowing and 70% at tillering. Other nutrients were applied prior to sowing each crop.

Root pruning of Inga hedgerows. Pruning was done by slicing vertically downwards with a square end machete at a distance of 25 cm from the base of the hedgerow. As the length of the blade to the handle is approximately 25 cm, effective root pruning to a depth of 20 cm was assumed. For each crop, hedgerow roots were pruned immediately after sowing the crop and again just prior to tillering (rice) or flowering (cowpea). More specifically, the first pruning was done at 11 mo during Cowpea 3, the second and third prunings at 14 and 18 mo during Rice 4 and 5, and the fourth pruning at 22 mo during Cowpea 6.

Plots were 7 m long by 4 m wide. As each hedgerow comprised two rows of trees, prunings for each plot were obtained from the tree row immediately adjacent to the plot. Crops were harvested by row in order to estimate the regression relationship between crop yields and distance from hedgerow. The remainder of the within-plot variation was used as an error term to test the polynomial components of the row to row variation.

Experiment and Crop Management

Two weeks after burning, upland rice was planted and 3 wk after crop emergence, *Inga* hedgerows were established using bare-root plants (8 wk old) that had been raised from seed in a cleared area adjacent to the site. At sowing, *Inga* seeds were immersed for 12 h in a slurry of soil, fine roots and nodules collected from mature *Inga* trees in the area. The slurry served as a source of inoculum of native rhizobia and Vesicular Arbuscular mycorrhizal (VAM) fungi. At field planting, all *Inga* seedlings were profusely nodulated and an examination of the nodules revealed that most had leghemoglobin as evidenced by a reddish pink coloration.

The hedgerows in all blocks had an east–west orientation to minimize shading of crops in the alleys. Hedgerows were inspected weekly for a month after establishment and dead trees were replaced with healthy seedlings of the same age. The first rice crop after burning (subsequently referred to as Rice 1) was harvested in March 1988. The mean grain yield harvested from the site was 1.5 Mg ha^{-1} while 2.3 Mg ha^{-1} of dry crop residue were left in place. The next six crops were referred to as Rice 2, Cowpea 3, Rice 4, Rice 5, Cowpea 6, and Rice 7. Crop residues were retained on the plots for all crops.

Crops were hand planted using a planting stick. Crop inter-row spacing was 0.5 m in all plots including the nonalley cropped controls. Within-row spacing was 25 cm for rice and 20 cm for cowpea. Experimental treatments commenced with the sowing of Rice 2 in March 1988. As hedgerow trees were considered too young (24 wk) to prune, *Inga* mulch was imported for Rice 2. Mulch for subsequent crops was obtained from plot hedgerows. All field operations (planting, weeding, hedgerow shoot and root pruning, harvesting and fertilizer applications) were carried out by hand. No tillage or herbicide was used. Leaf cutter ants (*Atta cephalotes*) were controlled by placing poisoned baits at entrances to identified nesting sites.

Pruning Measurements

The hedgerows were pruned only once per crop to minimize the requirement for labor. Hedgerows were first pruned (October 1988) 11 mo after field establishment. A pruning height of between 0.65 and 0.75 m was selected, as this is a "comfortable" height to prune while using machetes or pruning shears. Varying the pruning height up or down by 5 to 10 cm at each pruning appears to be important for *I. edulis* as observations indicated a marked decrease in vigor and resprouting if the stem was repeatedly pruned at the same height.

Pruning yields were estimated by weighing fresh prunings immediately after pruning and converting to DM by determining moisture content of fresh prunings from a subsample. Hedgerow yields for each plot were determined by bulking yields of the tree rows immediately adjacent to the plot. Composite samples of prunings from each plot were obtained for nutrient analyses.

Measurement of Crop Yields

Crop yields from the nonalley cropped plots were estimated by weighing fresh grain and stover from randomly chosen 5-m^2 areas. In the alleys, crop grain and stover was harvested by row from the central 5-m portion of the plot, thereby allowing a 2-m border between adjacent plots. Subsamples of stover were dried to constant weight in a 70 °C forced air drier and moisture content determined. The remaining stover was distributed evenly over the plot from which it was harvested. Grain also was dried in a forced air drier and its moisture content measured with a moisture meter. Calculations of crop yields per hectare included the area occupied by the hedgerows and weights were adjusted to 14% moisture content for both rice and cowpea.

Measurement of Weed Biomass

The biomass of all grasses and broadleaved weeds in each plot was sampled approximately 4 wk after crop germination and again immediately after harvesting each crop. In each plot, weeds within four randomly placed 0.25-m^2 quadrats were clipped at ground level and separated into grasses and broadleaves. All harvested material was dried at 70 °C to constant weight in a forced air drier and weighed.

Soil Sampling and Analyses

Composite samples (20 soil cores per 100 m^2) were collected from the site in October 1987, prior to clearing the secondary forest, at depths of 0 to 15, 15 to 30 and 30 to 45 cm. Samples were air dried and ground to pass a 2-mm sieve. Subsamples were extracted with either the modified Olsen or 1 M KCl extractants (Hunter, 1979). The modified Olsen extract was used in the colorimetric molybdate blue determination of P (Murphy & Riley, 1962) and in the determination of exchangeable K by atomic absorption spectroscopy. Exchangeable Al, Ca, and Mg were measured in the KCl soil extracts. Levels of exchangeable Al + H also were measured in the KCl soil extract by titration with NaOH to the phenolphthalein endpoint (Hunter, 1979). Exchangeable Ca and Mg were measured by atomic absorption spectroscopy. Soil pH values were measured in a stirred soil/water (1:2.5 v/v) slurry. Organic C was determined by the Walkley-Black method (Nelson & Sommers, 1982).

Plant Analyses

For each crop, subsamples of crop grain, stover and *Inga* prunings per treatment were oven dried at 70 °C and then ground to pass a 1-mm mesh in a Wiley mill. Nitrogen was analyzed by standard microKjeldahl methods. Phosphorus, K, Ca, and Mg were analyzed by digestion in concentrated H_2SO_4 and extraction by 30% H_2O_2. Phosphorus was measured colorimetrically by the molybdate blue method (Murphy & Riley, 1962) while K, Ca, and Mg were measured by atomic absorption spectroscopy.

RESULTS AND DISCUSSION

Upland Rice and Cowpea Test Crop Yields

Analysis per Crop

Mulch significantly ($P = 0.002$) increased rice grain yields relative to nonmulched plots only for Rice 2 (Fig. 7–2). The positive mulch effect was presumably due to a very dry period in January at around tillering and flowering of the rice crop. The reverse occurred for Cowpea 3 with an apparent negative effect ($P = 0.01$) of mulch on grain yield. Cowpea is normally grown in the dry season and it is likely that the high rainfall in October and November contributed to the lowering of yields in mulched plots, apparently caused by increased fungal deterioration of the pods relative to nonmulched plots. Grain yields of rice crops in fertilized plots were significantly higher relative to nonfertilized plots for Rice 4 ($P = 0.03$), Rice 5 ($P = 0.0001$), Rice 7 ($P = 0.002$) and Cowpea 6 ($P = 0.0001$). Hedgerow root pruning, however, resulted in significantly higher grain yields only for Rice 5, Cowpea 6 ($P =$

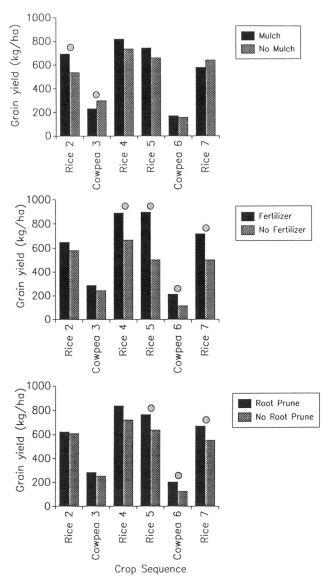

Fig. 7–2. Mulch, fertilizer, and hedgerow root pruning effects on rice and cowpea grain yield (● = significantly different at $P < 0.05$).

0.001) and Rice 7 even though plots with hedgerow root pruning consistently yielded more than nonroot pruned plots for all crops.

Analysis over Time

Trends of alley cropped rice grain yields with time relative to nonalley cropped plants were obtained by plotting mean yields from fertilized and

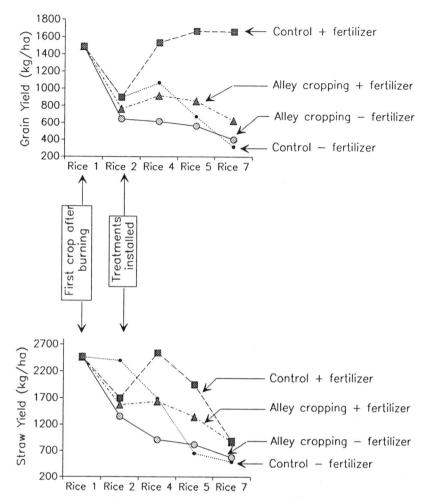

Fig. 7-3. Mean rice grain and straw yields for alley cropped and nonalley cropped (fertilized and nonfertilized) controls for five consecutive rice crops.

nonfertilized, nonalley cropped controls and alley-cropped plots receiving only mulch or mulch and fertilizer (Fig. 7-3). Rice 1 was the first crop after burning of the fallow vegetation. Treatments were imposed from Rice 2 onwards. Grain and straw yields from alley cropped plots and nonfertilized, nonalley cropped plots declined with time. Only grain yields from fertilized, nonalley cropped plots increased and then stabilized with time. While grain yield was maintained at a moderate level in the (control plus fertilizer) treatment from crop Rice 4 onward, the straw yield decreased. There was no obvious cause for this divergent pattern in yields.

Analysis of variance on grain yields for rice Crops 2, 4, 5 and 7 over time revealed highly significant ($P = 0.0001$) increases in both grain and straw yields for fertilized compared to nonfertilized plots and also a significant time × fertilizer interaction. An evaluation of the means suggests a cumulative effect of fertilizer with time.

Crop-Tree Competition

Rice Crops

The interaction of time × crop row was established by plotting. A study of time × crop row means of grain yields showed that Time 2 (corresponding to rice Crop 2) had a very different pattern from those for the other times. Figure 7–4 was then prepared to show these relationships with Times 4, 5 and 7 (corresponding to rice Crops 4, 5 and 7) combined into a single curve each for the root pruning and no root pruning treatments. Grain yields at Time 2 did not have a significant trend with distance from hedgerow whereas the regression was strongly quadratic for Time 4, 5 and 7 with a peaking in yield at the center of the alley. Equations for the two curves of rice grain regressed on row number and (row number)^2 are shown in Fig. 7–4.

After combining all rice grain data, except for Time 2, the hedgerow root pruning and no root pruning curves were parallel and the vertical distance between these was significantly different at the 0.05 level (one-tailed test). Thus for all times other than Time 2, root pruning significantly increased yields of rice grain. The average increase in grain yields due to root pruning was 30% in the row closest to the hedgerow and 15% in the center of the alley. The overall effect was a 20% increase in crop yield due to root pruning. At Time 2, however, there was no significant effect of root pruning at any crop row position in the alley. Grain yields for root pruning and no root pruning treatments from Cowpea 3 (hedgerow 11 mo old) and Cowpea 6 (hedgerow 22 mo old) showed that a significant competitive effect of hedgerow roots was present 11 mo after hedgerow establishment.

Competition between the hedgerows and crops was never completely alleviated by two root prunings and a single shoot pruning per crop as shown by a convex yield curve for root pruning treatments. Obviously, part of the crop yield decline close to the hedgerow is due to shading by the hedgerow. Root pruning did not result in a decrease in aboveground hedgerow biomass production but it did reduce root competion, as evidenced by increased yields in the root-pruned plots. Other effects such as soil compaction close to the hedgerows due to trampling during hedgerow pruning were not evaluated. It also was observed that grain-eating birds often took refuge in the hedgerows. These birds were more likely to feed on crop rows nearest to the hedgerows.

A time × crop row response surface for grain yield of rice Crops 2, 4, 5 and 7 illustrates the competitive effects of hedgerows on grain yields with distance from hedgerow over time (Fig. 7–5). Despite the steady increase in grain yields in the center of the alley (crop Rows 3 and 4), grain yield next to the hedgerow dropped by almost 50% by Time 7 (Rice 7). The data from Fig. 7–4 and 7–5 support our hypothesis of poor crop yields from alley cropping due to increasing root competition between hedgerows and crops with

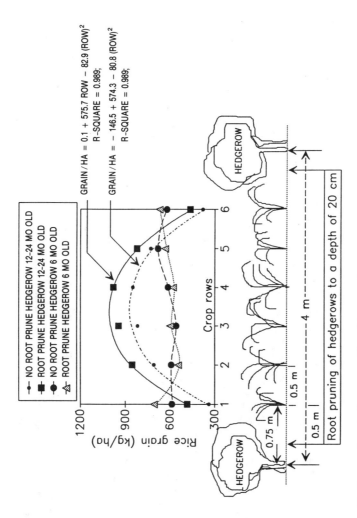

Fig. 7-4. Rice grain yield with distance from hedgerow as affected by age of hedgerow and pruning of hedgerow roots.

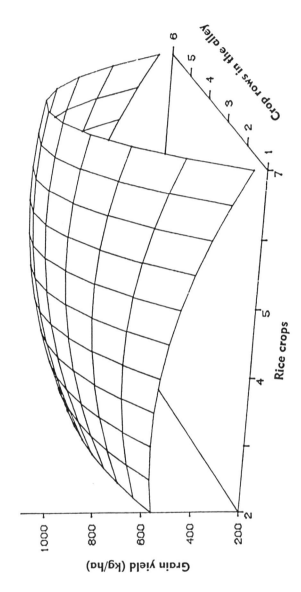

Fig. 7-5. Rice grain yields by crop row for rice Crops 2, 4, 5, and 7 in alley cropping on a pH 4.3 Ultisol at Yarimaguas, Peru. Crop Rows 1 and 6 were closest to the hedgerows.

time. The application of prunings to the alleys probably results in a progressively increasing zone of nutrient depletion starting at the base of the hedgerow and extending into the alley. The resulting fertility gradient will promote the growth of hedgerow roots into the alleys.

The declining yields close to hedgerows emphasize the importance of measuring and reporting yields across the entire alley (Huxley et al., 1989). In this study, measuring yields in the central portion of the alley would have led to the conclusion that alley cropping increases yields with time (as shown in Fig. 7-5).

It is likely that belowground competition is more intense in acid soils relative to nonacid soils because the high Al saturation (80–90%) of the subsoil restricts fine root growth of hedgerows to the same rooting zone as the agronomic crops (0–20 cm). Cannell (1983) proposed the hypothesis that frequent shoot pruning of hedgerows results in significant root death and thus minimizes root competition between hedgerows and crops in the alleys. Our observations indicate that only the fine roots (< 2-mm diam.) may die and that new fine roots begin to emerge from secondary and tertiary laterals within 8 to 16 d after shoot pruning.

Root competition between hedgerows and crops for nutrients released from prunings applied as mulch could be minimized by the use of fast decomposing mulches where the bulk of the nutrients are released during the period following shoot pruning when tree fine roots die and new root regrowth commences. It is hypothesized, however, that shoot pruning alone becomes less effective in controlling belowground competition for soil nutrients and water as hedgerow plants get older because accumulated carbohydrates and nutrients in the larger hedgerow stems and coarse roots are sufficient to sustain rapid regrowth of new shoots and fine roots after shoot pruning.

Hedgerow Biomass and Nutrient Contents

Mean pruning yields (oven-dry basis) of *Inga* hedgerows for the different treatments over time, ranged from 2.69 to 3.72 Mg ha^{-1} pruning $^{-1}$ at a pruning frequency of approximately 4 mo which is equivalent to between 8.07 and 11.2 Mg ha^{-1} yr^{-1}. The biomass of prunings was significantly greater in hedgerows next to fertilized plots than nonfertilized plots even though fertilizers were only applied to the alleys. Fertilizer application to alleys increased hedgerow biomass by 18% after four cuttings. Hedgerow root pruning did not have a significant effect on hedgerow biomass production during the period of measurement. It is important to note that tree rows were only root pruned on the side immediately adjacent to the alley.

There was a trend (not significant) of higher hedgerow biomass yields (7% higher after 4 cuttings) from mulched as compared to nonmulched plots despite the fact that plots not receiving mulch had the hedgerow prunings placed at the base of the hedgerow to minimize export of nutrients from the site. The trend of higher hedgerow yields from mulched plots (alleys) may be due to better conditions for root growth caused by the mulch application to the alleys. Mulching may promote exploitation of a larger soil volume by the hedgerow roots than when mulch was placed within the hedgerow itself. A significant time × mulch × root prune interaction appears to support this hypothesis. With the exception of the third mulch application, mean hedgerow biomass yields of mulched plots that had been root pruned were lower than hedgerow yields from mulched plots that were not pruned. The surface application of slowly decomposing prunings to alleys may promote hedgerow root growth into the alleys and thus increase root competition between hedgerows and crops. Palm (1988) reported slow decomposition of *Inga* leaves with close to 70% of original leaf mass remaining after 30 wk of decomposition. After 20 wk, the decomposing *Inga* leaves still had close to 50% of original N and P remaining.

Nutrient Content of Hedgerow Biomass

There was a significant fertilizer and time effect for all the major nutrients in hedgerow biomass. In general, prunings from hedgerows next to fertilized plots had significantly greater nutrient contents than hedgerows next to nonfertilized plots. The prunings contained 82 to 147 kg N, 4 to 10 kg P, 33 to 60 kg K, 14 to 25 kg Ca, and 3 to 8 kg Mg ha^{-1} pruning^{-1}. On average, the prunings from hedgerows adjacent to fertilized alleys had 25% more N, 29% more P, 20% more K, 24% more Ca, and 21% more Mg than prunings from hedgerows near nonfertilized alleys. The data provide evidence of the ability of hedgerow trees to capture and recycle nutrients. The N content of prunings exhibited a significant time × mulch interaction. With the exception of the first pruning (11 mo after hedgerow establishment), hedgerow prunings from mulched plots had higher N contents than prunings from nonmulched plots.

Phosphorus contents of prunings also showed a significant time × fertilizer interaction and time × mulch × fertilizer interaction. An examination of the means indicated that with time, P content of prunings increased in plots with both mulch and fertilizer as compared to plots with only mulch or fertilizer. Higher P contents of prunings from plots with mulch and fertilizer may be due to (i) a greater utilization efficiency of added P via the stimulation of VAM by the moderate amounts of added P, (ii) enhanced fine root growth of the hedgerows in mulched plots, and (iii) faster decomposition of prunings and liberation of P from recently added and old prunings due to added fertilizer P.

Nutrient concentrations of hedgerow prunings showed trends similar to those observed for nutrient contents (Table 7–2). Fertilizer and time were the most important significant effects with time × fertilizer interaction also significant for N and P. The mean concentrations (4 prunings) of micronutrients in *Inga* mulch were: Mn, 112; Cu, 13; Zn, 35; Fe, 95; and Al, 35 mg kg^{-1}. The low concentration of Al in *Inga* mulch indicates that *Inga* is not an Al accumulator, addition of 3.33 Mg DM ha^{-1} of prunings as mulch per crop resulted in 0.12 kg Al ha^{-1} being added to the surface soil. The levels of Al in leaves of some accumulator species were found to range between 4 000 to 14 000 mg kg^{-1} (Haridasan, 1982).

Table 7-2. Effect of mulch (M), fertilizer (F), root prune (R), and control (O) on nutrient concentration and content of the prunings of *Inga* hedgerows in alley cropping on an Ultisol.†

Treat	Hedge biomass	N	P	K	Ca	Mg	N	P	K	Ca	Mg
	kg ha^{-1}			%					kg ha^{-1} pruning^{-1}		
O	2687	3.42	0.24	1.37	0.54	0.18	88.8	5.8	34.9	15.0	4.7
M	3228	3.51	0.23	1.46	0.57	0.16	108.8	6.8	43.8	19.3	5.0
F	3372	3.54	0.26	1.47	0.58	0.17	112.6	7.8	46.0	20.0	5.5
R	3182	3.39	0.21	1.47	0.59	0.17	102.7	6.0	43.2	18.9	4.9
MFR	3669	3.60	0.23	1.43	0.58	0.17	124.8	7.8	49.1	21.9	6.0

† Data are the means of four prunings.

Weed Biomass

There were significant crop, mulch and fertilizer effects for total weed biomass. Cowpea crops had a lower weed biomass than rice crops. Cowpea has a spreading growth habit and rapidly establishes full ground cover thus reducing weed growth. Grass weeds became more predominant with time as seen in Fig. 7-6. The decline of broadleaf species with continued cropping implies that technological alternatives to slash and burn agriculture that prolong the cropping period prior to abandonment of the site will adversely affect the subsequent regeneration of forest fallow when the site is abandoned. In the case of alley cropping, the hedgerows will fulfill the role of the woody fallow. It is not yet clear whether such a single species fallow would be as efficient at regenerating soil fertility as the traditional multispecies forest vegetation.

Mulched plots for both rice and cowpea had significantly lower weed biomass as compared to nonmulched plots. Fertilized plots showed the reverse trend. Plots with both mulch and fertilizer had lower (though not significant) total weed biomass than plots receiving only fertilizer. Total weed biomass was not significantly different in root pruned vs. nonroot pruned plots.

The reduction of weed biomass in mulched relative to nonmulched plots did not result in significantly better grain yields for mulched plots. The lack of better yields is probably due to the timely weeding carried out for each crop and hence the main advantage in this case was the time saved for weeding mulched plots relative to nonmulched plots. On average, mulching reduced weed biomass (and hence weeding time) by 40% relative to nonmulched plots while fertilizer increased weed biomass by 38% relative to nonfertilized plots.

Estimates of weeding time for 7- by 4-m plots and the weed biomass (20–130 g dry weight m^{-2}) encountered in our plots indicated a weeding rate of 70 g of weeds (dry weight) person minute^{-1}. The weed biomass for nonmulched plots averaged for seven crops was 90 g m^{-2} per crop. The estimated labor requirement for weeding 1 m^2 is 1.3 person minutes. Hedgerow shoot pruning and mulch applications to the alley were estimated on the basis of 32 plots. Each 7- by 4-m plot required the pruning of 14 m of hedgerow. The time required to prune the hedgerows and apply mulch to 1 m^2

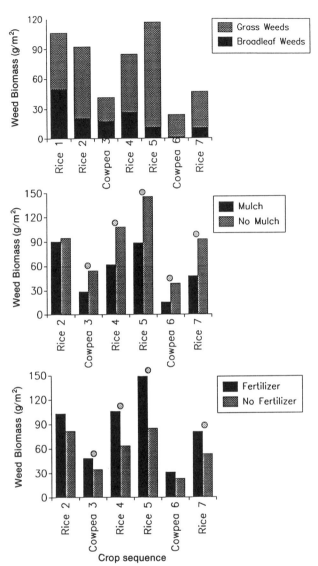

Fig. 7-6. Biomass of grass and broadleaf weeds by crop, and the effects of mulch and fertilizer on total weed biomass (• = significantly different at $P < 0.05$).

of alley was 0.6 person minutes and the time to weed the mulched plot was 0.8 person minutes resulting in a total labor requirement of 1.4 person minutes.

Under our experimental conditions and a 4-m alley width, the 40% saving in weeding time due to mulch was offset by the extra labor required to prune the hedgerows and apply the mulch. It is important to note, however, that the demand for the extra labor required for hedgerow pruning and mulching occurred after the crop had been sown and before the next major labor demand for crop weeding. There is, hence, the possibility for minimizing labor requirements during a peak period (at weeding), by pruning hedgerows and mulching at a time of low labor demand.

Suggested Technical Specifications for Alley Cropping on Acid Soils

Our data suggest that for alley widths of 4 m or less, root competition between *Inga* hedges and crops is a major factor involved in low crop yields of upland rice in alley cropping on acid soils. The following specifications are suggested to ameliorate tree–crop competition in alley cropping. Two major options are available for minimizing hedgerow–crop competition on acid soils—increased alley widths and/or intensive management of the hedgerows.

1. Alley widths greater than 4 m will decrease the tree–crop interface and reduce the competition between hedgerows and crops in the alley. On relatively flat land, farmers intuitively prefer wider alleys (6–8 m) as reported by Yamoah and Burleigh (1990). As alley widths are increased, hedgerow mulch production per unit area of alley could be maintained by planting wider hedgerows (e.g., 3–4 rows of trees).

2. On slopes, farmers may have to use alley widths of 4 m or less to minimize erosion. Hedgerows are normally pruned for the first time around 10 to 12 mo after establishment at which time root competition effects can already be detected. Earlier pruning of hedgerows is likely to reduce the competition due to decreased tree growth. Alternatively, pruning at 10 or 11 mo but at a pruning height of 25 cm also is likely to minimize competition between hedgerows and crops (Duguma et al., 1988). Low pruning heights (i.e., more intensive pruning) could substitute for the frequent, labor-intensive prunings recommended to reduce crop hedgerow competition (Evensen, 1989).

The intensive management of hedgerows suggested above will result in a decline in hedgerow biomass production and could result in higher mortality rates of trees. Visual observations indicate that even mild pruning regimes result in the death of around 15% of trees within 12 mo after pruning commences. On sloping lands, some farmers may use hedgerows along the contours to initiate terrace formation and then resort to intensive pruning to kill the hedgerows after 1 or 2 yr of growth in order to avoid the adverse effects of tree–crop competition.

3. The use of slowly decomposing mulches appears to increase root competition between hedgerows and crops probably due to the positive effect of the mulch on soil moisture and hedgerow root growth and the slow release of nutrients at the soil surface. To minimize the competition for nutrients released by the prunings, it may be preferable to use hedgerow species with prunings that decompose rapidly. Careful timing of pruning and mulch application in relation to stage of crop growth will be required to minimize competition and loss of nutrients in runoff or leaching. Data from pot experiments with seedlings of *Inga* indicated that shoot pruning intensities of

25 to 50% resulted in an initial increase of fine root biomass between 8 and 16 d after shoot pruning followed by a decline (Fernandes, 1990). Shoot pruning of hedgerows about 10 d prior to crop sowing could thus minimize root competition between hedgerows and germinating crops.

4. Incorporation of prunings with the soil by tillage can improve synchrony of nutrient release from prunings with demand by crop and help reduce root competition between hedgerows and crops by cutting the coarse and fine roots of hedgerows. Clearly, factors such as soil texture, structure, and erosivity (as affected by slope and aggregate stability) will be important determinants in the decision to forego the potential soil cover benefits of mulch. The use of two or more hedgerow species (with different mulch decomposition characteristics) could help with synchrony as well as provide soil protection.

5. Additions of moderate amounts of fertilizers can significantly improve crop yields from alley cropping on acid soils presumably due to a partial alleviation of nutrient competition. When grain yields exceed 1 Mg ha^{-1}, P is probably the most important nutrient since the P content of prunings generally does not compensate for removals in grain. Nitrogen-fixing hedgerow species require P both for nodulation and N fixation and are likely to compete strongly for P with associated crops. Yamoah and Burleigh (1990) found that addition of 30 kg ha^{-1} of N and P in addition to hedgerow prunings gave significantly higher yields of bean than plots receiving prunings alone on an Oxisol. The addition of 20 to 30 kg P ha^{-1} could enhance the effect of VAM fungi on P uptake and N$_2$ fixation by the hedgerow species and result in more N in the mulch (Fernandes, 1990). The addition of fertilizer nutrients to the hedgerow rather than alleys in order to compensate for the nutrients removed in prunings may be one way to minimize hedgerow–crop competition in alley cropping on acid soils.

6. The use of fast-decomposing prunings, incorporation of prunings, and/or the use of fertilizers is likely to increase the effects of weeds. One strategy for minimizing weeds would be to increase crop planting density within alleys. Significant increases in crop yields were reported by Mt. Pleasant (1987) when plant spacing was reduced from 50 by 35 cm to 35 by 35 cm for rice and from 50 by 35 cm to 35 by 25 cm for cowpea.

7. Upland rice is particularly susceptible to competition with weeds and hedgerows. Perhaps the production of other crops such as cassava, plantain, banana (*Musa* × *paradisiaca* L.), and acid-tolerant varieties of sorghum [*Sorghum bicolor* (L.) Moench] may prove more successful under alley cropping in acid soils.

CONCLUSIONS

The application of fertilizers to alleys and root pruning of hedgerows in alley cropping on a pH 4.3 Ultisol significantly improved yields of upland rice relative to nonfertilized and nonroot pruned hedgerows. The current model of *Inga* hedgerows spaced 4 m apart with upland rice as the crop in the alleys and only a single shoot pruning of *I. edulis* hedgerows per crop was not superior to conventional, cropping following slashing and burning of a secondary forest.

Reduced crop yields due to root competition between hedgerows and crops in the alleys were detected at 11 mo after hedgerow establishment and competition increased with age of hedgerow as measured by steadily declining yields of the crop rows closest to the hedgerows. Fertilizer application to alleys significantly increased hedgerow biomass production relative to hedgerows adjacent to nonfertilized alleys. Placing hedgerow prunings in the alleys (mulched plot) vs. at the base of the hedgerow (nonmulched plot) did not have a significant effect on hedgerow yields. Root pruning of hedgerows on mulched plots, however, resulted in significantly lower hedgerow yields than for nonroot pruned hedgerows suggesting that hedgerows compete strongly for the nutrients in prunings.

Fertilizer application to alleys significantly increased the contents of N, P, K, Ca, and Mg and concentrations of N and P in hedgerow prunings relative to nonfertilized alleys providing more evidence of the ability of hedgerow trees to capture and recycle nutrients at the expense of the crop in the alley.

Although hedgerow prunings accumulated large quantities of nutrients, and application of the mulch to the alleys significantly reduced weed biomass, there was no significant improvement in crop yields relative to nonmulched plots probably due to asynchrony between nutrient release from decomposing prunings and nutrient demand by crops.

REFERENCES

Atta-Krah, A.N., J.E. Sumberg, and L. Reynolds. 1986. Leguminous fodder trees in the farming system: An overview of research at the humid zone program of ILCA in southwestern Nigeria. p. 307–329. *In* I. Haque et al. (ed.) Potentials of forage legumes in the farming systems of sub-saharan Africa. ILCA, Addis Ababa, Ethiopia.

Cannell, M.G.R. 1983. Plant management in agroforestry: Manipulation of trees, population densities and mixtures of trees and herbaceous crops. p. 455–487. *In* P.A. Huxley (ed.) Plant research and agroforestry. ICRAF, Nairobi.

Duguma, B., B.T. Kang and D.U.U. Okali. 1988. Effect of pruning intensities of three woody leguminous species grown in alley cropping with maize and cowpea on an Alfisol. Agrofor. Systems 6:19–35.

Evensen, C.L.I. 1989. Alley cropping and green manuring for upland crop production in west Sumatra. Ph.D. diss. Univ. Hawaii.

Fernandes, E.C.M. 1990. Alley cropping on acid soils. Ph.D. diss. North Carolina State Univ., Raleigh (Diss. Abstr. 91-21993).

Ghuman, B.S., and R. Lal. 1990. Nutrient additions into soil by leaves of *Cassia siamea* and *Gliricidia sepium* grown on an ultisol in Southern Nigeria. Agrofor. Systems 10:131–133.

Haridasan, M. 1982. Aluminum accumulation by some cerrado native species of central Brazil. Plant Soil 65:265–273.

Hunter, A.H. 1979. Suggested soil and plant analytical techniques for tropical soils research program laboratories. Agro Serv. Int., Orange City, FL.

Huxley, P., T. Darnhoffer, A. Pinney, E. Akunda, and D. Gatama. 1989. The tree/crop interface: A project designed to generate experimental methodology. Agrofor. Abstr. 2:127–145.

Kang, B.T., G.F. Wilson, and L. Sipkens. 1981. Alley croping maize (*Zea mays* L.) and leucaena (*Leucaena leucocephala* Lam.) in southern Nigeria. Plant Soil 63:165–179.

Kang, B.T., H. Grimme, and T.L. Lawson. 1985. Alley cropping sequentially cropped maize and cow pea with Leucaena on a sandy soil in southern Nigeria. Plant Soil 85:267–277.

Kang, B.T., L. Reynolds, and A.N. Atta-Krah. 1990. Alley farming. p. 315–359. Adv. in agron. Vol. 43. Academic Press, San Diego, CA.

Lal, R. 1989. Agroforestry systems and soil surface management of a tropical Alfisol: II. Water runoff, soil erosion, and nutrient loss. Agrofor. Systems 8:7–29.

Mt. Pleasant, J. 1987. Weed-control measures for short-cycle food crops under humid-tropical environments in developing countries. Ph.D. diss. North Carolina State Univ., Raleigh (Diss. Abstr. 88-04816).

Murphy, J., and J.P. Riley. 1962. A modified single solution method for the determination of phosphate in natural waters. Anal. Chim. Acta 27:31–36.

Nelson, D.W., and L.E. Sommers. 1982. Total carbon, organic carbon and organic matter. p. 539–580. In A.L. Miller et al. (ed.) Methods of soil analysis. Part 2. 2nd ed. Agron. Monogr. 27. ASA, Madison, WI.

Palm, C.A. 1988. Mulch quality and nitrogen dynamics in an alley cropping system in the Peruvian Amazon. Ph.D. diss. North Carolina State Univ., Raleigh (Diss. Abstr. 88-21773).

Sanchez, P.A. 1987. Soil productivity and sustainability in agroforestry systems. p. 205–223. In H.A. Steppler and P.K.R. Nair (ed.) Agroforestry: A decade of development. Int. Council for Res. in Agrofor., Nairobi, Kenya.

Sanchez, P.A., and J.R. Benites. 1987. Low input cropping for acid soils of the humid tropics. Science (Washington, DC) 238:1521–1527.

Singh, R.P., R.J. Van den Beldt, D. Hocking, and G.R. Korwar. 1989a. Alley farming in the semi-arid regions of India. p. 108–122. In B.T. Kang and L. Reynolds (ed.) Proc. of Int. Workshop. Ibadan, Nigeria.10–14 Mar. 1986. IDRC, Otawa, Canada.

Singh, R.P., C.K. Ong, and N.N. Saharan. 1989b. Above and below ground interactions in alley cropping in semi-arid India. Agrofor. Systems 9:259–274.

Szott, L.T. 1987. Improving the productivity of shifting cultivation in the Amazon basin of Peru through the use of leguminous vegetation. Ph.D. diss. North Carolina State Univ., Raleigh (Diss. Abstr. 88-04827).

Szott, L.T., E.C.M. Fernandes, and P.A. Sanchez. 1991. Soil-plant interactions in agroforestry systems. For. Ecol. Manag. 45:127–152.

Tyler, E.J. 1975. Genesis of the soils within a detailed soil survey area in the upper Amazon Basin, Yurimaguas, Peru. Ph.D. diss. North Carolina State Univ. Raleigh.

van Eijk-Bos, C., and L.A. Moreno. 1986. Barreras vivas de Gliricidia sepium (Jacq.) Steud. (mataraton) y su efecto sobre la perdida de suelo en terrenos de colinas bajas-Uraba (Colombia). Conif-Informa, Bogota, Colombia.

Wiersum, K.F. 1984. Surface erosion under various tropical agroforestry systems. p. 231–239. In C.L. O'Loughlin and A.J. Pearce (ed.) Symp. on Effects of Forest Land Use on Erosion and Slope Stability, Honolulu, Hawaii. 7 to 11 May. East–West Center, Honolulu, HI.

Yamoah, C.F., A. Agboola, and K. Mulongoy. 1986. Decomposition, nitrogen release and weed control by prunings of selected alley croping shrubs. Agrofor. Systems 3:238–245.

Yamoah, C.F., and J.R. Burleigh. 1990. Alley cropping Sesbania sesban (L.) Merill with food crops in the highland region of Rwanda. Agrofor. Systems 10:169–181.

8 Nutrient Cycling by *Acacia Albida* (syn. *Faidherbia albida*) in Agroforestry Systems

Ray R. Weil

Agronomy Department
University of Maryland
College Park, Maryland

S. K. Mughogho

Crop Science Department
Bunda College Agriculture
University of Malawi
Lilongwe, Malawi

Agroforestry provides a number of land-use options that can enhance agricultural sustainability by increasing the diversity and quantity of useful outputs from a land area while helping to conserve and protect the area's natural resources. Agroforestry practices, under study and implementation in the Tropics, range in management complexity from simply mulching crops with leaf litter collected from nearby forested areas to intensive intercropping systems using alley or multistoried arrangements (see for example, Nair, 1990; Attah-Krah & Kang, 1993).

Effective intercropping involves a careful balance between competitive and complementary processes. Because trees may have root systems that extend far more deeply than those of annual crops, agroforestry may enhance opportunities for complementary use of soil resources, especially nutrients that may be recycled from below annual crop root zones. Further, N-fixing trees may create a positive balance with respect to that nutrient. However, intense competition for water, light and nutrients may occur during the growing season and result in significantly reduced crop production in the vicinity of trees not carefully managed (e.g., Fernandes et al., 1990). Therefore, there would be a great advantage to an agroforestry system in which the periods of peak resource use by the tree and crop components were largely not overlapping. The cultivation of agronomic crops under the canopies of *Acacia albida* (Del.) [syn. *Faidherbia albida* (DEl.) A. Chev.] appears to be such a system.

CULTURE AND USE OF *ACACIA ALBIDA*

Acacia albida [also known as *Faidherbia albida* (French), Kad (Ouolof), Gao (Haussa), Msangu (Chewa), apple ring acacia, and winter thorn] is commonly found throughout the subhumid and semiarid zones of sub-Sahara Africa at elevations below 2000 m (Wickens, 1969; Miehe, 1986) as well as in parts of the Middle East (Halevy, 1971). It is a thorny, leguminous tree that can grow to 25 m in height and more than 1 m in diameter. It is most common on alluvial soils or other areas with relatively shallow perennial water tables, but also can be found on upland slopes if the soils have sufficient water-holding capacity (Miehe, 1986; Radwansky & Wickens, 1967). Vandenbelt (1991) describes two distinct ecotypes, Ecotype A adapted to deep-water tables and severe drought in the western Africa plateaus, and Ecotype B of eastern and southern Africa requiring riparian environments. The most unique characteristic of this acacia species is its deciduous behavior known as "reverse foliation" (NAS, 1984). In most environments, the mature tree retains its leaves during the dry season and sheds them early in the rainy season. This behavior results in a temporal pattern of growth and resource use nearly opposite to that of most agronomic crops.

Although not normally used for human food, *A. albida* provides many valuable products including fuelwood, wood for building and tool-making, thorny branches for fencing, mulch, fixed N, proteinaceous forage (NAS, 1979) from its pods and leafy branches, and shade for livestock and people. The latter two products are especially valuable because they are produced during the dry season when forage and shade are scarce (Miehe, 1986).

In addition to the above benefits, scientists and traditional farmers (Felker, 1978; Sturmheit, 1990) generally recognize that crops grow better in the soil under *A. albida* canopies than in areas outside the trees' influence. Unlike in other agroforestry systems, there appears to be little direct competition between *A. albida* and associated annual crops. In fact, Vandenbeldt (1990) stated that, "There are no definitive reports of species other than *Prosopis cinceraria* (in India) and *Acacia albida* (in Africa) that exhibit the ability to increase crop yields under their canopies."

Because of these unique beneficial effects, efforts have been made to propagate *A. albida* and systematically plant it on cropland. Several large-scale efforts in Francophone West Africa prior to 1965 met with failure because of difficulties with seed germination and seedling survival after transplanting (Felker, 1978). Subsequent work has developed methods of mechanical and acid scarification, nursery methods and transplanting techniques using 30 cm long polyethylene tubes and precise timing that result in 80 to 90% germination and 60 to 90% survival after 1 yr (Poschen, 1986; Felker, 1978; Sandiford, 1988). *Acacia albida* is slow to moderate in early growth and there is great variability among individual trees produced from seed with respect to growth rate and vigor. The trees also may be highly sensitive to variability in soil fertility and physical properties. Some recent studies indicate that, on sandy soils of Niger, the higher fertility conditions often found under the trees may, in part, precede the establishment of the tree (Geiger et al., 1993; Vandenbeldt & Geiger, 1991). In these studies, improved soil properties such as increased subsoil clay content and decreased exchangeable acidity were associated with the occurrence of relic termite (*Macrotermes* spp.) mounds that correlated with enhanced growth and survival of newly planted trees. It may therefore be possible to improve planting success on poor soils by observing the growth of an indicator crop (eg., millet, *Pennisetum typhoideum*) to identify the best microsites, and then concentrating *A. albida* planting on those sites the following season. This hypothesis is currently being tested at the ICRISAT Sahelian Center in Niger (J. Brouwer, 1992, personal communication).

INTERCROPPING WITH *ACACIA ALBIDA*

By far the greatest agroforestry use of *A. albida* is the opportunistic protection of naturally occurring trees on cropland and cultivation of crops under their canopies. Nair (1985, 1990) classifies this type of use as an "agri-silvicultural system of multipurpose trees on farmland." This is a traditional technique practiced by African farmers for centuries. *Acadia albida* is mentioned with respect in African proverbs from Malawi to Senegal, and the Sultans of Zinder in the Sahal are said to have cut off the hands of people who illegally felled one of these trees (NAS, 1984). Despite efforts to protect the trees, *A. albida* on farmland are usually very unevenly spaced, leaving much of a given field without the beneficial effects of the tree.

There is little agreement in the literature as to what is the optimum spacing or number of *A. albida* per hectare (recomendations range from 12–100). Some of this confusion may result from different assumptions about the age and size of the trees. In Malawi, mature *A. albida* are often 20 to 25 m high with canopy radii of approximately 8 to 12 m and trunk diameters of 0.5 to 1.3 m. Approximately 20 such trees could give nearly complete coverage of 1 ha. This is the number of trees per hectare recommended in a report by NAS (1984). While this density of mature trees does occur in a few localized areas in eastern and southern Africa, in many areas only 6 to 10 mature trees per hectare have been oberved, and scattered individual trees at densities of 1 or 2 per hectare are most common in this region.

Effects of *A. albida* on crops grown under them have been well documented in several parts of the Sahel and in the Sudanese and Ethiopian highlands (Charreau & Vidal, 1965; Poschen, 1986). Millet, groundnut (*Arachis hypogaea* L.), and sorghum [*Sorghum bicolor* (L.) Moench] have been reported to yield 1.5 to 3 times as much under *A. albida* canopies as outside the canopies. However, such yield enhancement does not occur in all cases. Poschen (1986), for example, found a 50% or greater yield increase in 20 out of 40 paired samples, but in 10 of the pairs sorghum or maize (*Zea mays* L.) yielded about the same or even less under the *A. albida*. Bunderson et al. (1990, unpublished data) reported similar inconsistent effects on the yield of maize at several locations in Malawi. The reasons for these inconsistencies have not been well defined. Perhaps there are genetic differences in the trees and/or crops at different sites. Or perhaps *A. albida* is only beneficial where certain environmental factors are limiting crop growth. The reports do not indicate whether there was any relationship between the vigor of the *A. albida*, soil physical properties in the tree's microsite, and observed effects on crop growth. It is clear from several studies that reported crop densities and yield components that the observed yield increases resulted principally from more vigorous growth and not from increased crop populations under the *A. albida* (Dancette & Poulain, 1969).

Relatively little has been published on microclimate factors under *A. albida* in attempts to explain the enhanced crop growth. Dancette and Poulain (1969) found soil moisture to be enhanced under *A. albida* during heavy rains, but less moisture under the trees during light rains. They also posited, but did not measure, reduced evaporative demand and lower maximum temperatures under the tree canopies. Felker (1978) cites one report of greater water-holding capacity of soil under *A. albida*, possibly as a result of increased soil organic matter.

SOIL FERTILITY UNDER *ACACIA ALBIDA*

Several authors have studied soil properties in paired soil samples taken under and away from *A. albida*. Charreau and Vidal (1965) analyzed soil in the vicinity of four *A. albida* trees near Bambey, Senegal, and reported approximately twice as great a concentration of organic C, total N, and exchangeable cations and extractable (but not total) P under the *A. albida* canopies. Dancette and Poulain (1969) found less dramatic, but still significant, increases in all these properties except extractable P. The organic C levels in both studies were very low (0.3–0.6%) and C/N ratios were just under 10:1. In the Sudan, Radwanski and Wickens (1969) compared two slopes, one dominated by *A. albida* and the other by nonleguminous trees. They reported that the soil with *A. albida* had approximately two times the organic C (1.3%) and six times the total N (0.43%) giving an unexplicably narrow C/N ratio.

Workers at the ICRISAT Sahelian center in Niger have shown that the greater fertility under *A. albida* in young plantations near Niamey is largely an artifact of the better survival and growth of the trees in pre-existing microsites with better soil conditions, especially increased clay content, associated with termite activity (Brouwer et al., 1991). Also, in some places droppings from cattle attracted by the shade in the dry season and increased dust accumulation due to reduced wind velocity may contribute to the greater soil fertility under *A. albida*. However, the leaf litter probably has sufficient nutrient content to account for the observed effects through nutrient cycling. Charreau and Vidal (1965) determined that leaf litter from the relatively small (approximately 100-m^2 canopy) *A. albida* they studied contained per tree 1.13 kg N, 0.4 kg P, and 2.75 kg Ca. Jung (1970), studying larger *A. albida* (approximately 230-m^2 canopy) reported leaf litter containing 4.3 kg N, 0.09 kg P, and 5.1 kg Ca per tree.

It is the conclusion of most workers who have studied the effects of *A. albida* on crop productivity that N plays the key role. Dancette and Poulain (1969) stated that "without denying the improvement of other fertility factors, nitrogen appears the essential factor in soil productivity under *Acacia albida*." The study of Radwanski and Wickens (1969) comparing *A. albida* to nonleguminous trees would seem to support this point of view. *Acacia albida* seedlings have been shown by Habish and Khairi (1968) to form effective (red) nodules with rhizobium isolated from *A. albida*, but not from 20 other legumes species. Attempts to observe nodules on *A. albida* in the field under natural conditions have been unsuccessful (e.g., Jung, 1970), possibly due to high temperature and low moisture in surface soils during the dry season. Although N fixation by *A. albida* has not been demonstrated in the field (NAS, 1975), the high N content of the pods and the accumulation or maintenance of soil N under *A. albida* even where cattle are not present and grain crops are grown and removed, suggest that N fixation does take place.

Nutrient uptake from deep soil layers may be another important mechanism of nutrient cycling and enrichment of surface soils under *A. albida* with both cations and anions. This mechanism is likely to be especially important with regard to S which is present partly as an anion relatively mobile in sandy surface soils but subject to adsorption in Fe- and clay-enriched subsoils (Parfitt, 1978). Although S is known to be present in many African soils in amounts insufficient for optimal crop production, no information is available on the effects of *A. albida* on soil S. The only data available on S in *A. albida* systems are some leaf tissue analyses given by Charreau and Vidal (1969). They reported that S concentration in *A. albida* leaves was 0.9% and that the S concentration of Millet leaf tissue was significantly lower under *A. albida* (0.31 vs. 0.56%).

CURRENT RESEARCH IN MALAWI

The Malawi Ministry of Agriculture has recently formed an Agroforestry Commodity Team to conduct research and collaborate with the Ministry of Forestry and Natural Resources and the Agricultural Development Districts. This team plans to promote "systematic planting of *Acacia albida* trees on smallholder farms to improve the productivity and sustainability of crops and livestock, and to reduce the need for expensive fossil fuel-based fertilizers" (Saka et al., 1990). Homestead and boundary plantings also are being encouraged. Saka et al. (1990) report that interest among farmers and extension workers is high. The program began in 1990–1991 with the production of 5 000 to 10 000 seedlings in each interested Agricultural Development Division.

To further investigate nutrient cycling by these trees, we studied the soil and maize crops in farmers' fields associated with nine individual mature *A. albida* trees: three east of the town of Bolero in Northern Malawi and six west of Salima in east central Malawi (see Fig. 8–1). The soils near Sali-

Fig. 8–1. Maize growing in association with mature *A. albida* near Salima, Malawi (Site S8) at the time soil samples were obtained.

ma (Tropepts) were developed on sandy and loamy lacustrine deposits. Those near Bolero (Fluvents and Tropepts) were on old sandy alluvium. Annual rainfall in both areas is 800 to 1000 mm coming between December and May. Each site met these criteria: (i) *A. albida* trunk circumference was greater than 175 cm, (ii) maize was planted and uniformly managed under the tree's canopy and to at least 10 m beyond, (iii) soil and topography appeared to be uniform under and to at least 10 m beyond the tree's canopy, (iv) no fertilizer had been applied to the site in the current or previous cropping season. The area covered by the *A. albida* canopies ranged from 300 to 650 m^2. The *A. albida* in these areas are distributed non-uniformly with densities ranging from 2 to 12 trees per hectare.

At every site studied, maize was visibly taller and more vigorous under rather than beyond the *A. albida* canopy. Maize plant heights at the 10- to 12-leaf stage for the sites near Salima are shown in Fig. 8–2. It is worth noting that under the *A. albida*, not only the maize grew more vigorously, the weeds did too. Optimum crop production and valid yield comparisons involving intercropping with *A. albida* may require more frequent and timely efforts to manage weeds.

When the maize was in the 10- to 12-leaf growth stage and again in the tasseling stage, leaves from 20 random plants in two roughly concentric sampling areas were collected and analyzed, the one 5 m beyond the edge of the *A. albida* canopy and the other under the *A. albida* midway between its trunk and the edge of its canopy. Concentrations of N, S, P, and Zn in these leaves are shown in Table 8–1. The concentrations of N, S, and Zn were significantly higher under the *A. albida* canopy, but only for S was the effect highly significant at both growth stages. Foliar symptoms of S deficiency in maize were observed at most of these sites on maize beyond, but not under the *A. albida* canopy. Phosphorus was more concentrated away from the tree at maize tasseling stage, possibly because of greater dilution of P taken up by the larger plants under the tree. Concentrations of other

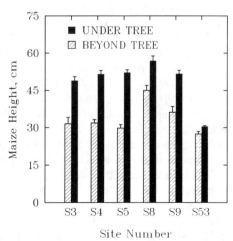

Fig. 8–2. Effect of *A. albida* on the height of maize in the 10- to 12-leaf growth stage at six sites near Salima, Malawi. Error bars are standard errors of the means.

Table 8-1. Concentrations of selected elements in maize leaf tissue at two growth stages as affected by association with *A. albida*, with means of nine sites in Malawi.

Element	Seedling stage			Tasseling stage		
	Under tree		Beyond tree	Under tree		Beyond tree
N (g/kg)	26.3	†	22.0	22.9	**	16.6
S (g/kg)	1.35	**	1.05	1.09	**	0.74
P (g/kg)	2.62	NS	3.05	2.15	*	3.13
Zn (mg/kg)	13.2	**	11.3	12.0	†	11.4

*,**,† indicate adjacent means are different by paired *t*-test at the 0.05, 0.01, and 0.1 levels of probability, respectively.

nutrients (K, Ca, Mg, Fe and Cu, not shown) were not significantly affected by the *A. albida*.

Figure 8-3 shows the association of *A. albida* with some properties of the surface (0-15 cm) soil at the nine Malawi sites. The effect of the *A. albida* canopy on total soil N ranged from 200% increase to 30% decrease. The effect on total soil C followed a similar pattern, but the differences were not as great. Across all nine sites, the probabilities of greater *t* values in paired *t* tests of the effect of *A. albida* on soil N and C were 0.09 and 0.45, respectively. Also, there was no consistent nor statistically significant effect of *A. albida* on extractable soil P (Bray P1 soil test). Only at site S4 did extractable P increase dramatically in the manner reported by Charreau and Vidal (1965) for their sites in Senegal.

In contrast, total soil S was higher under the *A. albida* at eight of the study sites and unaffected at one. Across all sites, the effect of *A. albida* on total soil S was highly significant. Sulfate-S in the upper 15 cm of the soil also was higher under the *A. albida*, except at the very sandy sites in northern Malawi (Sites A1-A3). At the latter sites, the soil samples were collected at the end of the growing season instead of early in the season as at the Salima sites. Figure 8-4 indicates that sulfate S had leached down in these very sandy profiles. The effect of *A. albida* in increasing the sulfate S was much more apparent (and highly significant) in the samples from 30 to 45 cm deep than in the surface soil. A series of on-farm fertilizer trials currently being conducted at similar sites in Malawi indicate that S is often yield limiting for maize in these soils, especially where N fertilizer is used.

The source of the increased nutrients under *A. albida* is not certain (Dunham, 1989). Although manure from cattle seeking dry season shade and forage under *A. albida* is considered to play a role in nutrient concentration under these trees in parts of the Sahel, the local scarcity of cattle and use of communal grazing lands makes this an unlikely mechanism of nutrient enrichment under *A. albida* in the Malawi study areas. Much of the *A. albida*'s N is most probably biologically fixed from atmospheric N_2 gas, but the proportion that derives from N dissolved in rainfall, from dust accumulation, and from increased efficiency of N cycling within the soil profile is unknown. Even less is known of the sources of the enhanced S under *A. albida*. In addition to uptake of S released from mineral weathering or sorp-

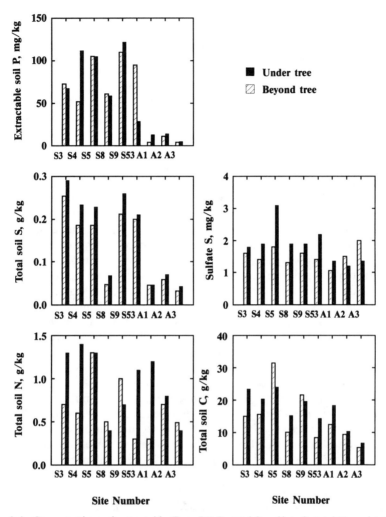

Fig. 8-3. Concentrations of extractable (Bray P1) P, total S, sulfate S, total N, and total C in the upper 15 cm of soil under *A. albida* canopy and beyond the canopy at nine sites in Malawi.

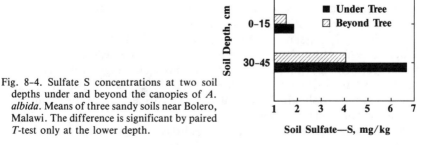

Fig. 8-4. Sulfate S concentrations at two soil depths under and beyond the canopies of *A. albida*. Means of three sandy soils near Bolero, Malawi. The difference is significant by paired *T*-test only at the lower depth.

tion deep in the soil profile, the *A. albida* foliage may absorb SO_2 from the atmosphere during the dry season when savannah vegetation and crop residues are burned over wide areas. This type of atmospheric S input may not be included in the measurements of S in rainfall. While no data are available for Malawi, Bromfield (1974) estimated that rainfall annually added 0.5 to 1.9 kg/ha S to soils of the Nigerian savannah, an amount insufficient to replenish soil S removed by even the modest grain yields common on unfertilized savannah soils.

RESEARCH NEEDS

Existing plant and soil data taken together would suggest that *A. albida* is important in enhancing soil fertility both by N fixation and by nutrient cycling from deep soil and atmospheric sources. The unique aspect of *A. albida* in agroforestry is that the tree apparently provides so little competition during the growing season that crops growing under its canopy can take advantage of the soil fertility enhancement without suffering from overriding shading or moisture stress that results from close association with most other tree species. This, coupled with the fact that African farmers already appreciate the benefits of growing crops in association with *A. albida* (Saka et al., 1990; Sturmheit, 1990), makes it very likely that successful programs can be implemented to more intensively and systematically take advantage of *A. albida* in agroforestry systems. To do this, however, considerable research remains to be done in a number of key areas.

Much work is needed to better understand the dynamics and mechanisms involved in the ability of *A. albida* to enhance soil fertility. The rate of N_2 fixation and the sources and rates of accumulation of S and other nutrients must be identified in order to predict what level of crop production the system can sustain in the long term. Nutrient balance models should be developed for the principal crop production limiting nutrients in each major *A. albida* region. This will most likely require measurements of atmospheric inputs and changes in deep soil profile concentrations over many years. Soil physical properties unlikely to be affected by vegetation, such as clay content, should be determined to evaluate the possible role of termite activity or other pedogenic processes that may create soil variability independently of the trees. The effect of *A. albida* on local water tables also should be evaluated before large-scale planting programs are implemented.

Progress has been made in understanding the factors that affect the success of mass propagation of *A. albida* (eg., Bahuguna et al., 1987; Favre & Traore, 1987; Gassama & Duhoux, 1987; Harsh & Muthuna, 1985; Sary & Some, 1990), but nursery and propagation techniques will continue to need further refinement. More critical, perhaps, are technologies for maximizing the survival and early growth of *A. albida* seedlings planted out on semiarid and subhumid cropland. Considerable work is needed to accurately identify the geographic and ecological range of adaptation for intercropping with *A. albida*. There are reports of *A. albida* established on upland and mountainous terrain (eg., Poschen, 1986; Miehe, 1986), but comparative studies have not been conducted to determine the relative potential of *A. albida* intercropping in these diverse environments.

Other questions as yet inadequately answered include: How much of the regional variability in size and vigor of *A. albida* is environmental and how much is genetic? Can *A. albida* be established in even stands, and not just as scattered individual trees, on upland soils with no shallow water table? In a specific environment, what is the optimum density of *A. albida* in order to maximize the beneficial effects on crop yields while encouraging the long-term survival and growth of the *A. albida*? How can the benefits of lopping *A. albida* for firewood be balanced with obtaining optimum soil fertility enhancement in an intercropped field? What is the sustainable yield potential of *A. albida* intercropping under improved crop management and are supplemental inputs needed to optimize the benefits of this agroforestry system? Finding answers to these critical questions will require both research experiments and carefull on-farm observation of existing *A. albida*-associated cropping systems.

ACKNOWLEDGMENT

Partial funding for this work was provided by USAID under contract DPE-5542-6-55-8034-00.

REFERENCES

Atta-Krah, A.N., and B.T. Kang. 1993. Alley farming as a potential agricultural system for the humid and subhumid tropics. p. 67–76. *In* J. Ragland and R. Lal (ed.) Technologies for sustainable agriculture in the tropics. ASA Spec. Publ. ASA, CSSA, and SSSA, Madison, WI.

Bahuguna, V.K., G.P. Maithani, U.D. Dhaundiyal, K.P. Unnikrishnan. 1987. Standardization of nursery techniques of *Acacia albida* Del. under north Indian moist climatic condition. Ind. For. 113:95–100.

Bromfield, A.R. 1974. The deposition of sulphur in the rainwater in northern Nigeria. Tellus 26:408–411.

Brouwer, J., S.C. Geiger, and R.J. Vandenbeldt. 1991. Variability in the growth of *Acacia albida*: The termite connection? Paper presented at Regional A. albida Workshop, Niamey Niger. 22 to 26 April.

Charreau, C., and P. Vidal. 1965. Influence de l'*Acacia albida* Del. sur le sol, nutrition minerale et rendements des mils *Pennisetum* au Senegal. Agron. Trop. (Paris) 6–7:600–626.

Dancette, C., and J.F. Poulain. 1969. Influence of *Acacia albida* on pedoclimatic factors and crop yield. Afr. Soils 14:143–184.

Dunham, K.M. 1989. Litterfall, nutrient-fall and production in an *Acacia albida* woodland in Zimbabwe. J. Trop. Ecol. (Cambridge) 5:227–238.

Favre, J.M., and M. Traore. 1987. Compared root morphogenesis in seedlings of *Acacia albida* Del. and two othe rleguminous trees (Prosopis juliflora D.C. and Robinia pseudoacacia L.). 37:131–137.

Felker, P. 1978. State of the art: *Acacia albida* a a complimentary permanent intercrop with annual crops. Grant no. AID/afr-C-1361. Univ. of California, Riverside.

Fernandes, E.C.M., C.B. Davey, and P.A. Sanchez. 1990. Alley cropping on an Ultisol: Mulch, fertilizer and hedgerow root pruning effects. p. 56. *In* Agronomy abstracts. ASA, Madison, WI.

Gassama, Y.K., and E. Duhoux. 1987. Micropropagation of adult Kad trees, *Acacia albida* (Leguminosae). p. 216–224. *In* Les arbres fixateurs d'azote. L'Amelioration biologique de la fertilite du sol: Actes des seminaires, Dakar, Senegal. 17–25 Mars 1986. Cah. OR-STOM, Ser. Pedol.

Geiger, S.C., R.J. Vandenbeldt, and A. Manu. 1992. Preexisting soil fertility and the variable growth of *Faidherbia albida*. p. 121–125. *In* R.J. Vandenbeldt (ed.) *Fairherbia albida* in the West African semi-arid tropics. Proc. Workshop, Niamey, Niger. 22–26 Aor, 1991, ICRAF, Patancheru, India, and Nairobi, Kenya.

Habish, H.A., and S.M. Khairi. 1968. Nodulation of legumes in the Sudan: Cross inoculation groups and the associated rhizobium strains. Expl. Agric. 4:227–234.

Halevy, G. 1971. A study of *Acacia albida* in Israel. La-yaaran 21:52–63.

Harsh, L.N., and K.D. Muthana. 1985. Rooting media for vegetative propagation of *Acacia albida*. J. Trop. For. 1:275–276.

Jung, G. 1970. Seasonal variations of the microbiological characteristics of a weakly leached tropical ferruginous soil (Dior), submitted or not to the influence of *Acacia albida* (Del.). Oecol. Plant 5:113–136.

Miehe, S. 1986. *Acacia albida* and other multipurpose trees on the Fur farmlands in the Jebel Marra highlands, Western Darfur, Sudan. Agrofor. Systems 4:89–119.

Nair, P.K.R. 1985. Classification of agroforestry systems. Agrofor. Systems 3:97–128.

Nair, P.K. 1990. Classification of agroforestry systems. p. 31–97. *In* K. MacDicken and N. Vergara (ed.) Agroforestry: Classification and management. John Wiley & Sons, New York.

National Academy of Sciences. 1975. Underexploited tropical plants with promising economic value. NAS, Washington, DC.

National Academy of Sciences. 1979. Other forage acacias. p. 142–145. *In* Tropical legumes: Resources for the future. NAS, Washington, DC.

National Academy of Sciences. 1984. Agroforestry in the West African Sahel. Natl. Acad. Press, Washington, DC.

Parfitt, R.L. 1978. Anion adsorption by soil and soil materials. Adv. Agron. 30:1–50.

Parvez, A. 1987. Vegetative propagation of *Acacia albida*: A promising species for farm forestry in arid areas. Ind. For. 113:459–465.

Poschen, P. 1986. An evaluation of *Acacia albida*-based agroforestry practices in the Hararghe highlands of Eastern Ethiopia. Agrofor. Systems 4:129–143.

Radwanski, S.A., and G.E. Wickens. 1967. The ecology of *Acacia albida* on mantle soils in Zalingei, Jebel Marra, Sudan. J. Appl. Ecol. 4:569–579.

Saka, A.R., W.T. Bunderson, O.A. Itimu, Y. Mbekeani, and H.S.K. Phombeya. 1990. The national agroforestry research programme of Malawi: Goals, strategies and achievements from 1986–1990. Natl. Agroforestry Symp., Blantyre, Malawi, 4–10 November. Agroforestry Commodity Team, Chitedze Agric. Res. Stn., Lilongwe, Malawi.

Sandiford, M. 1988. Burnt offerings: An evaluation of the hot-wire seed scarifier. Common. For. Rev. 67:285–292.

Sary, H., and L.M. Some. 1990. Attacks by spermatophagous insects on the seed of three acacias-A. albida, A. nilotica var. adansonnii and A. senegal. p. 21–24. *In* J.W. Turnbull (ed.) Tropical tree seed research. Proc. of Int. Workshop at Forestry Training Centre, Gympie Bld, Australia. 21 to 24 Aug. 1989. ACIAR Proceedings Series 1990. Vol. 28. Australian Centre for Int. Agric. Res., Canberra, Australia.

Sturmheit, P. 1990. Agroforestry and soil conservation needs of smallholders in southern Zambia. Agrofor. Systems 10:265–289.

Vandenbeldt, R.J. 1990. Agroforestry in the semiarid Tropics. p. 150–194. *In* K. MacDicken and N. Vergara (ed.) Agroforestry: Classification and management. John Wiley & Sons, New York.

Vandenbeldt, R.J. 1991. Rooting systems of western and southern African *Faidherbis albida* (Del.) A. Chev. (syn. *Acacia albida* Del.)—A comparative analysis with biogeographic implications. Agrofor. Systems 14:233–244.

Vandenbeldt, R.J., and S.C. Geiger. 1991. Does soil fertility under *Faidherbia albida* precede the tree? Nitrogen Fixing Tree Res. Rep. 9:105–106.

Wickens, G.E. 1969. A study of *Acacia albida* Del. (Mimosoideae). Kew Bull. 23:181–202.

9   Vetiver Grass—A Hedge Against Erosion[1]

J. W. Smyle and W. B. Magrath

The World Bank
Washington, District of Columbia

Over the last 5 yr the World Bank has been promoting the use of *Vetiveria zizanioides* (Linn.) Nash or vetiver grass as a plant species which is particularly well suited for use as a contour vegetative barrier or hedgerow. Vetiver grass is a species which is distributed pantropically. It has a long history of traditional use as a soil conservation plant in Asia, Africa and the Caribbean. Vetiver grass has the potential to perform a key role in developing and maintaining sustainable upland farming systems.

THE PROBLEM

Soil erosion in the Tropics and sub-Tropics is perceived as being one of the major environmental problems of our times. Following from this perception, millions, if not billions, of dollars has been spent in the last three decades on programs to control soil losses and stabilize hillslope agriculture. Despite massive expenditures, which have gone mainly to engineered soil conservation systems, the judgment remains that the magnitude of the problem has increased. One result has been the growing awareness that a major problem with the extension of soil conservation has been the frequent inappropriateness of the technologies in the light of government's ability to deliver and the farmer's desires, capabilities and resources. Farmer acceptance has been low for technologies that require significant changes in land use, farming practices, labor inputs, and whose benefits are predicated almost exclusively on soil conservation. Today, on-farm soil conservation is losing its status as a free-standing pursuit or project component. In its place, the focus is shifting to how farmers can manage their farms and maintain (or restore) the productivity of their soils. Soil conservation can then be the

[1] The findings, interpretations and conclusions expressed here are entirely those of the authors and should not be attributed in any manner to the World Bank.

additional benefit of good farming practices that, if adopted, increase farm productivity and conserve resources.

Agroforestry, sloping agricultural land technology, and alley cropping have become popularized as promising soil conservative components that are compatible or complementary with a farming systems approach. One attraction is their potential to supply a number of farm inputs from within the farm boundaries while concurrently protecting the farm's resource base and, ideally, sustaining and/or increasing yields. From a strict soil conservation program point of view, they avoid the high per unit costs associated with the construction and maintenance of engineered systems—costs which few farmers can afford unless heavily subsidized.

CONTOUR HEDGEROW SYSTEMS

As a generalization, Sloping Agricultural Land Technology, alley cropping and some agroforestry systems are contour technologies based on hedgerows. The main soil conservation functions of contour hedgerows are to: (i) break the slope length, reduce runoff velocity and increase infiltration opportunity time; (ii) reduce the erosivity and transport capacity of surface runoff; (iii) cause the deposition of eroson products, trap sorbed nutrients and induce terracing; (iv) cause cultivation and planting operations to be carried out on the contour and (v) stabilize contour cultivated areas on steeper slopes.

The other, nonsoil conservation, benefits of hedgerow systems can be numerous. Several typologies have been proposed to classify hedgerow systems based on their composition, architecture and the products which they produce—such as fruit, fodder, green manure, fuelwood, cash or subsistence (16,31).

Hedgerow technologies, however, are still emerging systems that remain to be sufficiently characterized. Nonetheless, a large body of anecdotal information and grey literature suggests that they are effective for soil conservation. What research exists tends to confirm that contour vegetative barriers can be very effective at reducing erosion and soil loss (1, p. 1–7,2,6,9,33,35) and in reducing surface runoff (1, p.1–7,33,35,57). Several studies, some on 30% slopes, have made direct comparisons of the effectiveness on contour vegetative barriers vs. terracing (9,33,62). In the aggregate they find vegetative barriers more or less as effective as terraces in reducing soil loss. Combined with mulching, manuring and certain tillage or cultivation practices, there is an ample basis to suggest that they should be as effective as within field structures (14,38). There is also evidence that the so-called "induced or controlled erosion terraces" that form behind contour barriers, show superior stability (65) and crop yields (12,19).

The management and benefits of hedgerow systems are not well defined or understood. Most available research and project data are on tree hedgerows. As yet this data do not support the broad assumption that annual crops benefit from hedgerows (4,6,16,26,27,28,35,51,58,66). In cases where positive benefits are demonstrated, both species and management dependency on agroecological zones (31) restricts wider applicability. Further, information on the management and integrated benefits of hedgerows for fruit, fuel and/or fodder, crop productivity and soil conservation is still largely unavailable.

Problems have also been associated with smallholder adoption of hedgerows: (i) they often represent a radical change from present practices; (ii) they can occupy significant percentages of annual cropping areas; (iii) they can have high establishment and management labor requirements; (iv) they often need protection from grazing; (v) competition effects cause reduction in the associated annual crops and; (vi) the hedgerows can act as hosts to rats, birds, insects, diseases and weeds. Despite these difficulties and unknowns, contour hedgerow systems have not been ignored. Indeed, they are in use all over the world, put in place by farmers without the aid of research or subsidies.

CONTOUR HEDGEROWS OF *VETIVERIA ZIZANIOIDES* (LINN.) NASH

In 1985, John Greenfield encountered vetiver grass being grown in the semiarid zone of India. He had worked with vetiver in Fiji, a humid zone, during the 1950s using it as a soil conservation measure for hillslope farming of sugarcane (*Saccharum officinarum* L.). Greenfield recommended that trials be initiated with vetiver grass under three World Bank-assisted watershed management projects in the semiarid zone of India. In the 7 yr since he began that work, vetiver grass has been found to be both a technically sound and economically practical basis for a hedgerow/contour cultivation system. It has recently been incorporated into both on- and off-farm conservation programs in a number of countries including China, India, Malaysia, Nepal, Philippines, Sri Lanka, Thailand, Madagascar, Nigeria, South Africa, Zimbabwe, Bolivia, Costa Rica, Guatemala, and Honduras.

Vetiver grass has a number of features which make it desirable, both from a farmer and a project viewpoint: (i) it is a densely tufted, perennial clump grass, the leaf blades of which are relatively stiff. The foliage is mostly basal with the leaf sheaths closely overlapping, strongly compressed and keeled (45) which creates a physical barrier of great density at the ground surface; (ii) adaption to a wide range of climatic and edaphic conditions; (iii) ease of propagation and establishment; (iv) low maintenance requirements; (v) minimal space requirements; (vi) ability to withstand grazing or burning; and (vii) to date it has exhibited little or no competition effects with associated crop plants.

Because of vetiver's physical characteristics, it has a number of distinct advantages. First, it can be utilized to create dense, vegetative contour barriers under almost any situation in which annual or perennial cropping is carried out in the Tropics or sub-Tropics. Because of the density of the barrier, vetiver hedgerows can control runoff and act as a water spreading system, increasing plant available moisture. It is this function which can provide short-term benefits to the farmer in the form of crop yield. And, second, the hedges of vetiver do not channel runoff, as do engineered systems. They allow surface runoff to percolate slowly through the hedge. Besides the obvious implications of not needing to convert significant portions of the field to drainageways, there is the significant advantage that there is no need for detailed layouts and alignments. Vetiver hedgerows can be planted along an averaged contour and still function well. This makes it possible to reach much wider areas and allows farmers a great deal of latitude in where and how they wish to incorporate the hedgerows. Third, vetiver hedgerows can perform their function while occupying a narrow strip of less than 50 cm width per hedgerow and with the hedgerow kept pruned down below the level of the crop. This allows the introduction of hedgerow systems into the field with the minimum possible change in current farming practices.

In effect, *V. zizanioides* has the potential to provide a baseline strategy for the adoption of contour hedgerow systems over a broader area. The concepts of these systems have held promise for many years, however the extension of the concept requires successful species. *V. zizanioides* is a successful species.

HISTORICAL USE OF *VETIVERIA ZIZANIOIDES*

While attention has only recently been focused on this species as a soil conservation plant, it is not correct to assume that there is little experience or knowledge of *V. zizanioides*. The plant has been traditionally used as mulch, as a thatch grass, as a mattress stuffing, for making brooms and baskets, as animal bedding, for paper pulp, as an ornamental plant and as a cut-and-carry fodder plant. There are several references (13,39,64) of *V. zizanioides* planted on field or garden boundaries to prevent the extension of weeds such as *Cynadon dactylon* (L.) Pers. and *Desmostachya bipinnata*; this function also has been mentioned by farmers in south India and Zimbabwe (18). The roots of vetiver grass have been used for centuries to weave as mats, screens or fans, powdered for use in religious ceremonies, taken for medicinal purposes and used in insect repelling sachets. The major commercial usage is and has been the extraction of an exceedingly complex, volatile oil from the roots of vetiver grass used in perfumes and cosmetics. It is believed that the aromatic characteristics of the root account for the original spread of *V. zizanioides* germplasm throughout most of the Tropics and parts of the sub-Tropics, it is now reported to exist in 86 countries (Table 9-1). Interest in *V. zizanioides* as an essential oil crop also has made it the subject of a number of cytological (8,32,36,48,49,54), adaptability (41,46,53) and growth (5,40,55,56,60) studies.

Vetiver grass has been known in this century as a soil conservation grass though little formal mention is made of it. Prior to present interest, a number of authors throughout this century have noted its soil conservation function (3,17,21,22,25,37,39,44,45,47, p. 297–298,50,59) and the grass also has

Table 9–1. Countries where *Vetiveria zizanioides* (Linn.) Nash is currently known to exist.

Africa	Asia	America	Caribbean	Pacific	Miscellaneous
Algiers	Bangladesh	Argentina	Antigua	Australia	France
Angola	Bhutan	Bolivia	Barbados	Cook Islands	Isreal
Burkina Faso	China	Brazil	Cuba	Fiji	Italy
Burundi	Hong Kong	Canada	Dominican Republic	New Caledonia	Russia
Cape Verde	India	Chile	Haiti	New Guinea	Spain
Comoro	Indonesia	Colombia	Jamaica	New Zealand	United Kingdom
Central African Republic	Japan	Costa Rica	Martinique	Samoa	
Ethiopia	Laos	Ecuador	Puerto Rico	Tonga	
Gabon	Malaysia	Guatemala	St. Lucia		
Ghana	Myanmar	Guyana	St. Vincent		
Kenya	Nepal	Honduras	Trinidad		
Madagascar	Pakistan	Mexico			
Malawi	Philippines	Nicaragua			
Mauritius	Singapore	Paraguay			
Nigeria	Sri Lanka	Suriname			
Rwanda	Thailand	USA			
Reunion	Vietnam				
Seychelles					
Somalia					
S. Africa					
Sudan					
Tanzania					
Togo					
Tunisia					
Uganda					
Zaire					
Zambia					
Zimbabwe					

Table 9-2. Currently reported locations and traditional uses (for other than oil production) of *Vetiveria zizanioides* (Linn.) Nash for soil conservation and/or other purposes.

Location	Stabilizing Canals/ditches/ embankments	Roads/ paths	Field/ boundary hedge	Miscellaneous
Africa				
Burundi	X			
Central African Republic				Mattresses
Ethiopia	X	X		Weed barrier/mulch/ mattresses/thatch
Gabon	X	X	X	
Ghana			X	
Kenya			X	
Madagascar			X	
Malawi			X	
Mauritius			X	
Rwanda	X			
South Africa	X	X	X	
Tanzania			X	Fodder
Tunisia		X		
Zaire	X			
Zambia			X	
Zimbabwe			X	
Asia				
Myanmar	X			
China			X	
India	X		X	Fodder/weed barrier/ paper mats/thatch/ medicinal
Malaysia			X	
Nepal	X			
Pakistan			X	
Sri Lanka	X			
Thailand	X			
Pacific				
Fiji	X		X	
New Guinea				Repel insects
Samoa				Weed barrier
Latin America				
Costa Rica		X	X	
Guatemala				Animal bedding
Caribbean				
Barabados			X	
Haiti		X	X	
St. Lucia		X	X	
St. Vincent		X		
Trinidad		X	X	
Martinique			X	
North America				
USA				Sachet/repel insects/ ornamental

been promoted by government agencies 35 or more years ago in Kenya, St. Vincent, Trinidad and Fiji as a contour hedgerow species. Aside from official promotion, vetiver grass has been functioning as a soil conservation species for more than a decade to well over 100 yr in more than 30 countries worldwide (Table 9-2). As Purseglove stated in his 1972 edition of *Tropical Crops*

> The grass (*V. zizanioides*) is widely used throughout the tropics for planting on the contour as an anti-erosion measure, for protective partitions in terraced fields, and as a border for roads and gardens.

THE USE OF *VETIVERIA ZIZANIOIDES* (LINN.) NASH

As a hedgerow species, the usefulness of vetiver grass has been established in the Tropics and sub-Tropics from sea level to 2 000 m in Uttar Pradesh, India; over a temperature range from 38 °C mean dry season temperature/45 °C maximum in Andhra Pradesh, India, to a 5 °C mean winter temperature, minus 9 °C minimum and 10 frost days in Cantalice, Italy (V. Branscheid, 1989, personal communication); in soils with pH ranging from 4.5 in Ethiopia (S. Kebede, 1990, personal communication) and China (Wang Zi Song, 1990, personal communication) to 10.5 in India (29,53); across rainfall zones of 600 mm yr^{-1} with an 8- to 9-mo dry season in Andhra Pradesh, India to about 6 000 mm yr^{-1} in Sri Lanka; and in China on slopes between 35 to 40%. The notable exceptions to vetiver's usefulness are in areas of low temperatures—though the extent of the grass's cold tolerance is not yet established; and in saline soil—the literature classifies *V. zizanioides* as a glycophate (63) and 75% reductions in growth rates have been reported at electrical conductivities (EC) of 4.0 (S. Miyamoto, 1990, personal communication)[2]. Other potential problem areas for *V. zizanioides* may be in areas with rainfall much less than 600 mm yr^{-1} and/or dry seasons longer than 9 or 10 mo and in areas where the commercial market for oil distilled from the roots has resulted in uprooting of plantings, as in parts of Haiti and Indonesia.[3]

Beyond its wide range of adaptability, other characteristics make *V. zizanioides* particularly suitable as a hedgerow species.

1. *V. zizanioides* hedges stop surface runoff from concentrating and create greater spatial and temporal opportunity for infiltration. As the major effect of erosion on productivity is due to the reduction in plant-available water, this function of vetiver hedgerows is perhaps the most important.

---

[2] Since this paper was written in 1990, Dr. P. Truong and colleagues from the Queensland Department of Primary Industries, Brisbane, Australia have completed both field and pot studies of salinity effects on vetiver. They have concluded that vetiver has a fairly high salt tolerance, especially in mature plants, almost as high as Rhoda grass (*Chloris gayava* Kunth.). According to their results, 50% yield reduction can be expected when soil salinity (EC) reaches 16 mS cm^{-1} (60).

[3] Since 1990, it has been found that the combination of moisture stress under semiarid conditions, termite infestation and heavy pressure on the young shoots can result in severe plant mortality rates which negate the value of the hedgerow.

2. *V. zizanioides* is not known to exhibit invasive behavior, based on empirical evidence from years to decades of observation in dozens of countries. Vetiver grass does not spread by rhizomes, nor has it been reported to have widely spread from seed. It has been reported as an escape in Malaysia (15), Fiji (44), Louisiana (23), the Caribbean (24) and Costa Rica (45). However, there is no evidence that it has become a problem. What is documented regarding this question is that two types of *V. zizanioides* are recognized; (i) a wild type from north India which flowers regularly and sets fertile seed and (ii) a south India type that is reported to be either a nonflowering type (11) or a nonseeding type (49,60). This latter type has existed under cultivation and been vegetatively propagated for a number of centuries and it has been noted that some accessions of vetiver, particularly those of south Indian origin, can only be maintained by vegetative methods as they do not produce seed under natural or hand-pollinated conditions (48). The south India type is currently believed to be the more widely distributed of the two. It should be noted, however, that some accessions of *V. zizanioides* do produce large quantities of viable seed and it is as yet unclear what the conditions are that favor fertile seed and seed germination.[4]

3. Vetiver grass has been grown in close association with crop plants for decades in Africa, Asia and the Caribbean, there are no reports of *V. zizanioides* as a host fo diseases, insects or pests that are of practical or economical concern to agriculture. Vetiver's susceptibility to some diseases (11,63) and insects or pests (11,34) are known, but they are not of concern in terms of vetiver's ability to function as a hedgerow. There also are common reports from Asia and Africa that rats are either repelled by the grass or tend not to create new burrows in or under it. Field and plot border hedges of vetiver grass are used in several countries to exclude rhizomatous weeds.

4. Competition effects between vetiver and adjacent crop plants has to date proven to be minimal. Preliminary research data from acidic and slightly alkaline, drought prone soils show no yield reduction in either finger millet [*Eleusine coracana* (L.) Gaertn.] (20,30), castor bean (*Ricinus communis* L.) (43) or peanut (*Arachis hypogaea* L.) (30) when planted next to vetiver hedgerows. Under the same conditions, maize (*Zea mays* L.) yield reductions were noted in the two rows closest to the hedge (20). Results on nonacidic soils under semiarid conditions show no yield reductions in green gram (*Phaseolus mungo* L.), pearl millet (*Pennisetum typhoideum* Rich.), sorghum (*Sorghum vulgare* Pers. 'R-73') or safflower (*Carthamus tinctoria* L.) (7). Aside from these crops, interviews with farmers have supported that competition effects are minor or unnoticeable in cotton, rice and pigeon pea [*Cajanus cajan* (L.) Millsp.]. It is hypothesized that the minimal competition effects are a function of root morphology—the extensive root system tends to move

---

[4] Initial DNA fingerprinting has shown that the North India seeds type to be quite heterozygous and genetically distinguishable from the South India materials. The clones tested thus far have shown unusual hymozygosity among these South India types. It is increasingly clear that the vetivers outside South Asia are overwhelmingly infertile, although this should be demonstrated on a case by case basis (M. Dafforn, 1993, personal communication).

vertically rather than horizontally, improved moisture and nutrient status near the hedgerow and the low nutrient requirements of the grass (52).

5. Vetiver hedges persist with little or no maintenance. Examples of intact, functional hedgerows ranging in age from 12 yr to over 100 yr may be seen in the Caribbean, East and South Africa, Fiji and India. Importantly, these hedges have not dried out in the center and split up into individual clumps. Instances of mortality within the centers of vetiver clumps has, however, recently been associated with termites under semiarid conditions in Andhra Pradesh, India (see footnote 4).

6. *V. zizanioides* is easily and inexpensively propagated and established. Normal production costs in India and China are on the order of $8 to $18 (U.S. dollars) for sufficient material to plant 1000 m of hedgerow. Approximately 10-man days are necessary to plant 1000 m of hedgerow. One hectare of vetiver nursery will supply sufficient material to establish from about 50 to 125 km of hedgerow.

7. Depending on climate and soils it will take 9 mo to 4 yr for the vetiver hedgerow to close completely. Once established, the hedgerow is a dense barrier that can be maintained at a width of 50 cm and still perform its function. Prior to closure the hedgerow would still be as effective as hedgerows of woody perennials that are spaced 10 cm and less apart. Also, the grass can be kept low pruned (to about 40 cm) so there is no light competition between the hedge and the crop. Pruning also will increase tillering and the rate of hedge closure. Because of the minimal amount of space occupied by the hedgerow, spatial opportunities exist for the inclusion of other plant species (such as leguminous trees) along the hedgerow, if the farmer so desires.

8. *V. zizanioides* is generally nonpalatable to grazing animals and thus once established, can be maintained in free grazing areas without resorting to fencing or requiring a change in grazing habits. The young leaves of the plant, however, are palatable and the hedges can be pruned at short, regular intervals to provide cut-and-carry fodder. The fodder value of the young leaves falls between that of Napier grass (*P. purpureum* Schumach.) and fresh maize stover (10,11,42) (Table 9–3).

9. The crown of *V. zizanioides* occurs below the ground, making it resistant to fire, mechanical and grazing damage. Its fire resistance allows farmers to continue burning to their fields.

Since 1987 a number of research station and farmer's field trials have been carried out in south India to compare the impacts of vetiver hedgerows on surface runoff, soil losses and crop yields. In total the trials represent 27 plot years of data from two areas and two soil groups. The data is preliminary, and is not statistically significant.

On slopes under 5%, contour hedgerows of *V. zizanioides*, planted at 1-m vertical intervals, have reduced surface runoff an average of 30% (±23%) and 47% (±9%) compared to conventional practices of graded banks (7,30) and across slope cultivation (7). Compared to *Leucaena* spp. hedgerows they have reduced surface runoff an average of 24% (±14%) (7,30).

Table 9-3. A comparison of the fodder value of *V. zizanioides* (Linn.) Nash—managed for fodder—with the fodder value of some other hedgerow grasses and major sources of ruminant forage.

Name	Crude protein	Ether extract	Crude fiber	Total ash	Calcium	Phosphorus
				%		
Vetiveria zizanioides† (fresh, young leaves(	6.1–6.7	1.1–2.1	38–42	5.3–9.0	0.28–0.31	0.05–0.60
Zea mays indentata‡ (Maize; stover)	6.60	1.3	34	7.2	0.49	0.08
Sorghum bicolor (Sorghum; stover)	5.20	1.7	34	11.0	0.52	0.13
Pennisteum purpureum (Napier; fresh, late bloom)	7.80	1.1	39	5.3	0.44	0.35
Paspalum notatum (Bahia; fresh)	8.90	1.6	30	11.1	0.46	0.22
Pasture§ (grass dominant)		5.10			0.17	0.07
Maize (cobs maturing)		3.20			0.17	0.07
Millet		3.00			0.10	0.05
Sorghum (heading)		2.50			0.11	0.05

† Source is CSIR (1976).
‡ Source is NAS (1982).
§ Source is Cochrane et al. (1983).

Plots and fields with vetiver hedgerows have shown a reduction in sediment yields of an average of 74% (± 5%) compared to across slope cultivated areas (7) and 43% (± 19%) compared to areas with graded banks (7,30). Compared to hedgerows of *Leucaena*, vetiver hedgerows have reduced sediment yields an average of 54% (± 4%) (7,30).

Crop yield data comparing the conventional practices of graded banks and across slope cultivation to areas with contour hedgerows of vetiver grass shows that yields averaged 6% (± 10%) and 26% (± 20%) (7,30) higher from the areas with the vetiver hedgerows, respectively. Compared to *Leucaena* hedgerows yields with hedgerows of *V. zizanioides* averaged 10% (± 9%) higher.

Research opportunities exist in a number of areas, for example, the ecology and symbioses, physiology, morphology and hedge function and maintenance over time. Some of the outstanding questions associated with *V. zizanioides* may be resolvable through existing knowledge and field observations.

ECONOMIC ASPECTS

A review of soil conservation investment programs in developing countries (14,38) shows the extent to which conventional practice relies on high cost, physical and engineering approaches to soil conservation. Investment costs for bunding (contour or graded banks), bench terracing and associated waterway construction typically exceeds $500 (U.S. dollars) ha^{-1}. Considering the benefits of moisture conservation and soil productivity loss prevented, the internal rate of return (IRR) for vetiver hedgerow establishment in India approaches 95%. Even with generous assumptions, engineering treatments showed IRR's of approximately 35% (14,38).

Farmer reaction to vetiver grass hedgerows is not well understood, nor for hedgerow technologies in general. As noted, there are areas with records of long use of vetiver grass hedgerows and there have been periods in several countries in which vetiver grass hedgerows have been promoted by government. On the other hand, the use of vetiver grass is not nearly as widespread as its prospective benefits would suggest. A possible explanation may be the emphasis that government agencies and research institutions have placed on engineering systems and the widespread use of subsidies and coercion to encourage their construction. Recent experience with vetiver grass in China and India is encouraging. The Chinese are testing vetiver grass hedgerows in nine provinces with the main field testing and demonstrations in Fujian and Jiangxi. In south India, under four World Bank-supported rainfed agriculture projects more than 20 000 ha of land have been treated with vetiver grass hedgerows as of the end of the 1990 rainy season.

CONCLUSIONS

Vetiver grass hedgerows can provide the soil and moisture conservation function that is one of the key elements for sustainable farming. Vetiver hedgerows can perform this function with a minimum of technical assistance, expenditure and required change in current farming practices. While contour hedgerows of vetiver grass can provide short-term benefits to the farmer, as a baseline intervention, they can provide the long-term stability that allows other management techniques to become effective.

Information on this species was formally compiled for publication by the Board of Science and Technology for International Development (BOSTID) in the National Academy of Sciences. Once their efforts began in 1989, it became apparent that *V. zizanioides* was probably the most widely researched "unknown" species with which they have dealt. In 1993, BOSTID published their scientific audit entitled, "Vetiver Grass—A thin green line against erosion." The BOSTID's panel, headed by Dr. Norman Borlaug, concluded that "the accumulated experiences with vetiver adds up to a compelling case that vetiver is one practical, and powerful solution to soil erosion for many locations thoughout the warmer parts of the world." They also concluded that: "vetiver is unique," and it "brings new advantages" and though "not a panacea" it "adds another technique that seems to have notable benefits...(including) the most important feature...is its compatibility with all other (erosion control) techniques."

REFERENCES

1. Abujamin, S., A. Abdurachman, and Suwardjo. 1985. Contour grass strips as a low cost conservation practice. ASPAC Ext. Bull. 221.
2. Abujamin, S., A. Abdurachman, and U. Kurnia. 1983. Strip rumput permanen sebagai salah satu cara konservasi tanah. Pemberitaan Penelitian Tanah dan Pupuk 1:16-20.
3. Allan, J.A. 1957. The grasses of Barbados. London.
4. Anonymous. 1990. By no means just a "side effect": Results of the field buffer strip programme in Hessen. pests suppressed, and beneficial animals promoted. Appl. Sci. 24(5):7-9.
5. Bajpai, P.N., I. Singh, L.P. Tiwari, O.P. Chaturvedi, and J.P. Singh. undated. Varietal performance of khus (*Vetiveria zizanioides* Stapf.). Punjab Hort. J. 208-211.
6. Basri, I.H., A.R. Mercado, and D.P. Garrity. 1990. Upland rice cultivation using leguminous tree hedgerows on strongly acid soils. IRRI Saturday Seminar, 31 March. IRRI. Los Baños, Laguna, Philippines.
7. Bharad, G.M., and B.C. Bathkal. 1990. Role of vetiver grass in soil and moisture conservation. *In* Proc. of the Colloquium on the Use fo Vetiver in Sediment Control. 25 April. Watershed Management Directorate, Dehra Dun, India.
8. Celarier, R.P. 1959. Cytotaxonomy of the Andropogoneae. IV. Subtribe Sorgheae. Cytologia 24:285-303.
9. Chan, C. 1981. Evaluation of soil loss factors on cultivated slopelands of Taiwan. ASPAC Tech. Bull. 55.
10. Cochrane, M., B. Bartsch, and S. Valentine. 1983. Feed composition tables. Fact Sheet no. 29/83. Dep. of Agric., South Australia.
11. Council of Scientific and Industrial Research. 1976. *Vetiveria* p. 451-457. *In* The wealth of India. Vol. 10. Public. & Inform. Directorate, CSIR, New Delhi, India.
12. Central Soil and Water Conservation Research and Training Institute. 1989. 25 years of research on soil and water conservation in southern hilly, high rainfall regions. *In* K.G. Tejwani (ed.) White Cover Rep.—Draft Material. World Bank Internal Document, Washington, DC.
13. Dalziel, J.M. 1937. Useful plants of west tropical Africa. Crown Agents for the Colonies, London.
14. Doolette, J.B, and J.W. Smyle. 1990. Soil and moisture conservation technologies: Review of the literature. p. 35-71. *In* J.B. Doolette and W.G. Magrath (ed.) Watershed development in Asia.
15. Elatler, E., and C. McCann. 1982. Revision of the flora of the Bombay presidency. Gramineae. J. Bombay Nat. Hist. Soc. 32:408-410.
16. Garrity, D.P. 1989. Hedgerow systems for sustainable food crop production on sloping lands: Elements of a strategy. *In* Symp. Workshop on the Development of Upland Hilly Areas in Region XII, Bansalan, Davao del Sur. 20-22 December 1989.
17. Gilliland, H.B. 1971. A revised flora of Malaya. p. 231-232. *In* Grasses of Malaya Vol. 3. Singapore Botanic Gardens, Singapore.
18. Greenfield, J.C. 1989. Vetiver grass (*Vetiveria* spp.): The ideal plant for vegetative soil and moisture conservation. Asia Tech. Dep. Publ., Agric. Div., The World Bank, Washington, DC.
19. Gupta, S.K., D.C. Das, K.G. Tejwani, S. Chittaranjan, and Srinivas. 1971. Mechanical measures of erosion control. p. 146-182. *In* K.G. Tejwani et al. (ed.) Soil and water conservation research 1956-1971. Indian Council of Agric. Res., New Delhi, India.
20. Hedge, B.R., K.T. Krishnegowda, K.M. Bojappa, P. Ramanagowda, and Panduranga. 1989. Studies on the feasibility of farm ponds for individual holdings. AICRP Ann. Rep. GKVK Univ. of Agricultural Sciences, Bangalore, India.
21. Henty, E.E. 1969. A manual of grasses of New Guinea. Div. of Botany, Dep. of Forests, Lae, New Guinea.
22. Hill, W.G. undated. Note on *Vetiveria zizanioides* specimen 5901 in Kew Herbarium as reported by Frances Cook. Econ. and Conserv. Sect., Royal Botanic Gardens, Kew Richmond, Surrey, England.
23. Hitchcock, A.S. 1980. Manual of grasses of the United States. U.S. Gov. Print. Office, Washington, DC.
24. Hitchcock, A.S., and A. Chase. 1917. Grasses of the West Indies. Contributions from the United States National Herbarium. Volume 18. Part 7. U.S. Gov. Print. Office, Washington, DC.

25. Hooker, J.D. 1975. Flora of British India. Bishen Singh Mahendra Pal Singh, Dehra Dun, India.
26. Kang, B.T., and L. Reynolds. 1986. Alley farming in the humid and sub-humid tropics. IITA, Ibadan, Nigeria.
27. Kang, B.T., H. Grimme, and T.L. Lawson. 1985. Alley cropping sequentially cropped maize and cowpea with Leaucaena on a sandy soil in southern Nigeria. Plant Soil 85:267–276.
28. Kerkhof, P. 1990. Agroforestry in Africa: A survey of project experience. PANOS, Angel House, London.
29. Khoshoo, T.N. 1987. Ecodevelopment of alkaline land: Banthra—a case study. Natl. Botanical Res. Inst., CSIR, Lucknow, India.
30. Krishnappa, A.M. 1989. Report on the Kabbanala watershed project. Operational Res. Project, Kabbanala, Karnataka, India.
31. Kuchelmeister, G. 1987. State of knowledge report on tropical and subtropical hedgerows. Version 1. GTZ, G. Kuchelmeister, Tiefenbach, Germany.
32. Kumar, S. 192. Studies on morphological and genetic variability in Vetiveria zizanioides (Linn.) Nash. Masters thesis Indian Agric. Res. Inst., New Delhi, India.
33. Liao, M.C. 1981. Soil conservation measures for steep orchards in Taiwan. p. 301–310. In South-East Asian Regional Symp. Problems of Soil Erosion and Sedimentation.
34. Lal, A., and V.K. Mathur. 1982. Occurrence of Heterodera zeae on Vetiveria zizanioides. Maize cyst nematode—A new host. Indian J. Nematol. 12:405–407.
35. Lal, R. 1989. Potential of agroforestry as a sustainable alternative to shifting cultivation: Concluding remarks. Agrofor. Systems 8:239–242.
36. Lavania, U.C. 1985. Nuclear DNA and karyomorphological studies in vetiver (Vetiveria zizanioides (L.) Nash). Cytologia 50:177–185.
37. Lazarides, M. 1980. The tropical grasses of southeast Asia. J. Cramer, Vaduz.
38. Magrath, W.B. 1990. Economic analysis of soil conservation technologies. p. 71–97. In J.B. Doolette and W.B. Magrath (ed.) Watershed development in Asia. World Bank Technical Paper no. 127. Washington, DC.
39. McIntosh, A.G.S., and J.A. Allan. undated. Note on Vetiveria zizanioides specimen 440 in Kew Herbarium as reported by Frances Cook. Econ. and Conserv. Sec., Royal Botanic Gardens, Kew Richmond, Surrey, England.
40. Nair, E.V.G., N.P. Channamma, and R.P. Kumari. 1982. Review of the work done on vetiver (Vetiveria zizanioides Linn.) at the Lemongrass Research Station, Odakkali. p. 427–430. In C.K. Atal and B.M. Kapur (ed.) Cultivation and utilization of aromatic plants. Regional Res. Lab., CSIR, Jammu-Tawi, India.
41. Nair, E.V.G., K.C. Rajan, N.P. Chinnamma, and A. Kurian. 1983. Screening of different vetiver hybrids under kerala conditions. Indian Perfum. 27(2):88–90.
42. National Academy of Sciences. 1982. United States-canadian tables of feed composition. NAS Press, Washington, DC.
43. Padmaraju, A., and M. Singa Rao. 1989. Soil and water losses under different management practices involving castor. Maheshwaram Watershed Development Project, Spec. Res. Rep. 1989-90. Andhra Pradesh Agricultural Univ., Pahadi Sharif, Rangareddy District, India.
44. Parham, J.W. 1955. The grasses of Fiji. Fiji Gov. Press, Suva.
45. Pohl, R.W. 1980. Flora Costaricensis, Family 15, Gramineae Fieldiana Botany, New Series no. 4.
46. Punia, M.S., P.K. Verma, and G.D. Sharma. 1989. Heritability estimates and performance of vetiver (Vetiveria zizanioides) hybrids. Indian J. Agric. Sci. 59:71.
47. Purseglove, J.W. 1972. Tropical crops; Monocotyledons. 1. p. 297–298. Halsted Press Div., John Wiley & Sons, New York.
48. Ramanujam, S., and S. Kumar. 1963. Irregular meiosis associated with pollen sterility in Vetiveria zizanioides (Linn.) Nash. Cytologia 28:242–247.
49. Ramanujam, S., and S. Kumar. 1963. Preliminary studies on the mode of reproduction of Vetiveria zizanioides (Linn.) Nash. Proc. Indian Sci. Cong. 50th session. Agric. Sec., Delhi, India.
50. Raponda-Walker, A., and R. Sillans. 1961. Les plantes utiles du Gabon. Editions Paul Lechevalier, Paris, France.
51. Sarrantonio, M. 1990. An assessment of current activities on the use of legumes for soil improvement in world cropping systems. Rodale Press, Emmaus, PA.

52. Sethi, K.L. 1982. Breeding and cultivation of new khus hybrid clones. Indian Perfum. 26(2-4):54-61.

53. Sethi, K.L., V. Chandra, and A. Singh. 1976. Adaptability of vetiver hybrid clones to saline-alkali soils. p. 166-169. *In* Proc. of 2nd Workshop on Medicinal and Aromatic Plants. Gujarat Agricultural Univ., Anand, India.

54. Sethi, K.L., M.L. Maheshwari, V.K. Srivastava, and R. Gupta. 1986. Natural variability in *Vetiveria zizanioides* collections from Bharatpur. Part I. Indian Perfum. 30(2-3):377-380.

55. Singh, U.N., and R.S. Ambasht. 1980. Floristic composition and phytosociological analysis of three grass stands in Naugarh Forest of Varanasi Division. Indian J. For. 3:143-147.

56. Sreedharan, A., and K.C. Nair. 1975. Effect of fertilizers on the yield of root and oil of vetiver, *Vetiveria zizanioides*. Agric. Res. J. Karala 13:197-199.

57. Subagyono, K. 1988. Pengaruh penanaman berbagai tanaman penutup tanah secara strip terhadap erosi dan limpasan permukaan pada Mediteran Desa Jatikerta Sumberpucung Malang. Tesis Fakultas Pertanian Univ. Brawijaya, Malang.

58. Szott, L.T. 1987. Improving the productivity of shifting cultivation in the Amazon Basin of Peru through the use of leguminous vegetation. Ph.D. diss. North Carolina State Univ. Raleigh.

59. Trochain, J. 1940. Contribution a l'etude de la vegetation du Senegal. Mem. de l'Instit. Francais d'Afrique Noir.

60. Virmani, O.P., and S.C. Datta. 1975. *Vetiveria zizanioides* (Linn.) Nash. Indian Perfum. 19:35-73.

61. Truong, P.N., I.J. Gordon, and M.G. McDowell. 1991. Effects of soil salinity on the establishment and growth of *Vetiveria zizanioides*. Asia Tech. Dep. Agric. Div., World Bank, Washington, DC.

62. Wang. 1969. Soil and water conservation experiments in a banana plantation. Fengshan trop. Hort. Exp. Stn. Publ. no. 72.

63. Watson, L., and M.J. Dallwitz. 1988. Grass genera of the world. Australia Natl. Univ. Print. Serv., Canberra, Australia.

64. Whistler, A. 1976. Note on *Vetiveria zizanioides* specimen W 3271 in Kew Herbarium as reported by Frances Cook. Econ. and Conserv. Sec. Royal Botanic Gardens, Kew Richmond, Surrey, England.

65. Williams, L.S., and B.J. Walter. 1987. Controlled-erosion terraces in Venezuela. p. 177-187. *In* W.C. Moldenhauer and N.W. Hudson (ed.) Conservation farming on steep lands. Soil and Water Conserv. Soc. and World Assoc. of Soil and Water Conserv. Workshop, San Juan, Puerto Rico. 22-27 March.

66. Yamoah, C.F., and J.R. Burleigh. 1990. Alley cropping *Sesbania sesban* (L.) Merrill with food crops in the highland region of Rwanda. Agrofor. Systems 10:169-181.

10 Decision Support Systems for Sustainable Agriculture

James W. Jones

Agricultural Engineering Department
University of Florida
Gainesville, Florida

W. T. Bowen

International Fertilizer Development Center
Muscle Shoals, Alabama

W. G. Boggess

Department of Food and Resources Economics
University of Florida
Gainesville, Florida

J. T. Ritchie

Department of Crop and Soil Sciences
Michigan State University
East Lansing, Michigan

Recent literature has focused attention on many of the issues that jeopardize our ability to meet human food, fuel and fiber needs for the future. World population is continuing to increase, thereby creating increasing demands on limited resources. Many agricultural practices threaten the sustainability of our resource base and sometimes cause unacceptable degradation of the environment. Clearly, agriculture is both affected by, and affects, other components of our planet, hence a framework for developing or improving agriculture must be based on long-term and broad-based perspectives.

A concept of "sustainable agriculture" is emerging as a philosophy to guide agricultural research, development, and technology transfer. Embodied in this concept is a recognition that agriculture is a hierarchy of systems operating in and interacting with economic, ecological, social, and political components of the earth. This hierarchy ranges from a field managed by a single farmer to regional, national, and global scales where policies and decisions are made that can influence production, resource use, economics, and ecology at each of these levels. Because it is such a holistic philosophy, agricultural researchers who wish to develop sustainable agricultural systems and policy makers who attempt to influence agriculture are confronted with many difficulties. There are various goals embodied by the philosophy of sustainable agriculture, some of which may be in conflict. Solutions that work in one soil, climate, and socioeconomic setting may not in others. The multiplicity of problems that can prevent an agricultural system from being sustainable and the many variations of practices and policies that could be implemented create a situation in which rapid progress is difficult.

Indicators are needed to quantify the changes that occur in agricultural systems over time at different hierarchical levels for studying agricultural sustainability. These indicators can be grouped into four major categories: productivity, socioeconomic, resource, and ecological. More than one indicator will be necessary to characterize the sustainability of a particular system. Analysis of sustainability then will be based on measurement or prediction of the long-term changes in the indicators appropriately defined for the level of system under study.

Suggestions on how to improve the sustainability of agricultural systems include (i) reduce input use, (ii) improve efficiency of resource use, and (iii) increase use of natural processes such as biological N fixation, nutrient cycling, and integrated pest management. One approach to research is to implement long-term crop management experiments with crop yields, soil characteristics, environmental quality and other indicators periodically measured. Although long-term studies are needed, the impracticality of establishing enough field experiments to provide answers to all questions in all locations should be apparent. Methods are needed, in addition to long-term field experiments, to provide planners and policy makers with the capability to analyze various practices and policies for improving agricultural sustainability.

Modern computer technology can play a major role in the future development of sustainable agricultural systems and in deciding on policies and plans that help ensure their adoption. Decision support systems (DSS) integrate data bases, models, expert systems, analysis and graphics programs, and other information to assist users in planning and decision making. The purposes of this paper are to present design concepts for a decision support system for sustainable agriculture, to present an example of the use of an existing DSS for studying sustainability issues, and to discuss enhancements needed for existing models and data bases.

DESIGN OF A DECISION SUPPORT SYSTEM
FOR SUSTAINABLE AGRICULTURE

In agriculture, the terminology "Decision Support System" has been used to describe a wide range of computer software designed to aid various types of decision makers. Such systems have been developed for making site-specific recommendations for pest management (Michalski et al., 1983; Beck et al., 1989), fertilizer (Yost et al., 1988), farm financial planning (Boggess et al., 1989), and general crop management (Plant, 1989). These decision aids have been developed for use by farmers or by extension agents who are advising farmers. Many have been based on expert system concepts and the programming of rules and logic used by experts in making these decisions. Users provide site-specific information on crop and field conditions, then the DSS provides one or more recommendations. As such, these DSS are usually narrow in scope and may be applicable only within the region where they were developed.

Other DSS provide information and analysis capabilities to assist researchers and planners in determining the best management practices for various crops, soils, and climate conditions. For example, the Decision Support System for Agrotechnology Transfer (DSSAT) provides data bases, crop and soil models, and analysis programs integrated together to evaluate weather-related risks associated with user-specified combinations of crops, locations, and management practices (IBSNAT, 1986; Jones, 1989). This type of DSS has some capabilities for evaluating the sustainability of certain crop systems as will be demonstrated below. Existing DSS, however, lack the scope and functionality to address the range of problems that have to be considered in assessing the sustainability of agricultural systems. They have usually been site oriented, focused on either production or environmental quality, but not on both, and have not dealt with long-term changes in soils, weather, resources and economics.

It is impractical to attempt to design a single DSS that will address all of the issues being discussed in the sustainable agriculture literature or one that will meet the goals of every researcher, planner, or policy maker. However, based on the successes of recent systems, such as the DSSAT, and continuing improvements in technology, it is reasonable to expect that within the next few years considerable progress will be made in developing relevant and useful systems to assist researchers, planners, and policy makers in developing more sustainable production systems. In this section, the functional requirements, components, and preliminary design of DSS for sustainable agriculture (DSSSA) are presented.

Functional Requirements

The proposed DSSSA will have the capabilities needed by researchers, planners, and policy makers to identify agricultural systems that are not likely to be sustainable in a region and to analyze the potential benefits of alternative practices over long periods of time. Table 6-1 summarizes the general functional requirements of the system. It will provide access to information on soils, weather, ecology, and land use and provide maps of the distributions of various attributes over space in the region as well as statistics on their frequencies of occurrence. For example, all saline soils could be located, or all areas within 1 km of lakes could be identified. A Geographic Information System (GIS) will form the basis for storing data in the system and linking them to specific areas in the region. Through an interactive user

Table 10-1. General functional requirements of a DSSSA for use by researchers, planners, and policy makers.

1. Provide information to users on characteristics of a region under study, such as land use, soil types, weather characteristics, farm types, and socioeconomic conditions.
2. Assist users in identifying areas in which existing production systems are likely to be nonsustainable.
3. Suggest changes in practices that should be considered in fragile agroecosystems to improve the sustainability of agricultural production systems, natural resources, and the environment.
4. Analyze alternative practices to determine optimal practices to meet short- and long-term production, economic and environmental goals for a site or farm within the region.
5. Provide policy makers with the consequence analysis tool for evaluating the probable impacts of various policies and practices on productivity, stability and sustainability of agriculture in a region.
6. Provide methods for developing regional agricultural plans that meet desired economic, production, and environmental goals.

interface, DSSSA users wil be able to interactively select areas for analysis within a specified region as well as the type of information to display.

The proposed system also will be able to identify potential problem areas in a region, display them for users, and categorize them as to the type and probable magnitude of the problem. For example, users may wish to identify all nonagricultural areas in a region that would likely be highly erodible under intensive crop production practices or areas in which highly sensitive or endangered species of plants and animals exist. This part of the DSSSA will be based on expert system types of programs using knowledge bases collected from scientists in several fields. It will compare attributes of the landscape with rules and logic in the knowledge base using diagnostic operations and present interpretations to users using maps, graphics, and tables.

The proposed DSSSA also will contain simulation models for evaluating the effects of alternative practices on various indicators of sustainability over time and space. For example, it will contain crop/soil models to simulate the changes in crop yield, soil organic matter, and nutrients over long periods of time under specified practices. It also will contain models to simulate soil erosion and nutrient and pesticide contamination of ground and surface water in the region. These models will need to simulate the desired set of sustainability indicators for each combination of crop, management, soil, and weather condition in the region.

Finally, the DSSSA will provide an interactive tool for use in developing plans and policies that influence the mosaic of crops and practices over a region. For example, a goal may be set by planners to produce a certain crop yield in a region with a high probability level, but at the same time minimizing water pollution, soil loss, and destruction of natural forests.

To provide these functional requirements, the DSSSA will include three major classes of components: data bases, analysis tools, and planning or decision aids. The data bases will store characteristics of the region under study in the computer, and the analysis tools will use the data to predict various sustainability indicators. The planning or decision aids will provide interactive dialogue for users to access any of the data bases and analysis tools and perform any of the functions in the system. Figure 10-1 shows a schematic of the organization of the proposed DSSSA with its spatial and attribute data bases, analysis tools, and decision aids.

Data Bases

The DSSSA will require both spatial and attribute data bases. The spatial data bases will divide the region into areas that delineate soils, weather regimes, land use, socioeconomic conditions, ecology, hydrology, and transportation. Each of these data types are referred to as "layers" in a GIS. By overlaying the layers, every point in the region can be identified by its soil, land use, and other layer characteristics. These spatial data bases provide the information for creating maps of the region in which each area on the map can be colored based on its characteristics to provide users with a picture of their distributions over space.

The attribute data bases will store specific information about each type of soil, land use, farm type, and management practice in the region. For example, soil characteristics such as slope, color, depth, and percentage sand, silt, and clay by horizon would be stored in the soil attribute file for every soil in the region. The attribute files will be related to the spatial files so that the spatial distribution of any attribute, such as soil slope, can be plotted on maps. A GIS will provide the capabilities for storing both spatial and attribute data, creating maps, and performing many calculations that will be required to aggregate and summarize regional or subregional information.

Finally, a second type of attribute data base will need to store, on a temporary basis, results from various models as well as plans being considered for changes in land use or practices. The permanent attribute files will provide inputs to run various models, and the temporary attribute files will store simulated yields, soil loss, organic matter, economics, and other variables which can then be displayed on maps or printed in reports. The temporary attribute file also will store possible changes in land use and practices being analyzed during a session.

Analysis Tools

The main analysis tools of the proposed DSSSA will be simulation models and expert systems. These tools will make use of the data bases and perform the various diagnostic and analysis functions listed in Table 10-1. The major role of the expert system component will be to identify areas in a region where problems are likely to exist under current or proposed practices, and to suggest alternatives for additional analyses. There will be a need for several simulation models in the DSSSA, particularly site specific crop-soil simulation models and hydrology and water quality models. The crop-soil models must include the capabilities to simulate changes in soil properties for given cropping systems and the associated changes in crop production with time. There are many existing crop and soil models (Jones & Ritchie,

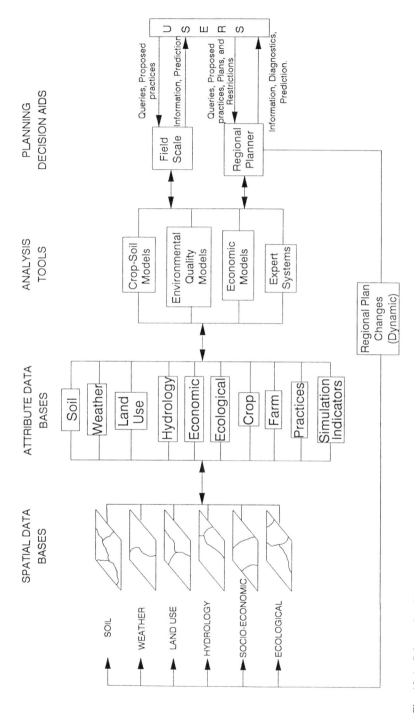

Fig. 10-1. Schematic of data bases, models, and analysis tools for the proposed DSSSA.

Table 10-2. Indicators of changes in field-level cropping systems that may be useful in studying agricultural sustainability.

Productivity	Socio-economic	Resource	Ecology
Yield	Cost of production	Chemical use	Chemical losses
Yield quality	Net returns	Water use	Water quality
Yield stability		Soil depth	Beneficial organisms
		Soil properties	Soil toxins
		Soil nutrients	
		Genetic material	

1990), but most of these currently lack the capabilities to simulate long-term changes in soil properties or to simulate the effects of various soil factors on growth and yield.

Table 10-2 provides a list of some of the major indicators of change for studying the sustainability of cropping systems at a field scale. The crop-soil models must be able to simulate these indicators in response to crop rotations and fallow periods, tillage, irrigation, residue management, fertilizer management, and pest management. The crop–soil models must be sensitive to weather variables, simulate the uptake of nutrients and water, and be sensitive to fertility and pests. Existing crop–soil models have the capabilities to perform some of the analyses required for studying sustainable cropping systems. An example will be presented later in which maize (*Zea mays* L.) and [*Glycine max* (L.) Merr.] rotations, with fallow periods between the crops, will be simulated taking into account irrigation, residue management, and N fertilizer management. However, existing models do not have the capability to simulate all of the indicators of change at the field scale nor the crop responses to them.

Sustainability indicators at watershed or regional scales would have broader scope than those in Table 10-2 for a field scale. For example, the regional indicators would include characteristics of lakes, wetlands, and natural forests. Various hydrology and water quality models exist (USDA, 1980; Beasley, 1977; Hornsby et al., 1990; Hahn et al., 1982). These models simulate nutrient or pesticide losses from fields or water sheds to predict changes in water quality and environmental contamination. By combining these types of models with the cropping system models in the DSSSA, users will have the capability to investigate the trade-offs between production and environmental quality under alternative practices or plans.

Some work has been initiated in which different types of simulation models are included in a regional DSS (Lal et al., 1990). This system, called the Agricultural and Environmental Geographic Information System (AEGIS), uses a GIS, crop–soil models, and an erosion model to assist researchers and regional agricultural planners in Puerto Rico and the Caribbean.

Economic models also will be needed to simulate the changes in farm and regional economies under various scenarios. Agricultural cropping systems must be profitable over time. Changes in supply and demand of inputs and outputs will affect prices and thus profits. Also, changes in agricultural or environmental policies can have dramatic effects on the profitability of cropping systems. Thus, the DSSSA will show profits varying in response to changes in prices and management practices. Some work has been initiated to link farm-scale economic models with crop–soil models in a GIS framework for a Guatemala study of bean- and maize-producing areas (Thornton and Hoogenboom, 1990). In a related study, the results of general circulation and crop–soil models have been used in a spatial equilibrium model to evaluate the economic effects of climate change on U.S. agriculture (Adams et al., 1990).

Planning/Decision Aids

A collection of programs in the DSSSA will be structured in a hierarchical manner, providing analysis and diagnostic capabilities at field, farm, and regional scales. They will provide information to the user, allow options to be selected from pop-up menus on the screen, and provide for flexible map, report, and graphic presentation of results. The site or field-scale user interface will allow users to select a site (soil and weather), specify initial conditions and management inputs, simulate the cropping system over a number of years, and provide graphical and numerical outputs of sustainable indicators and their variabilities. The existing DSSAT (IBSNAT, 1989) has many of the features needed for this field-scale analysis. An expansion of the capabilities of this system to analyze long-term trends can form the basis for this decision aid at the field scale.

The regional-scale planning decision aid also will provide information to users and allow for options to be selected from on-screen menus, but the options will be different from the field scale. Users will be able to select all or part of the region based on a wide range of options, select regional socioeconomic characteristics, select farm and cropping system characteristics and simulate over time and space the different sustainability indicators. It will provide a capability for "remembering" good combinations of land-use practices so that users can incrementally build a plan for the region with acceptable characteristics of any of the predicted attributes. This framework for spatial analysis and plan building is an extension of the one described by Lal et al. (1990) for a prototype regional agricultural planning system (AEGIS).

The development of the DSSSA will require a well-coordinated, multidisciplinary effort. It will require definitions for all of the data bases, their data structures, and methods for data collection. It will require considerably more detail in its design than is presented here before it can be further developed. In addition, it will require input from a wider range of scientific expertise (e.g., anthropologists, wildlife scientists, etc.) to design the system to ensure that appropriate knowledge, data, and analysis tools are included to provide users with credible planning tools.

The structure of the proposed DSSSA would lend itself for use across different regions by putting in data specific to a given region. The models and expert systems also must then be applicable to the region under study.

The models and expert systems should be tested, and the expert systems may need to be revised to include regional expertise.

The following section describes examples of the types of analyses that the crop/soil models will need to perform as a part of the overall system.

EXAMPLE USE OF CROP AND SOIL MODELS

The development of sustainable agricultural systems requires an understanding of the physical, chemical, and biological responses to management practices at a field production unit scale. The responses to existing and proposed practices form the basis for quantitative statements about the sustainability of production in the field itself and, by aggregation over space, some of the sustainability indicators in a region. These responses could be based on long-term experiments or on results from crop and soil simulation models. Resource requirements, production, and chemical losses from the fields in a region will be estimated by models and combined with spatial information on socioeconomics, ecology, and natural resources to estimate regional sustainability indicators. Several existing models simulate soil erosion and its effect on crop growth and yield. These two models (Williams et al., 1984; Shaffer & Larson, 1982) also simulate the dynamics of soil N and C under different tillage, fertilizer, and residue management. The CENTURY model (Parton et al., 1987) simulates long-term changes in soil organic C to study steady-state organic matter levels for grasslands under different climatic zones. Wolf et al. (1989) developed a model to study long-term crop response to fertilizer and soil N, taking into account various soil, crop, weather, and management conditions. The CERES-Maize model (Ritchie et al., 1989) contains a soil N and C dynamics component described by Godwin and Vlek (1985). Therefore, considerable experience has already been gained in modeling several of the important changes that may occur to soils and affect their long-term productivity.

Because of the critical role of crop–soil models for analyses at all scales in the DSSSA, two brief examples are presented here to demonstrate some of the features needed by these models. In the first example, models in an existing DSS are used to estimate changes in maize yield that may occur if practices result in significant soil erosion over some period of time. In a second example, existing maize and soybean models are combined to simulate the effects of long-term crop rotations on soil organic C and N and on crop yield under different N fertilizer levels. The first example is analogous to performing 10 yr of experiments at a site to measure maize response to N and then repeating this experiment 50 to 100 yr later after which time 15 cm of topsoil have been lost. The second example is analogous to a long-term, continuous experiment (60 yr) with two crop rotations at each of two N levels. Crop yield is the primary indicator of sustainability in both examples. In the second example, soil organic C is also simulated to show how different practices may affect this resource and affect the basic productivity of the site.

Table 10-3. Soil water-holding characteristics (θ_{LL} and θ_{DUL}, lower and drained upper limits, respectively), bulk density, pH, and initial soil organic C for the Oxisol from Brazil used in the simulations (From Bowen, 1990).

Depth	θ_{LL}	θ_{DUL}	Bulk density	Organic C	pH
			Mg/m^3	%	
0–15	0.170	0.250	0.93	1.70	5.6
15–30	0.237	0.337	1.03	1.35	5.4
30–45	0.250	0.330	1.01	1.10	5.3
45–60	0.256	0.329	0.95	0.93	5.2
60–75	0.259	0.325	0.93	0.81	5.1
75–90	0.250	0.319	0.91	0.76	4.9
90–105	0.250	0.322	0.91	0.69	5.0
105–120	0.250	0.326	0.91	0.63	5.0

Soil and Climate Characteristics of the Analysis Site

Weather and soil data for these hypothetical analyses represent a site in the Cerrado Region of Central Brazil. The climate in this region is characterized by wet summers and dry winters. Annual precipitation at the site averages about 1500 mm, 80% of which falls during the usual growing season from November to April. Monthly mean daily temperatures range from 19 to 23 °C at the site (Goedert, 1983).

Five years of weather data from a site in Cerrado (16°23′S, 49°8′W) were used with the WGEN weather generator (Richardson & Wright, 1984) contained in the DSSAT (IBSNAT, 1989) to simulate long-term daily weather data for the models.

The soil used in all simulation analyses was an Oxisol (clayey, oxidic, isothermic Typic Acrustoxes), having 65% clay, 20% silt, and 15% sand throughout the profile. The soil profile was 1.20 m in depth, and properties used as model inputs for water holding limits, organic C, pH, and bulk density (Table 10-3) were obtained from field measurements following several years of cultivation (Bowen et al., 1990).

Simulating Yield Changes for an Eroded Soil

The first simulation experiment was conducted using Version 2.1 of CERES-Maize contained in the DSSAT (Ritchie et al., 1989). In each of 10 yr, the maize cultivar Cargill 111 was planted on November 10 at a population of 6.2 plants/m². For each year, seven N fertilizer rates (0, 50, 100, 150, 200, 250, and 300 kg/ha) were simulated to obtain a yield response to applied N. The experiment was then repeated to estimate the N response if the soil was eroded. The eroded soil profile was created by assuming that 15 cm of the original topsoil were removed, organic C was reduced in the remaining soil by 11%, and soil water-holding capacities of the top two layers were decreased by 0.03 cm³/cm³ each, representing a decrease in available water of 0.9 cm.

Results of the 10 yr of simulated experiments were averaged and plotted in Fig. 10-2. When no N fertilizer was added, the yield loss was about 50% for the eroded soil, or about 0.65 t/ha. The response curves indicated

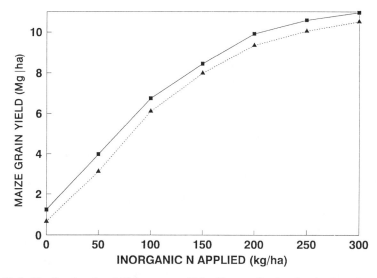

Fig. 10–2. Simulated maize yield response to N fertilizer application for the Cerrado Region of Brazil for the origional soil (■) and the same soil after 15 cm of topsoil erosion (▲).

15 to 20 kg/ha of N would be needed on the eroded soil to produce the same yield as no N on the noneroded soil. Under high levels of N, the yield decrease expected due to changes in soil properties by erosion remained about 0.65 t/ha. However, due to asymptotic yield responses at these high N levels, the amount of extra N needed by the eroded soil to offset erosion increased to 40 to 60 kg/ha. The lower yields of the eroded soil in this example were due to decreased N supply from organic matter in the soil as well as some increase in water stress due to its lower water-holding capacity.

Depending on the practices of the farmer, the magnitude of expected losses in this hypothetical example may or may not be of major consequence. If the normal practice is to apply 200 kg N/ha, then the yield loss would be about 6.5%. By increasing N to 250 kg/ha for the eroded soil, yield could be maintained. However, the resource-use efficiency (yield per unit of N in this case) would be lower. Furthermore, the cost of N may prevent the use of additional N to maintain yield.

If the farmer did not use N, the 50% loss in yield could be devastating, and it may not be possible to add 15 to 20 kg N to maintain the yield due to lack of availability or high cost of N. Clearly, the magnitude of the impact of this loss in soil productivity would depend on the socioeconomic setting. Results from this field-scale simulation provide information for computing productivity, resource, and socioeconomic indicators of sustainability at the field scale.

Simulating Losses in Soil Organic Matter and Crop Yield

In contrast to the previous example, this study simulates the long-term effect of various practices on soil properties for various practices as well as crop yield. In particular, losses in soil organic matter are simulated for 60-yr sequences of cropping practices.

Maize (CERES-Maize V2.1) and soybean (SOYGRO V5.42) models were modified to simulate long-term sequences of maize-fallow and maize-fallow-soybean-fallow crop rotations under two N levels (0 and 150 kg N/ha). The soybean model was modified by Hoogenboom et al. (1990) to simulate soil N processes, using submodels from CERES-Maize, and symbiotic N fixation. The soil N models in both CERES-Maize and SOYGRO were modified by incorporating the method reported by Parton et al. (1987) to represent soil C and N pools and their decomposition rates. Fresh organic matter dynamics as well as inorganic N transformation, transport and uptake remained as they were in the CERES-Maize model. The revised maize and soybean models were then modified to run in sequence with fallow periods (bare soil) between crop seasons.

The treatments were crop rotation (maize each year or soybean–maize, each with bare soil fallow during the dry season) and N fertilizer level for the maize crop (0 or 150 kg N/ha). The maize cultivar, row and plant spacings, and planting dates were the same as for the previous example. A tropical soybean cultivar, Jupiter, was planted each year on November 10 in 0.9 m rows at a density of 30 plants/m^2. Residue from each crop was assumed to be incorporated to 10 cm after harvest. The biomass residue and the C/N ratio depended on growth during each year. Neither of the crops was irrigated in any of the rotations.

Figure 10–3 shows the changes in maize yields for the different rotations and N treatments over the 60 yr of simulated experiments. Year-to-

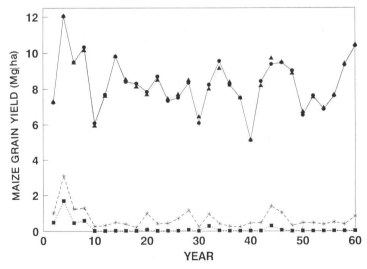

Fig. 10–3. Simulated maize yields for the Cerrado Region of Brazil on an Oxisol over 60 yr. Treatments were ● maize-fallow with 150 kg N/ha, ■ maize-fallow with no N applied, ▲ soybean-fallow-maize-fallow with 150 kg N/ha on maize, and ✱ soybean-fallow-maize-fallow with no N applied.

year variability in each treatment was due to differences in weather in each of the years. In the fertilized treatments, maize yields ranged from about 5.5 to 12.2 Mg/ha, with only a small effect of soybean in the rotation. There was no apparent trend in this sequence of maize yields under high N levels even though soil organic C declined during this time period (Fig. 10-4). The annual variability masked any apparent trends. However, additional analyses similar to the first example may show some loss in productivity.

Results for both crop rotations without N fertilizer showed significant yield declines during the first 10 to 15 yr in addition to annual variability. Without N, the yields of the maize-fallow sequence declined from an average of about 1.0 Mg/ha over the first 3 yr to no significant yield after 15 yr. This decline was attributed directly to the loss in soil organic C (Fig. 10-4). Over this same period, maize yield in the maize-fallow-soybean-fallow rotation treatment without N fertilizer declined, but remained about 0.5 to 1.0 Mg/ha above the yield of the maize-fallow rotation.

Organic C dropped for all treatment combinations (Fig. 10-4), but more so for the treatments without inorganic N fertilizer. Crop residue added back to the soil was considerably higher in the fertilized treatments, resulting in a less rapid decline. Such declines in soil organic C would not be expected to be continuous and equal under all circumstances. Other simulations with these and other models (Parton et al., 1987; Wolf et al., 1989) show that soil organic C reaches steady-state levels which depend on management practices, soil, and climate.

Results from this study estimate that a decline in productivity by maize will occur rather quickly without the replacement of N in this soil. Rotation with soybean supplied some N to sustain productivity. Further improvements in management could be investigated, such as timing of soybean residue incorporation or growing green manure crops prior to each maize crop as shown

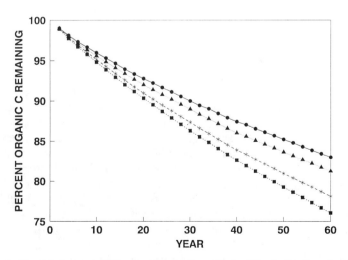

Fig. 10-4. Simulated organic C for the oxisol soil profile for 60 yr in the Cerrado Region of Brazil (See Fig. 10-3 for symbols definitions).

by Bowen et al. (1990). Other benefits of rotation may occur in cropping systems but were not considered in this example.

Crop–Soil Modeling Needs

For use in studying sustainable agriculture issues, crop models must respond to a wide range of soil conditions, such as water, nutrients, organic matter, acidity, salinity, and soil organisms. Currently, most crop models respond to weather and soil water, and some respond to soil N (Jones & Ritchie, 1991). Research is needed to better understand the effects of other soil conditions and how the plant responds to various combinations of stress conditions. Understanding the mechanisms of genetic differences in responses also is needed to simulate differences in cultivars in attempts to develop sustainable crop systems. In addition, more work is needed to incorporate these responses into existing crop models and to test them under a wide range of conditions.

Within a single crop season, changes in many soil properties are small enough to ignore them for predicting crop responses to soil, weather, and management over a season. Analysis of sustainable cropping systems, however, will require that the soil models simulate the slowly changing properties as well. Changes in soil organic C, soil depth, salinity and other variables must be included, and they must be coupled to crop models to properly account for the feedback effects between these two components. These models also must be tested under a range of conditions and over long time periods. It is for this reason that long-term experiments are so important. Monitoring of these long-term experiments should include a minimum date set required to run crop and soil models and to compare experimental and simulated indicators of sustainability.

DISCUSSION

In this chapter, a systems approach to study sustainability is proposed. This requires an explicit definition of the components to be considered, their interrelationships, variables that characterize their response to practices and environment, and external variables that act to influence the system under study. To illustrate this approach, emphasis was placed on an agricultural field, and indicators of the sustainability of a field production system were proposed. Crop and soil simulation models were used to show how sustainability at this scale would be affected by practices that contribute to soil erosion and soil organic matter loss. A point was made that existing crop/soil models need to be expanded to include other factors for broad application in a DSSSA, and that these models need to be tested over time and space.

This is only a starting point. The forces that prevent agricultural sustainability extend beyond the soil and climate changes that may occur at a field site. We do not live in a static world. Changes in populations, resource availability, economics, and policies at regional, national, and international scales create forces for change in agriculture. Therefore, planners and policy makers must be able to respond to these changes and make decisions that facilitate the development of sustainable agricultural systems that adapt to the changes. A systems approach is needed to assist decision makers understand the possible changes in agriculture due to changes that are occuring or are likely to occur in socioeconomic, ecological, and political conditions and to allow them to evaluate different policies and plans. Because the forces of change are variable over time and space, analysis tools in a DSSSA must allow for complete specification of these variables to create the appropriate setting in which analyses are to be performed.

Development of a DSSSA is not a trivial task. It will require a long-term commitment by interdisciplinary research teams to fully design the system and its needed models, data bases, and other analysis tools and to implement and test the various components. It will require an international effort in which scientists from different institutions and disciplines cooperate in the creation of such a system. The logistics of the communications required for accomplishing this goal far exceed those of any previous agricultural research program.

The payoffs of a large-scale effort to develop a DSSSA would more than justify the efforts and costs involved. It would provide a scientific framework for learning how to manage agricultural production in the larger contexts of its interactions with other components of the earth. It would provide context-sensitive analysis tools for planners and policy makers in different socioeconomic, political, agricultural, and ecological settings to help them understand the many consequences of their decisions. It also would provide for improved education across scientific disciplines, the general public, farmers, and policy makers.

ACKNOWLEDGMENTS

Approved for publication as Journal Series no. N-00797 by the Florida Agricultural Experiment Station. This research was supported by the Institute of Food and Agricultural Sciences, University of Florida, Gainesville, and by the IBSNAT program of the USAID, implemented by the University of Hawaii under contract no. AID/DAN-4054-A-00-7081-00.

REFERENCES

Adams, R.M., C. Rosenzweig, R.M. Peart, J.T. Ritchie, B.A. McCarl, J.D. Glyer, R.B. Curry, J.W. Jones, K.J. Boote, and L.H. Allen. 1990. Global climate change and U.S. agriculture. Nature (London) 345:219–224.

Beasley, D.B. 1977. ANSWERS: A mathematical model for simulating the effects of land use and management on water quality. Ph.D. diss. Purdue Univ., West Lafayette (Diss. Abstr. 77-30056).

Beck, H., P.H. Jones, and J.W. Jones. 1989. SOYBUG: An expert system for soybean insect pest management. Agric. Systems 30:269–286.

Boggess, W.G., P.J. van Blokland, and S.D. Moss. 1989. FINARS: A financial analysis review expert system. Agric. Systems 31:19–34.

Bowen, W.T., J.W. Jones, and U. Singh. 1990. Simulation of green manure nitrogen availability to subsequent maize crops. p. 140. *In* Agronomy abstracts. ASA, Madison, WI.

Godwin, D.C., and P.L.G. Vlek. 1985. Simulation of nitrogen dynamics in wheat cropping systems. *In* W. Day and R.K. Atkin (ed.) Wheat growth and modeling. Plenum Press, New York.

Goedert, W.J. 1983. Management of the Cerrado soils of Brazil: A review. J. Soil Sci. 34:405–428.

Hahn, C.T., J.P. Johnson, and D.L. Brakensilk (eds.). 1982. Hydrologic modeling of small watersheds. ASAE, St. Joseph, MI.

Hoogenboom, G., J.W. Jones, and K.J. Boote. 1989. Nitrogen fixation, uptake and remobilization in legumes: A modeling approach. p. 160. *In* Agronomy abstracts. ASA, Madison, WI.

Hornsby, A.G., P.S.C. Rao, J.G. Breath, P.V. Rao, K.D. Pennell, R.E. Jessup, and G.D. Means. 1990. Evaluation of models for predicting fate of pesticides. Project report. DER Contract no. WM-255. Soil Sci. Dep., Univ. of Florida, Gainesville, FL.

International Benchmark Sites Network for Agrotechnology Transfer. p. 1–5. Agrotechnol. Transfer 2:1–5.

IBSNAT. 1989. Decision support system for agrotechnology transfer (DSSAT). Version 2.1 Dep. of Agronomy and Soils, Univ. of Hawaii, Honolulu, HI.

Jones, J.W. 1989. Integrating models with data bases and expert systems for decision making. p. 194–211. *In* A. Weiss (ed.) Climate and agriculture: Systems approaches to decision making. Univ. of Nebraska, Lincoln, NE.

Jones, J.W., and J.T. Ritchie. 1990. The use of crop models in irrigation management. p. 65–89. *In* G.J. Hoffman et al. (ed.) Management of farm irrigation systems. ASAE, St. Joseph, MI. (in press.)

Lal, H., J.W. Jones, F.H. Beinroth, and R.M. Peart. 1990. Regional agricultural planning information and decision support system. p. 51–60. *In* Proc. of Application of GIS Simulation Models nad Knowledge-Based Systems for Landuse Management. Agric. Eng. Dep., Virginia Polytechnical and State University, Blacksburg, VA.

Michalski, R.S., J.H. Davis, V.S. Visht, and J.B. Sinclair. 1983. A computer-based advisory system for diagnosing soybean diseases in Illinois. Plant Dis. 67:459–463.

Parton, W.J., D.S. Schimel, C.V. Cole, and D.S. Ojima. 1987. Analysis of factors controlling soil organic matter levels in Great Plains grasslands. Soil Sci. Soc. Am. J. 51:1173–1179.

Plant, R.E. 1989. An integrated expert decision support system for agricultural management. Agric. Systems 29:49–66.

Ritchie, J.T., U. Singh, D. Godwin, and L. Hunt. 1989. A user's guide to CERES-Maize-V2.10. Int. Fert. Dev. Center, Muscle Shoals, AL.

Shaffer, M.J., and W.E. Larson (ed.). 1982. Nitrogen-tillage-residue management (NTRM) model: Technical documentation. Res. Rep. USDA-ARS, St. Paul, MN.

Thornton, P.K., and G. Hoogenboom. 1990. Agrotechnology transfer using biological and socioeconomic modeling in Guatemala. Res. Rep. Edinburgh School of Agriculture, Edinburgh, Scotland.

U.S. Department of Agriculture. 1980. CREAMS: A field-scale model for chemicals, runoff, and erosion from agricultural management systems. Conservation Res. Rep. 26. USDA, Washington, DC.

Williams, J.R., C.A. Jones, and P.T. Dyke. 1984. A modeling approach to determining the relationship between erosion and soil productivity. Trans. ASAE 27:129–144.

Wolf, J., C.T. de Wit, and H. van Keulen. 1989. Modeling long-term crop response to fertilizer and soil nitrogen. I. Model description and application. Plant Soil 120:11–22.

Yost, R.S., G. Uehara, M. Wade, M. Sudjadi, I.P.G. Widjaja-Adhi, and Z.-C. Li. 1988. Expert systems in agriculture: Determining lime recommendations for soils of the humid tropics. Res. Ext. Series 089-03.88. College of Tropical Agriculture and Human Resources, Univ. of Hawaii, Honolulu, HI.

11 Macroscale Economic and Policy Influences on the Sustainability of Tropical Agriculture

Clyde F. Kiker

Food and Resource Economics Department
University of Florida
Gainesville, Florida

The practice of agriculture in the tropics, as in all locations, is complex. Practice involves response to a large array of elements—climatic, biologic, soil, technologic, economic, social and political—which are highly variable in nature and which together create a context of substantial uncertainty for the farming community. Some elements make up the immediate local context, others are far removed. Still other elements are immediate in time, others off in the future. All, to varying degrees, influence the way agriculture is undertaken. The activities of the people are, in large part, responses to their understanding of these elements, and the sustainability of their agriculture is a function of their abilities to manage resources in face of these uncertainties in time and space.

There are many characteristics that set tropical agriculture apart from agriculture elsewhere. Much of this book deals with these characteristics and with specific technical aspects of the agriculture practiced in response to the biophysical environment. This chapter considers elements of importance to the practice of agriculture originating in the socioeconomic and political systems. Here the interest is in the decision-making environment that concerns opportunities and constraints, incentives and disincentives arising from the interaction of culture, public policy, and market signals as well as nonagricultural employment opportunities. Such influences stem from happenings at many levels ranging from the local community to global markets. Similarly, institutional and market inertia create both positive and negative effects that persist over long periods of time. Largely, an economic perspective is taken, since, to a substantial degree, the many macroscale influences are actually perceived by farmers through market signals. The intent of this chapter is only to introduce broad aspects of these microscale influences on the sustainability of tropical agriculture.

AN ECONOMIC FRAMEWORK

Throughout the Tropics rural people are linked to global economic forces. In many tropical regions, farmers produce for markets outside their immediate communities and purchase inputs and goods from outside. Through these activities, the farmers are connected to the vagaries of economic and policy forces far beyond their communities. People who primarily produce for the market are linked to world market forces by the prices they pay for their purchases and by the prices they receive for their products. This is the case whether these prices are established by market forces or shaped by government policy. Even agrarian people living near the outer frontiers and whose preoccupation is primarily with subsistence are joined to global markets through the few products they sell and purchase. The connections, for better or worse, are there.

While there is much discussion of the meaning of the term sustainability as applied to agriculture, in the following discussion three aspects of agriculture will be of primary concern. The first aspect is the degradation of the land quality through use in the present is such that future productivity is diminished. Second, is the potential of using inputs transported to the farm to alleviate the loss of the land's potential. The last aspect is the potential for losing control over the farm due to the lack of financial viability. Here land is used to encompass not only properties of soil—fertility, moisture retention capacity and other such soil characteristics—but also hydrologic and biologic influences of the broader surroundings. These three aspects of sustainability are interactive. The agriculture practiced by a farmer is considered nonsustainable when land productivity is not sufficient to continue market and/or subsistence levels of production, or when the farmer can no longer maintain financial obligations and gives up control over the farm. In the first case the land's productivity is no longer sufficient to provide the necessary product, while in the second case the land is still productive, but the farmer loses the ability to have benefit of the land's potential because control of the land is taken over by debt holders.

To understand the linkages between externally established forces—economic, social and political—and resource control and use by rural people, it is helpful to use a model. The following, based loosely on Perrings (1989) sets forth the basics of the model.

The Model

First, the level of output, either for direct consumption or for market sales, is a function of the inputs and management applied to the land, the inherent productivity of the land, and the environmental features of the region that encompasses the farm. The effects of agricultural activities on the productivity of the land (per unit area over time given fixed input levels) are characterized by first increasing returns, then diminishing returns following ultimately by negative returns. The implication is that overuse of the land can lead to declining productivity in future time periods. It is possible as part of the productive activity to purchase additional inputs and apply these to the land, thereby, ameliorating, to some degree, the loss of productivity. The potential of substituting is, however, dependent on having cash to purchase the inputs and the knowledge to effectively use them. Overall, the sustainable use of the land resource base will depend on both the people's knowledge base and the economic context.

Second, income is considered to be production used by the household, in-kind income, as well as cash income. The cash income of these rural people for a given period of time—for the present purpose, a year—is from net revenues derived from selling products or from wages for off-farm work in cases where family members work within the region or migrate seasonally to other regions. Similarly, remittances from kin who have permanently migrated serve as income.

Third, savings in a previous time period result from an income level (from production, net revenues, wages and remittances) that exceed the minimum consumption level set by the family. Savings can be in monetary terms or from stored production that can be consumed or marketed in a later time period. The ability to purchase inputs for use in production in a future time period is a function of the monetary savings and the potential for converting stored production into currency.

Fourth, when savings from the previous year's production are not sufficient, it may be possible to purchase inputs using credit. For this to happen year after year two conditions must exist: (i) A source of credit must be available. In many tropical countries credit for some groups may not be at all available, or if it is available, the terms may be so severe that for all practical purposes it is beyond reach. Within families or clans, credits may be provided to individual members who suffer setbacks, but where conditions are such that all members suffer the same setbacks, the savings of the group will be diminished. The only possibility remaining would be through members who have migrated to jobs and who have savings they are willing to loan. (ii) The principle plus interest must be repaid in a short period of time usually from that years' production. When the farmer's production and therefore income is not sufficient to repay, collateral, often land, will be lost.

Fifth, agriculture, from the perspective of the producing unit, is not sustainable when output from the land and off-farm labor are not sufficient to provide income—either monetary or in kind—necessary to continue production in future time periods. It is useful at this point to envision two groups of agricultural producers: one, an agrarian community very near a subsistence level of living with little savings and no credit available, and the second, a community of farmers producing at a commercial scale. For the group very near subsistence, agriculture is not sustainable when the area of land available cannot provide their fundamental needs, either directly or through revenues from sales. They will either have to find additional land or give up agriculture and attempt to find another way of sustaining themselves. The second group has been able in the past to obtain credit and depends on purchase of inputs and sales of product. For them agriculture is no longer sustainable when they lose decision control over the land. This occurs due to financial stress when the land is either taken through legal actions or the farmer sells the land, pays debts and leaves agriculture. In both cases, from the farmers' perspectives the agriculture they were practicing, given the environmental, economic and institutional context, could not be sustained.

Dynamics

Envisioning these two agricultural systems, it is possible to explore the dynamics of the systems and the actions of people in response to unfolding external conditions.

The hope of those attempting to foster agricultural development, and likely their expectation, has been that as agrarian people begin to produce and sell in the market, the land, given the technologies available, would allow production levels sufficient to continue to maintain an acceptable level of living while providing a surplus for marketing. The marketed surplus would give additional income for present consumption and for savings. The savings, in turn, would allow purchases of inputs for use in the next time period thereby increasing production and surplus. In each time period there would be greater production activities on the unit of land. Greater net return per unit of land would increase land values thereby increasing the potential of land as credit collateral. The hope has been that the form of agriculture would be sustainable over time and lift the people above mere subsistence. There is, of course, substantial uncertainty as to what the actual outcomes might be.

The range of possible outcomes is easiest seen by considering several cases (see Fig. 11-1). In each case the focus will be on the prices of inputs and product as the farmers move to greater participation in markets. Recall that these prices are established either in markets or by government, and that the farmers have virtually no control over them.

First, consider the most desirable case: product prices are high relative to input prices and stay that way. If production exceeds subsistence levels, the surplus is marketed and the net revenues saved for purchase of inputs, allowing even greater productive activities in the next time period. Given continued strong product prices and a propensity to save from the surplus, even greater productive activities can be undertaken in future time periods. A stable standard of living above subsistence could be achieved if the people recognize the limits of the land and hold the productive activities—given their technologies—to a level below the point where the land's productivity begins to decline. It may be possible if knowledge of new technologies is available to adopt technologies that provide greater productivity from the land. These new technologies, however, often commit the producers to greater dependence on purchased inputs. Since cash is necessary, sustained savings or credit also will be necessary. Ultimately, the productive potential of activities on the land will reach a limit, but the level of activity can be sufficient to sustain the people while maintaining the productivity of the land. A necessary condition is that product prices relative to input prices remain at a sufficiently high level to maintain the productive system from period to period.

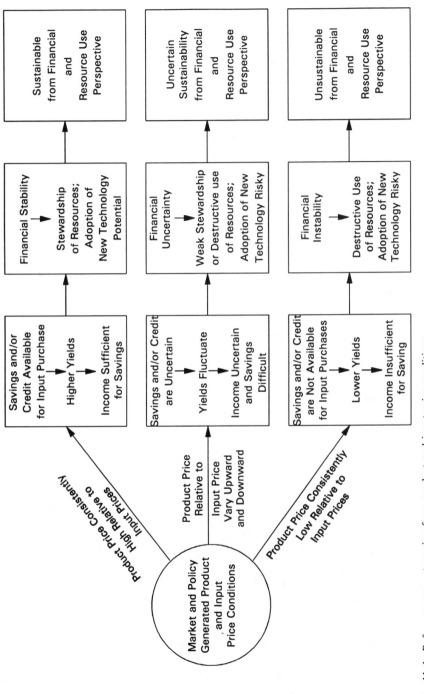

Fig. 11-1. Reference cases stemming from product and input price conditions.

As suggested, this is the most desirable case and will enhance the well being of both the farmers near subsistence and the commercial farmers.

The most likely case is that product prices relative to input prices will vary upward and downward creating uncertainty. For the agrarian people, obtaining a surplus of production that allows savings from one time period to the next generally will be difficult. If the input and product prices in one time period are such that the net income does not allow more than a subsistence level of living that year, savings for purchase of inputs in the next year will not be available, production will be yet lower and the land base will be degraded thereby lowering the following year's production potential. If the poor price situation continues for several years the production potential will deteriorate to the point where the people may not be able to maintain a subsistence level of living. The agricultural activities on the land are clearly non-sustainable and the people inevitably will look for additional land on which to continue these unsustainable practices.

The fate of commercial farmers under conditions of unfavorable relative prices will depend on their liquid financial assets and available credit to tide them over. But if the terms of credit are so harsh that it is unlikely that positive savings can result in years in which prices are good, the ultimate results will be similar: the land will be used in ways that cause a decline in productivity, and, where land was used as collateral, others will take control of the land.

What has been the actual situation in recent years? What signals are being sent to farmers in the Tropics from national governments, international organizations and global markets? The next section considers some aspects of the unfolding situation. Unfortunately, the information available is far from complete and the thought put forward represent at best only broad hypotheses.

ECONOMIC AND INSTITUTIONAL FORCES

Economic conditions, within which farmers in tropical countries produce, result from policy actions by their governments, global market forces, policies of the developed countries and actions by international organizations as well as immediate local market conditions. The policy actions by national and global organizations are discrete in nature, but are interrelated through local, national and global market forces. In fact, the conditions occurring in the broader context, in many ways, actually shape the local market conditions experienced directly by farmers. While it is not possible here to elaborate on the many forces and provide an array of cases, it is possible to illustrate several broad categories thereby giving insight into the relationship of farmers on the countryside to these national and global forces. Policies and actions originating from national governments, governments of developed countries and multilateral organizations will be discussed and then interrelated through a general discussion of agricultural prices.

National Policies

National goverments put into place policies and programs for a variety of reasons (see Olson, 1990). Macroeconomic policies dealing with fiscal, monetary, trade, exchange rate, investment and other issues, as well as policies targeted specifically to the agricultural sector, can either ameliorate or exacerbate conditions faced by farmers. It is clear that agricultural subsidy programs affect agricultural practices, but so do tax policies. Similarly rural development programs and actions can provide and improve basic infrastructure of rural areas thereby improving the farmers' positions. An all weather road can greatly improve farmers' income potential. A government program to assure agricultural credit can provide year to year stability where operating capital is very limited.

All to often, however, government policies in developing countries appear to work against certain groups of farmers, especially limited-resource farmers. In Africa many governments have purposely taken actions to keep food prices to urban dwellers low as other prices rise (Bates, 1991, p. 30-31). Additionally, some governments with the goal of increasing their revenues place export taxes on agricultural products, thereby reducing the price farmers receive. Similarly, they place taxes on imports, thereby increasing farmers' input prices. As Roe (p. 757, 1990) has pointed out "these policies made many countries vulnerable to the world market shocks."

Often groups in power use government policy to reward their supporters. Land policies and credit programs are directed to certain groups while others are left out. In many parts of the tropical world party functionaries and civil servants have become absentee farmers. The result, given the uncertainty of market prices, is that unfavored groups have difficulty maintaining viable enterprises and end up losing their land to the favored groups. Over the long-term very uneven distribution of productive resources occur causing the rural poor to become "informal sector" workers, to farm hillsides, to take game and fuelwood from biologically sensitive lands and to generally "mine" the margins of the rural country side. Some authors (Bunker, 1985; Hecht, 1985; Wood & Schmink, 1978; for example) contend that Brazilian policies have created a situation such that many landless workers moved into the Amazon region where agricultural potential was extremely limited.

The 1970s and 1980s were an especially unstable period. There was great change in export commodity prices and the oil crises sent shocks through many sectors of national economies. Governments attempted to put into place policies to ameliorate the resulting uncertainty stemming from these external shocks. Mexico, an oil exporting nation that would normally be in an advantageous position, can serve as an example. With the dramatic increase in oil revenues, Mexico undertook a number of programs, some directed at agriculture, most directed at broader development goals. The many policy initiatives had many unexpected impacts, and Villa-Issa (p. 744, 1990) puts the impact on agriculture in perspective: "...from 1970 to the present (1989), as a result of past and current policy conditions, the rural sector moved to a point where it was unable to contribute as effectively to the economy. Food imports increased...and the strength of agriculture generally deteriorated." In Mexico as in other nations, the rural economy is tied to the other internal sectors and to international markets. In times of increased external market pressures, the farmers demonstrate an increased tendency to exploit the land base. Long-term productivity drops and environmental degradation occurs.

Many countries not as fortunate as Mexico and other oil exporting countries experienced great difficulties in the late 1970s and 1980s. A pervasive problem arose as governments and large businesses in the 1970s and early 1980s borrowed heavily in the international capital markets. Many countries became highly indebted and this indebtedness lead to policies that worked against many groups within the countries. Devaluation of currency exchange rates was one result. While such devaluations can make export products of the country more competitive on the global market, this occurs as a result of a dramatic decrease in the relative value of indigenous resources—land and labor. Where return to land and labor was already very low, the devaluation can push them into unsustainable use patterns. Again, this can lead to the rural poor's "mining" of the land resources just to subsist. The long-term result is agriculture that is not sustainable in either human or biological terms.

Developed Countries' Policies

The policies of the governments of developed countries also can greatly influence conditions faced by farmers in tropical countries. These policies can be favorable or unfavorable to farmers. The banana (*Musa musaceae*) farmers of the Windward Islands, Jamaica and Belize are the recipients of positive benefits of a United Kingdom policy. As part of a preferential trade agreement the banana farmers receive a green banana wholesale price from England that is twice what it would be from the USA. This policy has been of great benefit to the small farmers of these countries. With the change in the European Economic Community that is presently occurring, the English may have to change this policy, again, greatly affecting the banana farmers.

Recent changes in the U.S. sugar policy can serve to illustrate negative effects on Third World farmers. During the 1980s the USA substantially decreased the quota of sugar allowed into the USA from Caribbean countries. The decrease in revenues have made it difficult for small sugarcane (*Saccharum officinarii*) in Jamaica, Dominican Republic and other countries (Messina, 1989). The U.S. trade agreements with Mexico similarly affect Mexican small farmers. If free trade occurs across the border, low cost U.S. farmers will increase the flow of feed and food grains to Mexico. Mexico's small, relatively high-cost farmers will be greatly affected through greater competition and lower product prices (Levy & van Wijnbergen, 1991).

Broader economic policies of the developed countries also can affect tropical countries. The U.S. government's persistent deficit along with expansionary economic policies affect global capital markets. One result is higher interest rates. For countries with already substantial debts, this makes it difficult to refinance and reduce the immediate debt service difficulties. The result is a decline in the currency exchange rate, and as explained above, this depreciates the relative value of indigenous resources.

Multilateral Organizations

In a way similar to the policies of the developed countries, multilateral organizations such as the World Bank, the International Monetary Fund (IMF) and regional development banks influence the policies of national governments that, in turn, ultimately affect the people of the country side. An example of such a policy area is the "structural adjustment policies" being promoted by the World Bank, the IMF and some developed nations. Under this broad rubric a number of actions are required of governments that seek access to these international organizations for assistance. Devaluation of the nation's currency, downsizing of government (Jamaica decreased their Ministry of Agriculture and Lands to one-third its previous size) and reducing government influence in the economy are typical requirements. Overall, the goal of these programs is to "get the macroeconomic framework right" and thereby to "get prices right," with the ultimate expectation that markets will operate freely and efficiently (Fischer & Thomas, 1990; Frenkel & Khan, 1990; Warford & Partow, 1990). Such changes can lead to greater exposure of farmers to globally established market product and input prices (Gladwin, 1991).

Lele (1990) in a recent article entitled, "Structural Adjustment, Agricultural Development and the Poor: Some Lessons from the Malawian Experience," points out many problems associated with these broad policies. Agriculture is Malawi's dominant economic sector but is characterized by extreme land pressure and poverty. The country's agricultural sector, like that of many other developing countries, can be divided into three subsectors: estates, commercial smallholders and subsistence smallholders. Lele indicates that the first three Structural Adjustment Loans ignored many aspects of the reality of Malawian agriculture, and "had a disappointing effect on growth, and an adverse effect on the poor" (p. 1207). An example is the lack of purchasing power and credit which may have prevented as many as 70 to 75% of smallholders from obtaining fertilizer. Lele suggests, "Broad-based growth in such a sector requires the adoption of an entire gamut of policies toward prices, taxes, subsidies, markets, and asset distribution involving all factors of production, and requiring a long time period to obtain a strong and sustained supply response. Compensatory measures are also needed to protect the poor from adverse effects of some of the adjustment measures" (p. 1207).

Malawi's situation is very little different than that of many developing countries. Farmers find themselves in a situation created at the very broadest levels of national and international policy that makes it very difficult to farm in a sustainable manner.

Agricultural Prices

Prices of products and prices of inputs are by far the most prevalent signals farmers receive from the world of commerce, trade and government beyond their communities. The policies of their governments, of the developed country governments and of the multilateral organizations ultimately affect prices, and these then signal to the farmers the reasonableness, or unreasonableness, of continuing certain agricultural practices. Where product prices are high relative to input prices, as explained in the first section, the farmers' viability improves. Where product prices are low and input prices are high, maintained production becomes difficult. Where product prices are low, input prices high and no savings or credit are available to tide the farm families over to a time of better prices, the primary alternative available to them is to use the land in a nonsustainable way, i.e., to essentially "mine" the land's future potential.

What have agricultural prices been in recent times? Rather than the favorable high product price relative to input price case, indications are that during the 1980s the opposite has been occurring: farm product prices have been low while input prices have been high. Indices of the value of selected agricultural exports for 45 tropical countries have been calculated (Capistrano, 1990; also specific data available from C.F. Kiker). The index is a unit value index for major agricultural exports and was set at 100 for 1980. A mean index for the period 1981 to 1985 was calculated, and the results were as follows: Central America and Caribbean, 36.6; South America, 26.9; Asia, 54.4; and Africa, 53.4. Clearly, during the 4-yr period the export prices to tropical farmers fell dramatically.

Input prices to tropical farmers are more difficult to obtain. For many tropical countries modern agricultural inputs must be imported (the Malawian case is an example). Fuels, fertilizers, pesticides and machinery are very often imported, and where they are not, the prices are substantially influenced by global demand. Not only are farm inputs affected, but also fuel and machinery costs affect the transport of agricultural products. In the USA and other developed countries input prices have been steadily rising; indications are that this also is the case in tropical countries. Since tropical countries must expend foreign exchange on imported inputs, the relative direction of their exchange rates against the dollar can give insight in the direction of input prices faced by farmers. Based on the same 45 tropical countries for the period 1981 to 1985, the average annual rates of change of real exchange rates relative to the U.S. dollar were calculated (Capistrano, 1990). All declined, that is, all have negative rates of change: Central America and Caribbean, −23.2%; South America, −84.2%; Asian, −12.7%; and Africa, −35.9%. These are mean annual declines for the aggregate regions; for many countries the declines were much greater. For example, Brazil, Peru and Mexico had mean annual rates of −191%, −115% and −71% respectively. Again, it is clear that farmers were experiencing dramatic increases in modern input and transportation prices.

Even in the developed countries similar situations occurred. In the USA a similar set of events in the 1980s lead to rapidly rising input prices relative to product prices. Many U.S. farmers were put in difficult positions. Melichar's analysis (1984) puts the situation in perspective. Farmers experienced a decline in total returns from equity from $143 billion in 1973 to $12 billion in 1983 (in 1983 dollars), resulting in at least one-third of the U.S. commercial-scale farmers experiencing a debt/asset ratio greater than 40%, a level indicating substantial financial instability. Fortunately, the U.S. had means of dealing with the agricultural sector's problems, although many farmers of the productive Midwest lost control of their land. The situation has been similar throughout the tropical parts of the world. Many farmers, especially subsistence farmers, either must use the land in an unsustainable manner or lose control of it.

CONCLUSIONS

Macroscale influences on the sustainability of farmers' activities in the Tropics will continue. Generally, events that generate the conditions which ultimately impact farmers lie well outside the realm of agriculture alone. These events have to do with a wide range of political and economic activities undertaken by national governments, governments of developed nations, and multilateral organizations. All too often actions are taken with very little information on the impacts to various groups, especially the rural poor and limited-resource farmers, and it is highly unlikely that this situation will change. Rather it is likely that tropical farmers will continue to face highly uncertain situations arising from powerful groups and organizations within their countries and from abroad.

In saying this, it is recognized that the farmers of the tropical country side will continue to be linked to the global forces through both local and global markets. For better or worse they will be a part of this global market. To help alleviate the problem, farmers and rural people must begin to create local isntitutions (i.e., ways of doing things, behavioral patterns, accords, etc.) that are mutually supportive and help spread the risk encountered during bad periods. These institutions must recognize that the "economic base" of a community must be first to produce for the local community and provide stability there, and then to pursue the monetary gains that come from exporting from the region. The recognition must be, as Daly and Cobb (1989) have noted, that "the real economic base of a community is not exports, but rather consists of all those things that make it an attractive place to live, work, or do business" (p. 135).

This does not mean that there is not a role for national governments and international organizations to bring about sustainable agriculture and communities in the tropics. Certainly, governments can do a great deal more. National governments can establish policies that work toward an equitable distribution of land. Their agricultural research organizations can work to develop agricultural technologies that fit the uncertain situation of many of their farmers. Rural infrastructure development will help get what is grown to markets in good condition and in a timely fashion so that the best prices may be obtained. And, the importance of affordable credit should not be overlooked. Similarly, international organizations, especially the members of the Consultative Group, can help national agricultural research and extension organizations develop agricultural technologies that fit into the objective of establishing a local community as an economic base for weathering the continued oscillation of global economic forces.

Macroscale influences on the sustainability of tropical agriculture are very real. They can, however, be greatly moderated by appropriate agricultural technologies, community organization and policies. But there is great need for assistance. It is highly unlikely that the people of the rural tropical countryside in many parts of the world, who are already at a severe disadvantage, can evolve the necessary appropriate technologies and social organization needed to moderate external shocks without the goodwill and help of the broader national and international communities.

ACKNOWLEDGMENTS

Recognition must be given to the assistance provided by Drs. Peter Hildebrand, Marianne Schmink, Steven Sanderson, Charles Wood, and other members of the Tropical Resources and Development Group at the University of Florida.

REFERENCES

Bates, R.H. 1981. Markets and states in Tropical Africa: The political basis of agricultural policies. Univ. of California Press, Berkeley, CA.

Bunker, S.G. 1985. Underdeveloping the Amazon. Univ. of Illinois Press, Urbana, IL.

Capistrano, A.D. 1990. Macroeconomic influences on tropical forest depletion: A cross-country analysis, 1967–1985. Ph.D. diss. Univ. of Florida, Gainesville, FL.

Daly, H.E., and J.B. Cobb. 1989. For the common good: Redirecting the economy toward community, the environment, and a sustainable future. Beacon Press, Boston.

Fischer, S., and V. Thomas. 1990. Policies for economic development. Am. J. Agric. Econ. 72:809–814.

Frenkel, J.A., and M.S. Khan. 1990. Adjusted policies and economic development. Am. J. Agric. Econ. 72:815–820.

Gladwin, C.H. (ed.). 1991. Structural adjustment and African women farmers. Univ. of Florida Press, Gainesville, FL.

Hecht, S.B. 1985. Environment, development and politics: Capital accummulation and the livestock sector in eastern Amazonia. World Dev. 13:663–684.

Lele, U. 1990. Structural adjustment, agricultural development and the poor: Some lessons from the Malawian experience. World Dev. 18:1207–1219.

Levy, S., and S. van Wijnbergen. 1991. Agriculture in the Mexico–U.S.A. Free Trade Agreement. Working Paper. World Bank, Washington, DC.

Melichar, E. 1984. A financial perspective on agriculture. Fed. Res. Bull. January, p. 1–13.

Messina, W.A. 1989. U.S. sugar policy: A welfare analysis of policy options under pending Caribbean Basin expansion act legislation. M.S. thesis, Univ. of Florida, Gainesville, FL.

Perrings, C. 1989. An optimal path to extinction? Poverty and resource degradation in the open agrarian economy. J. Dev. Econ. 30:1–24.

Olson, M. 1990. Agricultural exploitation and subsidization. Choices 5:8-11.

Roe, T.L. 1990. International financial markets and agricultural adjustment in North America: Discussion. Am. J. Agric. Econ. 72:755-757.

Villa-Issa, M.R. 1990. Performance of Mexican agriculture: The effects of economic and agricultural policy. Am. J. Agric. Econ. 72:744-748.

Warford, J.J., and Z. Partow. 1990. Natural resource management in the Third World: A policy and research agenda. Am. J. Agric. Econ. 72:1269-1273.

Wood, C.H., and M. Schmink. 1978. Blaming the victim: Small farmer production in an Amazon colonization project. Stud. Third World Soc. 7:77-93.

12

Farmer Participation for More Effective Research in Sustainable Agriculture[1]

Walter W. Stroup

Department of Biometry
University of Nebraska
Lincoln, Nebraska

Peter E. Hildebrand

Food and Resource Economics Department
University of Florida
Gainesville, Florida

Charles A. Francis

Department of Agronomy
University of Nebraska
Lincoln, Nebraska

If developing countries are to meet national food needs and alleviate rural poverty, millions of small farmers must become active participants in the agricultural research and development process (Whyte & Boynton, 1983).

Most crops and many predominant agricultural production systems are the result of empirical research, or trial and error, by generations of farmers working the land. Neolithic farmers knew much about 1500 different plant species used for food and medicine (Braidwood, 1967). Vestiges of their traditional subsistence system still exist in many regions (Francis, 1986; Plucknett & Smith, 1986). With the advent of scientifically based agriculture following World War II; however, farmers' influence on technology development became less and less.

In the late 1960s and early 1970s, the international development community began to see a need to reach the many small, resource-poor farmers who were being by-passed by the Green Revolution. Whyte and Boynton (1983) argued that this meant (i) an increased emphasis on on-farm research,

---

[1] Staff papers are circulated without formal review by the Food and Resource Economics Department. Contents are the sole responsibility of the authors.

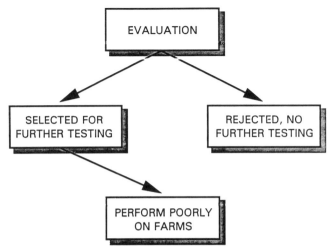

Fig. 12-1. Possible results of on-station testing.

(ii) greater inderdisciplinary collaboration, (iii) agricultural bureaucracies that are more responsive to the interests and needs of small farmers, and (iv) small farmers should no longer be treated simply as passive recipients of what the experts decide is good for them. Rhoades (1989) urged the participation of farmers in developing a research agenda.

To respond to this new clientele, a methodology was needed to efficiently find environment-specific technologies for large numbers of such farmers. This methodology had to not only reach farmers in widely varying and often difficult situations who lack the resources required to dominate the environment, but also (i) speed up the technology development, evaluation, delivery and adoption process; and (ii) efficiently use scarce institutional resources (those human, physical, and financial resources of national agricultural research and extension services in developing countries). In order to accomplish these needs, the methodology required an integrated, multidisciplinary approach that incorporated farmers, researchers and extension personnel (Gilbert et al., 1980; Hildebrand, 1988).

Internationally, over the last 20 yr, this "real world" or on-farm research for large numbers of farmers has come to be called Farming Systems Research and Extension (FSRE). In the broadest sense FSRE involves (i) rapid diagnosis of farm problems by multidisciplinary teams to provide the basis for, (ii) adaptive and descriptive biophysical on-farm research which is supported by, (iii) socioeconomic research on-farm and in the farm community, (iv) controlled biophysical research in laboratories and on-station, and (v) simultaneous dissemination and diffusion of results.

By incorporating farmers from the beginning of technology development—from problem diagnosis through adaptation and evaluation—FSRE methodology reduces the incidence of research results that *perform poorly on farms* (Fig. 12-1), or the rejection on-station of technologies which *might have performed well* on farms but were *never released* (Fig. 12-2).

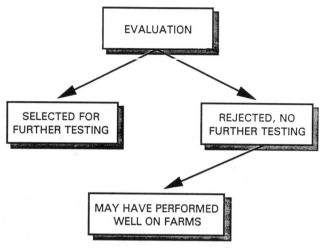

Fig. 12-2. Possible results of on-station testing.

As methods have developed over time, FSRE is not limited to small farms. Indeed, efficiency in the use of research resources is enhanced by incorporating farmers from multiple environments. In this chapter, methods for incorporating large numbers of limited resource farmers into on-farm research are discussed. Later in the chapter, means of including larger, more commercial farms are covered.

CONCEPTS AND METHODS

Diagnosis

Using FSRE methods, farm problems are diagnosed by rapid rural appraisal procedures (Chambers, 1981) or sondeos (Hildebrand, 1981), that incorporate farmers as active participants working with multidisciplinary research and extension teams. These methods are flexible and may or may not use formal questionnaires in the process. Problems encountered are elaborated and prioritized for research by several methods including those proposed by Tripp and Woolley (1989) from International Center for Improvement of Maize and Wheat (CIMMYT) and International Center for Tropical Agriculture (CIAT).

Research Domains

An earlier concept that sought homogeneous groups of farms (Hildebrand, 1981; Norman, 1980) has been modified to incorporate the concept of *a research domain* (Wotoweic et al., 1988) which recognizes the fact that farms and farmers are highly variable and *targets* this *variability*. Often research domains are chosen based on biophysical characteristics although

Fig. 12–3. Example of tobacco field on resource-poor farm in northern Florida.

they may be chosen politically. Research domains ideally contain a wide range of environments that are incorporated as early as possible in the technology screening process. Environments in this context can be associated with farms, fields or even portions of fields. The use of socioeconomic considerations in the choice of environments within the research domain enhances efficiency in technology development and evaluation.

To comprehend a research domain, compare the environment for producing tobacco in the field on the small, resource-poor farm in northern Florida shown in Fig. 12–3—with the environment for raising tobacco in the field on a larger farm, in the same area, but which has enough resources that it can dominate the environment to a much greater extent, shown in Fig. 12–4—and to the environment for raising tobacco on an experiment station, also in the same area, where it is grown with few limitations, allowing most environmental factors to be dominated, Fig. 12–5.

All of these environments can be considered part of the same research domain and be incorporated simultaneously in an integrated technology development, evaluation and diffusion process for tobacco (*Nicotiana tobacum*) in northern Florida. The nature of on-farm research in research domains is *exploratory*, to answer the questions *what* and *where*, not why and when. Diverse environments, such as those shown in northern Florida, enhance the exploratory nature of on-farm research in research domains.

Fig. 12-4. Example of tobacco field on resource-sufficient farm in northern Florida.

Fig. 12-5. Example of tobacco field on agricultural experiment station.

Recommendation Domains

In a research domain, an integrated, multidisciplinary research and extension team conducts both biophysical and socioeconomic on-farm research and analyzes the results to (i) characterize the biophysical environments associated with each location, (ii) elicit farmers' evaluation criteria with respect to the technology being evaluated, and (iii) define recommendation domains. *A recommendation domain is a unique combination of these environmental characteristics and evaluation criteria.*

Recommendation domains, then, are one or more subsets of a research domain which *target for homogeneity* of (i) natural and farmer-created biophysical environments, and (ii) farmers' evaluation criteria for the technology being evaluated. Also modified from previous thinking that recommendation domains pertained to whole farms (Byerlee et al., 1982; Harrington & Tripp, 1984), or cropping or farming systems (Hildebrand, 1981) are that these logically can refer as well to individual fields on a farm, or even different locations in the same field. The most important concept is to consider recommendation domains as *environments* whose biophysical and socioeconomic characteristics can be identified.

The nature of on-farm research in recommendation domains is *validation*, to confirm answers as to (i) how each alternative (treatment) will respond, and (ii) where each alternative is best, as well as to refine the characterization of the recommendation domains and farmer evaluation criteria. At this stage, the number of treatments in on-farm trials is limited. Extension personnel can play an increasingly important role by expanding coverage for evaluation and enhancing exposure (diffusion) of the technology.

Diffusion Domains

Diffusion domains are informal *interpersonal communication networks* through which newly acquired knowledge of agriculture technology normally flows. Knowledge of these networks is important in helping research and extension personnel locate on-farm trials to *target for communication.*

The challenge of diagnosis and identification of these several domains is complicated by how information is collected, analyzed, and evaluated from on-farm trials. We need to clearly identify where research results can be applied, how broad the recommendations can be, and for whom these new technologies are appropriate. To be credible for farmers as well as rigorous from a statistical point of view, results from on-farm research must be analyzed and evaluated according to valid statistical methods.

ANALYTIC VS. ENUMERATIVE STATISTICAL METHODS

Over the past several decades, procedures for the design and analysis of experiments have been developed and utilized very effectively in agricultural research. Many of these procedures have become so institutionalized that it is easy to lose sight of the fact that they are only *specific* applications of statistical theory to *specific* experimental conditions—namely, those of the agricultural experiment station.

Are the requirements of on-farm trials identical to those of experiment station trials? There is no good reason to expect they should be. In fact, on-farm trials differ from their on-station counterparts in two very significant ways: (i) the objectives are typically quite different; and (ii) the variability of the data in an on-farm trial is typically more complex and must be addressed with greater sophistication than is normally required for an on-station trial.

How do the objectives of on-farm trials differ from on-station research? How does this in turn affect decisions regarding appropriate statistical methodology? Although probably not obvious to agricultural researchers, on-farm trials have many statistical similarities to quality improvement experimentation in manufacturing. Deming (1953, 1975), the statistician whose contributions to quality in Japanese industry is legendary, distinguishes between two approaches to statistical analysis: enumerative and analytic (Deming, 1975).

Enumerative

"The action to be taken on the frame depends purely on estimates or complete counts of one or more specific populations of the frame. The aim of the statistical study in an enumerative problem is descriptive." Virtually all classical statistical procedures—*t*-tests, *F* tests, analysis of variance (ANOVA), standard confidence intervals—are enumerative in nature.

Analytic

"In which action will be taken on the process or cause system that produced the frame studied, the aim being to improve the practice in the future." Only statistical procedures which involve *prediction* rather than estimation or hypothesis testing are analytic in nature.

Deming (1975) puts it another way: "A 100 percent sample in an enumerative problem provides the complete answer to the problem posed for an enumerative problem...In contrast, a 100 percent sample of a group of patients, or of a section of land, or of last week's product, industrial or agricultural, is still inconclusive in an analytic problem. This point, though fundamental in statistical information for business, has escaped many writers."

Clearly, most on-farm trials have *analytic* rather than *enumerative* objectives. Thus, the literal application of enumerative statistical procedures, many of which form the core of statistical tradition in agricultural research, is not appropriate for most on-farm trials. For example, the ANOVA can be very useful for interpreting data from on-farm trials. However, traditional ANOVA places much emphasis on hypothesis testing and significance levels. These are important in enumerative studies, but essentially irrelevant to analytic studies, where the emphasis is on prediction and taking action.

Ad hoc statistical procedures are common in analytic studies. While many of these procedures can be validly criticized using enumerative statistical arguments, these criticisms often miss the point. Analytic studies are usually conducted with less prior knowledge of and control over experimental conditions. The choice frequently is between *no* knowledge and *useful, if imperfect* knowledge; conditions of *optimality* characteristic of enumerative statistical procedures are simply not an option. Analytic studies typically sacrifice control over variability for a broadened *research domain*. This does not make them incorrect or invalid, it just means that the researcher must understand the trade-offs and choose statistical methods accordingly.

The complex variability in on-farm trials often troubles those trained in traditional statistical methods for agricultural research. In statistical jargon, these methods are examples of "ordinary least squares"; their main virtue is that they are easy (relatively) to do without a computer, which was a vital consideration in the 1920s and 1930s when they were developed. Their main drawbacks are the rigid structure and narrow, frequently unrealistic assumptions required of the data to permit legitimate interpretation. Since on-farm trials rarely satisfy these assumptions, many have concluded— falsely—that they are somehow "statistically improper." In truth, *traditional methods simply cannot accommodate the complexity of on-farm trials.*

"Ordinary least squares" theory has long since been supplanted by more versatile methods, *mixed linear model* methods (or "mixed model methods" as they will be referred to here) being of particular importance to on-farm trials. The virtue of mixed model methods is their flexibility, their drawback is that they generally require a computer. Thus, while mixed model theory has been around for nearly 50 yr, it did not become practical to use until the 1970s in developed countries and the 1980s in most developing countries. By then, more traditional methods were so deeply entrenched in statistics courses, on experiment stations, and in agricultural research journals that substantial re-education has either been required or, more correctly, is still required.

Recently, there has been a great deal of interest in applications of mixed model theory in agriculture. Henderson (1975) developed *best linear unbiased predictors* (or BLUPs) as an alternative to more enumerative-type estimators. Perceived at first as an *ad hoc* procedure, Harville (1976) put BLUP on sound theoretical footing. A regional publication of the Southern Research and Information Exchange Group in statistics (Southern Regional Bulletin, 1989) contained several examples of mixed model applications in agriculture. This publication also contained articles by McLean (1989) and Stroup (1989a) describing mixed model theory and methods.

In the following section, mixed linear models appropriate for on-farm trials are discussed. These models *superficially* resemble models used to evaluate on-station data. The goal of this section is to show how to use mixed model theory to understand the distinction between the various assumptions that can be made about these models, their effect on the resulting analysis, and their implications for the on-farm researcher. The larger objective is to empower the on-farm researcher with a *relevant* statistical perspective so that design and analysis choices appropriate to on-farm trials can be made.

THE "TYPICAL" ON-FARM TRIAL

On-farm trials are conducted in a variety of ways, but most have a common basic structure. The following is a generic description of the essential elements:

Suppose a number of treatments, V, are to be evaluated. Each treatment is observed at F different farms where the specific biophysical and socioeconomic characteristics of the specific site on the farm will be characterized. At each farm site, each treatment is "replicated" R times—the word "replicated" appears in quotes here because, as will become apparent later in this discussion, multiple observations on treatments within a farm site may not be true replications. Note that the term "farm," to be designated in what follows by the letter "F" is generic. The term more specifically should be interpreted as "environment." In specific trials, "field," "location," "village," etc., may apply equally.

Schematically, this trial can be represented as in Fig. 12-6. As a starting point for analysis of this trial, the following mathematical model can be used

$$y_{ijk} = \mu + f_i + r(f)_{ij} + v_k + vf_{ik} + e_{ijk}, \qquad [1]$$

where

y_{ijk} is the observation on the jth replication of the ith farm for the kth treatment,

μ is the overall mean,

f_i is the effect of the ith farm,

$r(f)_{ij}$ is the effect of the jth replication on the ith farm,

v_k is the effect of the kth treatment,

vf_{jk} is the interaction between the ith farm and kth treatment, and

e_{ijk} is residual variation not accounted for by the above effects.

The ANOVA implied by this model has the following *general form*

Source of Variation		Degrees of Freedom
FARM	F	$F - 1$
REP(FARM)	$R(F)$	$F(R - 1)$
TREATMENT	V	$V - 1$
FARM × TREATMENT	$V \times F$	$(F - 1)(V - 1)$
RESIDUAL	resid	$F(R - 1)(V - 1)$
TOTAL		$FRV - 1$

This ANOVA has several possible interpretations, depending on the specific objectives of a given on-farm trial and how the effects in the model are defined as a consequence. In order to make *appropriate* use of this

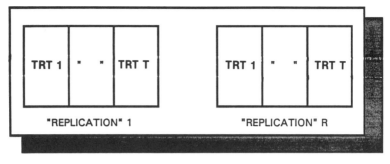

FARM 1

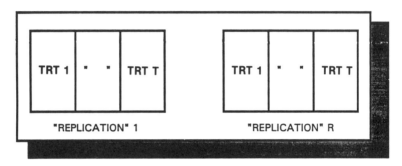

FARM F

Fig. 12-6. Representation of a typical on-farm trial.

ANOVA table, the researcher must be clear about the objectives of the trial and the nature of the effects being observed. Some useful definitions follow.

Population of Inference

The set of elements (e.g., farms) to which the results of the study are to be applied. This is similar to the concept of a research domain.

Prediction Space

Applications of study results from on-farm trials often take the form of recommendations. Recommendations are based on the predicted behavior of the treatments, either for the entire population or for various subpopulations. The set of elements (e.g., farms or environments) to which a prediction is intended to be applicable is called the prediction space. This is similar to the concept of a recommendation domain.

Random and Fixed Effects

Effects in the study—treatments, farms, "replications"—can be considered as *fixed* or *random* depending on (i) how they are chosen and (ii) what prediction space is appropriate to the objectives of the study. An effect is considered *fixed* if the levels of a particular factor are chosen deliberately in advance of the study. In this case identical levels would be used again were the study to be repeated based on the same prior knowledge, and prediction is limited to only those levels actually represented in the study. Typically, treatments such as tilage methods or fertilizer levels in a variety trial would be considered fixed effects. An effect is considered *random* if the levels actually observed in the study result from a random sample of a larger population—identical levels in a repeat of the study would be exceedingly unlikely. Prediction in this case is intended to apply to the population of which the levels observed are only representatives. The most blatant example of a random effect would be the effect of "replication" or of residual variation. Many effects are not clearly fixed or random—the effect of farm site or environment, for example. Whether an effect is fixed or random has a major impact on the analysis, as will be demonstrated below.

Most statistical methods texts, e.g., Steel and Torrie (1980) or Snedecor and Cochran (1980), contain discussions of fixed and random effects. Many texts on the design or planning of experiments discuss the population of inference, e.g., Cox (1958) or Mead (1988). The reader is referred to these texts for more detail.

IMPACT OF FIXED OR RANDOM EFFECTS
ON ANALYSIS OF VARIANCE

In the ANOVA for the on-farm trial given above, it is usually fairly clear that "treatments" are fixed effects and "replications" are random. Farms, however, are not so easily categorized. Different farms may have been selected quite intentionally based on certain criteria: size, income, technology level, soil type, climatic characteristics, etc. Or they may have been selected at random from a target population. Actually, these are extremes; usually, farms are selected using a combination of fixed and random effect tactics. That is, a spectrum of defined conditions must be represented, but some form of random sampling is done within each condition. Essentially, this amounts to stratified random sampling.

It follows that farms are not easily categorized as fixed or random. Usually, in fact, the "correct" analysis of the on-farm trial will involve some compromise between the analysis with farm as a fixed effect and the analysis with farms as random. Before examining this "compromise," it is instructive to look at the appropriate analyses with farms strictly fixed or strictly random.

If farms are fixed, then the only random components of Model [1] are $r(f)_{ij}$ and e_{ijk}. Denote the variance of $r(f)_{ijk}$ by σ_{rf}^2 and the variance of e_{ijk} by σ^2. Then the expected values of the mean squares of the ANOVA are as follows

Source of variation	Expected mean square
F	$\sigma^2 + V\sigma_{rf}^2 + RV\phi_f$
$R(F)$	$\sigma^2 + V\sigma_{rf}^2$
V	$\sigma^2 + FR\phi_v$
$V \times F$	$\sigma^2 + R\phi_{vf}$
residual	σ^2

where ϕ_f, ϕ_v, and ϕ_{vf} denote variation attributable to the fixed effects f_i, v_k, and vf_{ik}, respectively.

If farms are random, then the components f_i and vf_{ik} from Model [1] also are random. Denote their variances by σ_f^2 and σ_{vf}^2, respectively. Then the expected mean squares are

Source of variation	Expected mean square
F	$\sigma^2 + R\sigma_{vf}^2 + V\sigma_{rf}^2 + RV\sigma_f^2$
$R(F)$	$\sigma^2 + V\sigma_{rf}^2$
V	$\sigma^2 + R\sigma_{vf}^2 + FR\phi_v$
$V \times F$	$\sigma^2 + R\sigma_{vf}^2$
residual	σ^2

These two ANOVA tables imply very different approaches to inference. When farms are fixed, the data analyst's first concern must be the farm by treatment interaction ($V \times F$), the magnitude of which is assessed by the F ratio $MS(V \times F)/MS(\text{resid})$. If this F ratio indicates the existence of interaction, then effort must be focused on understanding its nature. Even if the interaction F ratio appears to be negligible, the data analyst would do well to partition $MS(V \times F)$ into meaningful components, e.g., using contrasts, since important interactions often are masked by the large number of degrees of freedom associated with the $V \times F$ effect (where MS = mean square and resid = residual [see Snedecor and Cochran (1980), p. 304–307]).

When farms are considered fixed, the treatment *main effect* is of interest only if the interaction effects are negligible, i.e., if it is clear that the same relationships among treatment means appear to hold for *every* farm in the population of inference. This is generally not true, but if it is then the treatment main effect can be evaluated using the F ratio $MS(V)/MS(\text{resid})$.

When farms are considered random, then test of farm by treatment interaction, which uses the same F ratio as above, has a far different interpretation. Specifically, it means that differences among treatments vary at random by farm. This is quite distinct from the fixed effect case, in which interaction implies that relative differences among treatments are affected by systematic, identifiable and repeatable farm characteristics (i.e., the characteristics that motivated the choice of the farms in the first place). In fact, the test for interaction is not particularly interesting if farms are random; if σ_{vf}^2 is not greater than zero, then the assumption of random farms is probably defective. Of interest is the treatment main effect. This is evaluated using the F ratio $MS(V)/MS(V \times F)$. Its purpose is to verify that differences among treatment means, substantial and consistent enough to be seen throughout the population of inference, over and above random differences among treatment by farm, actually exist.

To summarize, if farms are fixed, the F ratio of primary interest is that for the $V \times F$ interaction, $MS(V \times F)/MS(\text{resid})$, or, *if the $V \times F$ interaction is negligible* then the V main effect is assessed by $MS(V)/MS(\text{resid})$. If farms are random, the $V \times F$ test is of little intrinsic interest (except to verify the validity of the assumptions); of primary interest is the V main effect, which in this case has an F ratio $MS(V)/MS(V \times F)$.

In on-farm trials as they are actually conducted, farms are rarely purely fixed or purely random effects. The above ANOVAs, therefore, are useful as academic exercises to illustrate issues the farming systems researcher needs to understand, but neither, unmodified, is likely to be of much use in practice.

PARTITIONING THE FARM BY TREATMENT INTERACTION

In most on-farm trials, the population of inference includes a set of "types of environments," that the researcher wants to be represented. In the extreme fixed effects case, the number of types would be F, and thus only one environment per type would be observed. In the extreme random effects case, there would be exactly one type of environment (or so little would be known about the environments that typing could not be done prior to conducting the trial) and F randomly sampled environments per type. Usually on-farm researchers would reject either extreme, a more realistic design would be to randomly sample a number of environments from each of the several types in the population.

If the types of "farms" are very well defined, Model [1] could be modified as follows

$$y_{ijkl} = \mu + t_i + f(t)_{ij} + r(tf)_{ijk} + v_l + vt_{il} + vf(t)_{ijl} + e_{ijkl}, \quad [2]$$

where

t_i is the effect of farm type,
$f(t)_{ij}$ is the effect of farm within type,
vt_{il} is the farm type by treatment interaction,
and other terms follow by extension from Model [1].

In Model [2] type and treatment would be considered fixed, farm and replication random, and analysis would proceed accordingly based on the following ANOVA

Source of variation	df	Expected mean square
T	$T - 1$	$\sigma^2 + R\sigma^2_{vft} + V\sigma^2_{rtf} + RV\sigma^2_{ft} + FRV\phi_t$
$F(T)$	$T(F - 1)$	$\sigma^2 + R\sigma^2_{vft} + V\sigma^2_{rtf} + RV\sigma^2_{ft}$
$R(TF)$	$TF(R - 1)$	$\sigma^2 + V\sigma^2_{rtf}$
V	$V - 1$	$\sigma^2 + R\sigma^2_{vft} + TFR\phi_v$
$V \times T$	$(T - 1)(V - 1)$	$\sigma^2 + R\sigma^2_{vft} + FR\phi_{vt}$
$V \times F(T)$	$T(V - 1)(F - 1)$	$\sigma^2 + R\sigma^2_{vft}$
residual	$TF(R - 1)(V - 1)$	σ^2

The type by treatment $(V \times T)$ interaction would be of initial primary interest. Its F ratio is $MS[V \times T]/MS[V \times F(T)]$.

As before, partitioning $MS[V \times T]$ into meaningful contrasts would be strongly advisable. For example, suppose the farm types are (i) higher rainfall, mechanized; (ii) higher rainfall, nonmechanized; (iii) lower rainfall, mechanized; and (iv) lower rainfall, nonmechanized; and the treatments are (i) standard variety, no fertilizer; (ii) standard variety, with fertilizer; (iii) resistant variety, no fertilizer; and (iv) resistant variety, with fertilizer.

The type main effect could be partitioned into rainfall and mechanization main effects and a rainfall by mechanization interaction. The treatment main effect could be partitioned into variety and fertilizer main effects and a variety by fertilizer interaction. Then the interaction of any of the three type effects with any of the three treatment effects could be evaluated. For example, a rainfall by variety effect could be examined to see if the resistant variety is equally advantageous at lower and higher rainfall. In the unusual case that type by treatment interactions are negligible, the treatment main effect could be tested using $MS[V]/MS[V \times F(T)]$.

Predicted performance of treatments for particular farm types can be obtained using confidence intervals for the treatment × farm type means. Care should be taken to base the confidence interval on the correct standard error. Most statistical software packages are poorly suited to work with mixed linear models such as Model [2] without special attention. For a complete discussion of this issue, see McLean (1989) and Stroup (1989a). Predicted performance of specific farms within a given farm type for a particular treatment can be obtained by calculating BLUPs (Henderson, 1975). These are not the same as usual sample means. Again, see McLean (1989) and Stroup (1989a,b) for a full discussion of best linear unbiased prediction.

STABILITY ANALYSIS

A special case of the above analysis occurs when "environmental types" and their potential interactions with treatment are not well understood prior to conducting the on-farm trial. In such cases, the researcher makes an attempt to represent as wide a spectrum of types as possible within the population of inference but a "clean" partition of the variability among types and environments within types may not be possible. Indeed, one objective of the research may be to provide insight concerning which environments favor or disfavor certain treatments and what features are common to these environments. Various forms of "stability analysis" are important examples of this approach.

Excellent review articles on stability analysis are available [see Freeman (1973), Hill (1975), Westcott (1985)]. Hildebrand (1984) has adapted the approach for on-farm trials and its use is demonstrated in the following section. This discussion will be restricted to pointing out its relation to Model [2] above. In Hildebrand's modified stability analysis (MSA), an index for a given environment (EI) is defined as the mean response over all treatments at that farm site. A linear regression over EI's is obtained for each treatment and used as a basis for determining "recommendation domains," a notion loosely similar (but not identical) to the mixed model concept of prediction space. In terms of ANOVA, this could be expressed by modifying Model [2]

$$y_{ijk} = \mu + f_i + r(f)_{ij} + v_k + \beta_k (EI_i) + vf_{ik} + e_{ijk} \qquad [3]$$

where

EI_i is the index of the ith environment, and

β_k is the linear regression coefficient for the kth treatment.

In essence, EI in Model [3] replaces type in Model [2]. Also, f_i in Model [3] is equivalent to $t_i + f(t)_{ij}$ in Model [2] and vf_{ik} in Model [3] is equivalent to $vf(t)_{ijl}$ in Model [2]. Since environment (represented by "F") aside from EI, is a random effect, the ANOVA is

Source of variation	df	Expected mean square
F	$F - 1$	$\sigma^2 + R\sigma_{vf}^2 + V\sigma_{rf}^2 + RV\sigma_f^2$
$R(F)$	$F(R - 1)$	$\sigma^2 + V\sigma_{rf}^2$
V	$V - 1$	$\sigma^2 + R\sigma_{vf}^2 + FR\phi_v$
$V \times EI$	$V - 1$	$\sigma^2 + R\sigma_{vf}^2 + FR\phi_{EI}$
$V \times F$	$(V - 1)(F - 2)$	$\sigma^2 + R\sigma_{vf}^2$
residual	$F(V - 1)(R - 1)$	σ^2

Equality of the β_k can be tested using $MS[V \times EI]/MS[V \times F]$. A "significant" F ratio would imply that treatments respond unequally to EI (and thus to whatever environmental types the EI imply). This would in turn provide formal justification for predicting that different treatments are optimal for various "recommendation domains."

There is no reason why the use of environmental indices need to be limited to *linear* regression. For example, Model [3] can easily be extended to

$$y_{ijk} = \mu + f_i + r(f)_{ij} + v_k + \beta_{1k}(EI_i) + \beta_{2k}(EI_i)^2 + vf_{ik} + e_{ijk}, \qquad [4]$$

where

β_{1k} is the linear regression coefficient for the kth treatment, and

β_{2k} is the quadratic regression coefficient for the kth treatment.

The ANOVA for Model [4] would be identical to the ANOVA for Model [3] except that an additional line for $V \times EI^2$ (or $V \times EI \times EI$) with $V - 1$ degrees of freedom would appear immediately after $V \times EI$ and the remaining $V \times F$ term would have $(V - 1)(F - 3)$ degrees of freedom.

The F ratio $MS[V \times EI^2]/MS[V \times F]$ tests the equality of quadratic regression over EI for the various treatments. Pictorially, this can be visualized as in Fig. 12–7. Note that the quadratic regressions are quite different for the treatments, although their linear components are similar. Several authors have noted the limitations of linear-only regression over EI, e.g., Westcott (1985). However, Model [4] should make it clear that this restric-

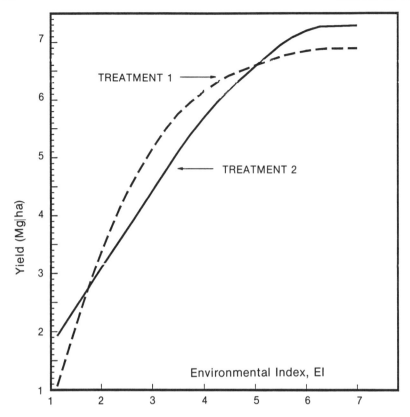

Fig. 12-7. Illustration of treatment by environment interaction.

tion is unnecessary. Indeed, Model [4] can be extended to more complex forms
of regression over *EI*.

   If there is only one "replication" per farm (a discussion of the advantages and disadvantages of this appears below) then the $R(F)$ and residual
terms in the ANOVA have no degrees of freedom and the result is the following simplified form

Source of variation	df	Expected mean square
F	$F - 1$	$\sigma_{vt}^2 + V\sigma_f^2$
V	$V - 1$	$\sigma_{vt}^2 + F\phi_v$
$V \times EI$	$V - 1$	$\sigma_{vt}^2 + F\phi_{EI}$
$V \times EI^2$	$V - 1$	$\sigma_{vt}^2 + F\phi_{EI}^2$
$V \times F$ (now the residual)	$(V - 1)(F - 3)$	σ_{vt}^2

Note that this has no impact on the F ratio used.

   The use of *EI* in stability analysis has been widely criticized because the
independent variable *EI* is in fact a function of the dependent variable. Westcott (1985) makes a case for greater use of independently determined "environmental variables." He also notes that "environmental measurements

are very seldom available." *Theoretical objections aside, the on-farm researcher often has but two alternatives: using EI or being unable to make useful recommendations within a reasonable period of time.* And, as McCullagh and Neider (1989) point out, "A first, though at first sight, not a very helpful principle, is that all models are wrong; some, though, are more useful than others and we should seek those." Critics often point to the weaknesses in formal statistical properties of analysis using *EI*. These difficulties clearly exist; however, a more compelling point is that the researcher often has the *EI* as the *only* objective guide to environmental quality. These criticisms would be severe problems if *formal, definitive* statistical inference were the objective—it is not. The more important use of this type of analysis is to obtain preliminary insight regarding the consistency of treatment performance, which fields, farms or groups of farms appear to be troublesome, what recommendations appear to be reasonable, etc. This sort of analysis always is a starting point, never an end in itself.

For the researcher to make the jump from finding a significant *EI* × treatment interaction from a model such as [3] or [4] to associating *EI* with *predictable* future environments or "recommendation domains" and making reliable treatment recommendations for them obviously requires a great deal of thought and care (and involves, to a large extent, nonstatistical questions, i.e., why are some *EI* low and others high). Predicted treatment performance for farms included in the trial can be made using well-known BLUP methods. The *EI*'s have no intrinsic meaning, so predictions for fields or farms not included in the trial are only as good as the researcher's ability to predict which fields or farms will be in which recommendation domain. The on-farm trial will not *by itself* generate data suitable for this purpose.

AN IMPORTANT NOTE ON DESIGNING ON-FARM TRIALS

Note that neither $MS[R(FT)]$ nor MS[resid] are ever used in the analysis of the "usual" on-farm trial, i.e., one described by some variation on Model [2]. The appropriate denominator term for all tests of interest is $MS[V \times F(T)]$. Why is this important? Both $MS[R(FT)]$ and MS(resid) require that R, the number of "replications" per farm, be at least two. However, neither of these terms has any role in the analysis of the standard on-farm trial. What would happen if only one replication per farm were observed? Neither $MS[R(FT)]$ nor MS(resid) could be calculated. However, since neither term plays any role in the analysis, this is no real disadvantage.

It *is* important to have as many farms per type as possible. This maximizes the degrees of freedom for $MS[V \times F(T)]$; since this is the denominator term for all F ratios of interest, this will maximize power and, consequently, the usable information available. Thus, it is the *farm* that is the true replication in an on-farm trial, not the "replication" within a farm (hence the motivation for the quotation marks!). This is important because on-farm researchers often have been advised to replicate within a farm, even for example in Hildebrand and Poey (1985)! From an ANOVA viewpoint, we now know this is clearly erroneous advice. Moreover, it is wasteful, the researcher would be better off observing more farms. Even worse, it abuses the hospitality of the farmer donating the space for the research to be conducted, the farmer should not have any more land out of ordinary production than absolutely necessary.

To repeat, in most on-farm trials, the number of farms observed should be maximized. Replication within a farm should not ordinarily be necessary and is usually wasteful. The *only* exception is for the purely "farms as fixed effect" case of Model [1] an unlikely though not unheard of, on-farm trial design.

MODIFIED STABILITY ANALYSIS

One method for managing research in such different environments as those shown above in northern Florida is with "stability analysis," *modified* to provide a *positive* rather than a *negative* interpretation to treatment by environment interaction (Hildebrand, 1990). Figure 12–6 shows hypothetical results of three varieties (as an example of three alternative technologies) that have been tested over an appropriately wide range of environments. In this hypothetical case, all three have the same overall mean yield and deviations from regression, $s^2d_1 = 0$. The linear regression coefficients are 1.5, 1.0 and 0.5 for Cultivars A, B and C, respectively. In the absence of other disqualifying characteristics, Cultivar B (the most generally adaptable according to Finlay and Wilkinson (1963), or the most stable according to Eberhart and Russell (1966)) would be selected based on the value of the regression coefficient. The argument against Cultivar A is that because it has a coefficient much higher than unity, it is too sensitive to environmental change and *does poorly in poor environments*. Cultivar C, because it has a coefficient much lower than unity, is *unable to exploit high-yielding environments*. Therefore, Cultivar B, which is *not superior in any environment*, is chosen as the best of the three.

Notice that the argument against Cultivar A with a high coefficient, moves from right to left or toward low environments (it does poorly in poor environments). The opposite is true of the argument against Cultivar C with a low coefficient, which moves from left to right or toward high environments (it is unable to exploit good environments). These are *negative* interpretations which lead to the selection of Cultivar B, Fig. 12–8.

If the emphasis regarding cultivar with a high regression coefficient were toward, rather than away from the best environments (which cultivar *can* exploit the better environments?), Cultivar A would be selected. Likewise, if for cultivar with a low coefficient, emphasis was toward (rather than away from) the poor environments (which cultivar can maintain yield even in poorer environments?), Cultivar C would be selected, Fig. 12–9. The difference is not one of analytical procedure, but of a positive rather than a negative philosophy, goal and/or attitude toward technology selection.

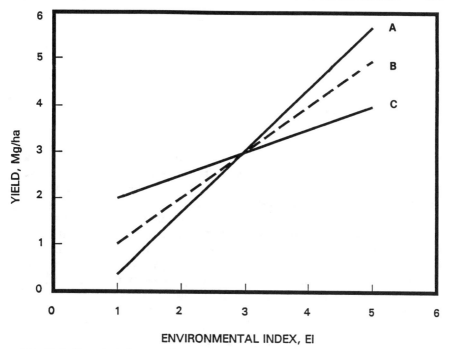

Fig. 12-8. Hypothetical results of variety testing over range of environments.

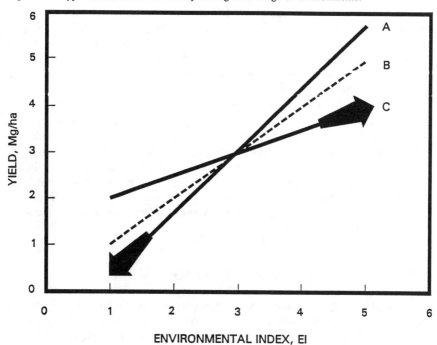

Fig. 12-9.Negative interpretation of the response of varieties to environment resulting in choice of Cultivar B for "broad adaptation."

The result of using this approach with MSA (Hildebrand, 1984) is to describe recommendation domains within which specific technologies excel (recommend Cultivar A for the better environments and Cultivar C for the poorer environments, in the above example, rather than Cultivar B for all environments).

Numbers of Locations (Environments)

Following Models [3] and [4], the number of environments required for estimation of treatment by environment response in research domains and verification in recommendation domains is not excessive. In order to have at least 20 degrees of freedom in the error term, and allowing for estimation of both linear and quadratic responses as in Model [4], if eight treatments are included in the trial, such as might be used in an exploratory trial in a research domain, six environments is an adequate number. For four treatments, 10 environments would be required, and in a verification trial with only two treatments (the recommended treatment and the farmer check, for example) 23 environments are adequate. These suggestions, of course, are approximate. The appropriate number of environments is a function of the variance and the required sensitivity—all case-by-case situations.

Number of Years

Experience has indicated that if three conditions are met, the estimates of environment by treatment response stabilize in 1 yr. These conditions are: (i) the range of environmental indices (EI) should be at least as great as the mean of the indices, (ii) the range of EI should approximate what would normally be expected over a period of years, and (iii) the distribution of environments should be reasonably uniform from good to poor.

However, it should be remembered that at least 2 yr of data will be available for estimates if both an exploratory trial (in a research domain) and a validation trial (in a recommendation domain) are carried out prior to making firm recommendations. Also, preliminary data often are available from on-station trials, conducted over previous years, as the technology is being developed. The treatments that are common from among these current and previous trials can be combined in a single MSA. The data from previous years also can help to verify whether the range of environments included in a current trial is adequate.

RECENT EXAMPLE

Singh (1990) reports on recent research conducted near Manaus, Brazil, that illustrates many of these concepts. The on-farm portion of his research was conducted in two small farming communities in the municipality of Rio Preto da Eva, Amazonas, Brazil, where the government was initiating a small watershed management program. The Brazilian national agricultural research institution (EMBRAPA) has a mandate to develop appropriate technology for different farming conditions in this relatively inaccessible area. Also collaborating in the research were EMATER (extension) and SEPA, the state development planning entity, TROPSOILS, and the University of Florida.

Secondary information regarding indigenous farming practices of the area was collected from published sources. A rapid appraisal of the area was conducted with a multidisciplinary team of individuals from EMBRAPA, SEPA and EMATER who visited the area on three different occasions. Farmers' knowledge of indigenous technology, agronomic practices, and land types being used were recorded. An extensive soil sampling program was carried out to understand soil physical and chemical characteristics and relate them to farmers' rationale for assigning a particular cropping pattern to a given land type.

Three treatments, based on previous on-station research, were selected for comparison with farmers' practices (FP) for growing maize (*Zea mays* L.) and cowpea (*Vigna unguiculata*). Only results from the cowpea are reported here. All three treatments with amendments received K (60 kg ha^{-1}) broadcast. Processed city waste (PCW), chicken manure (CM) and triple super phosphate (TSP) were applied in 25-cm bands. The cowpea cultivar IPEAN V-69 was planted in rows 60 cm apart. Plot size varied from 100- to 200 m^2. Land preparation and planting methods consisted of clearing the area by slash and burn, followed by manual land preparation and planting with sticks.

The project area is inhabited by subsistence farmers who clear land from primary forest (PF) or secondary forest (SF) and farm it up to 3 yr before abandoning it as waste land (WL). Cowpea trials were established on 13 locations. Eight were replicated and five were not. Yield results, averaged across replications where appropriate, and the environmental index for each location are shown in Table 12-1. Analysis of variance using the Model [4] with $R = 1$ is shown in Table 12-2.

For the criterion megagrams per hectare, the response of the four treatments to environment, using MSA, is shown in Fig. 12-10. It is clear that amendments are needed to maximize per hectare yield from these soils. In the poorer environments ($EI < 1.32$) CM produces the best results, and in the better environments ($EI > 1.32$) TSP is best.

The biophysical characteristics of the better and poorer environments closely follows the nature of the land being used. That is, the better environments ($EI > 1.32$) are all land taken from PF and in 1st or 2nd yr of use and SF in 1st yr of use. All other categories (PF$_3$, SF$_2$, SF$_3$, WL) are in the poorer environments. For farmers whose evaluation criterion is to maximize megagrams per hectare, the research domain can be divided into two recommendation domains. For farmers with PF$_1$ and PF$_2$, the recommendation is to use TSP. Farmers in all other cases should use CM.

Figures 12-11 and 12-12 show the MSA results for the alternate evaluation criterion of kilogram per dollar of cash cost (kg/$CC), a criterion usually of great importance to farmers in this area who have little cash to spend for agricultural inputs. Figure 12-11 shows that for the better environments (here

Table 12-1. Cowpea yield across environments, Rio Preto da Eva, Amazonas, Brazil, 1989.†

Location	FP‡	PCW	TSP	CM	EI
			Mg/ha		
8	0.10	0.20	1.30	1.65	0.81
13	0.00	0.00	1.30	2.00	0.82
6	0.15	0.50	1.35	1.35	0.84
9	0.20	0.40	1.20	1.70	0.88
2	0.50	0.65	1.10	1.50	0.94
5	0.15	0.50	2.10	2.05	1.20
4	0.60	1.20	1.60	2.25	1.41
1	0.70	0.90	2.30	1.80	1.42
11	1.20	1.50	2.20	1.90	1.70
12	1.50	1.80	2.10	1.70	1.78
3	1.45	1.95	2.50	1.90	1.95
10	2.20	1.90	2.60	1.40	2.02
7	1.70	1.65	2.65	2.15	2.04

† Source, Singh (1990).
‡ FP = farmers' practices, PCW = processed city waste, TSP = triple super phosphate, CM = chicken manure, EI = environment.

$EI > 1.25$, but covering the same soil situations) FP is by far the best practice of those tested. In the 1st or 2nd yr out of PF, none of the other tested treatments would be acceptable to farmers for whom cash is very scarce and therefore need to maximize kilograms per dollar of cash crop. For any other soil situation, however, either TSP or CM could be recommended even if the farmers had to use scarce cash to purchase the amendments.

The use of Fig. 12-13 narrows the choice somewhat in the poorer environments. The CM produces very stable results compared to TSP which could result in fewer kilograms per dollar of cash crop than FP. This leads to the recommendation of CM as the best choice of those treatments tested in cases where farmers use, or are forced by circumstances to use fields more than 1 yr out of SF or 2 yr out of PF.

Table 12-2. ANOVA, cowpea response to environment, Manaus, 1990.

Source of variation	df	Mean square	$Pr > F$
Location	12	0.9470	0.0001
Treatment	3	3.8127	0.0001
Environment × treatment	3	0.9923	0.0001
Environment × environment × treatment	3	0.1266	0.1840
Residual	30	0.0736	

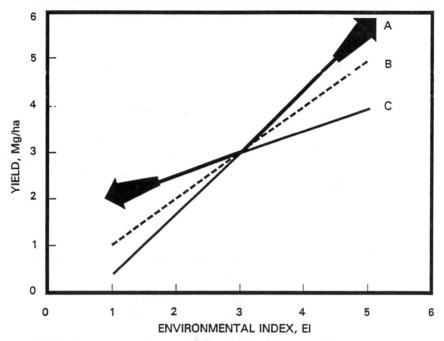

Fig. 12–10 Positive interpretation of the response of varieties to environment resulting in a choice of Cultivar A for the better environments and of Cultivar C for the poorer environments.

FOCUS ON INDIVIDUAL FARMERS, FARMS, AND FIELDS
IN COMMERCIAL AGRICULTURE

Farmers, themselves, continually fine-tune their systems to the specific resource and infrastructure conditions in which the farm and family are found. In addition to the crop cultivars mentioned earlier, a number of innovations in farm equipment originated with producers. Perhaps the agricultural machinery industry has been the sector most active in capturing the experience of farmers and putting this into commercial practices. Many of the current tillage, planting, and harvesting units have reached their present form through farmer modification of what was on the market, and then tested and adopted by industry for the next generation of commercial units. The ridge tillage planters and cultivators are currently going through this phase of farmer modification. A number of cropping system innovations likewise originated with farmers. Annual windbreaks have been proven useful to reduce transpiration in crops between the windbreaks, and perennial windbreaks used to break the wind and trap moisture as snow in the northern Great Plains. Alternating strips of different species, such as maize and soybean (*Glycine max*), have been used by a number of farmers in the western Corn Belt. Although there is a growing body of technical research on experiment stations to validate and quantify the effects of these practices, many of them, in fact, originated with farmers in the region. What has been diffi-

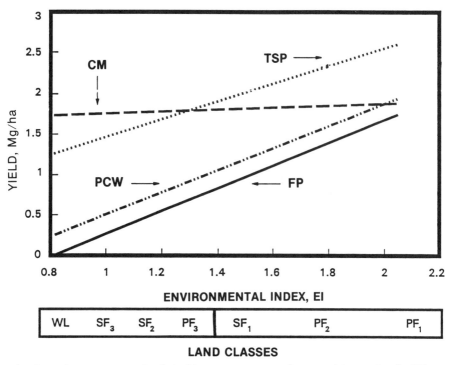

Fig. 12–11. Cowpea response (Mg/ha) of four treatments to environment, Manaus, Brazil, 1990 (Singh, 1990).

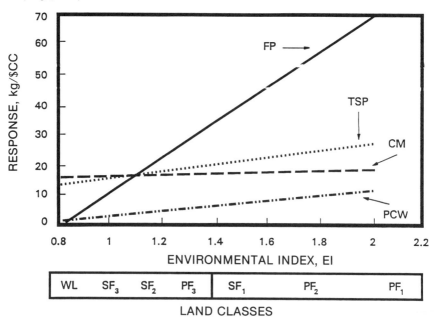

Fig. 12–12. Cowpea response (kg/$ CC) of four treatments to environment, Manaus, Brazil, 1990 (Singh, 1990).

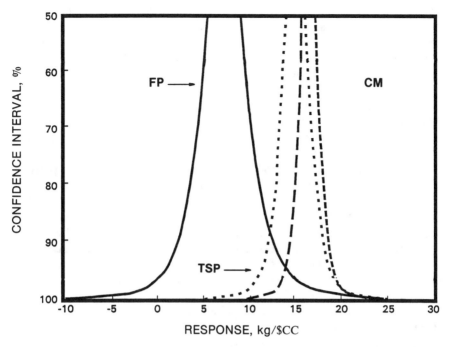

Fig. 12-13. Stability of three treatments in cowpea for poor environments, Manaus, Brazil, 1990 (Singh, 1990).

cult is the rationalization of different methods used by farmers to test their systems, and those used by scientists trained in a different research paradigm.

In many respects, on-farm research has a great deal in common with industrial statistical process control. In manufacturing, products are designed in the lab, then prototypes are produced and evaluated under "real world conditions." During the latter phase, problems are identified when typical workers, rather than research engineers, attempt to produce the product and when prospective consumers attempt to use the product. Invariably, they find ways to "break" the product that would never occur to lab workers. So it is with agricultural research. The experiment station or greenhouse can be thought of as the agronomic lab. The function of on-farm trials is identical in agriculture to "real world" testing in manufacturing. Real farmers will surface problems not encountered by experiment station workers, the research process is not complete until this is done.

DESIGNS FOR RESEARCH ON COMMERCIAL FARMS

In a recent symposium of ASA in San Antonio, there were many presentations about how research is being conducted on farms. The examples appeared to fall in one of two categories. First was the replicated trial with relatively small plots in which the university researcher developed an agenda, designed treatments and plots in the field, collected most of the data and interpreted the results. The farmer was a participant in providing land and some cultural operations during the season, but was not an active part of the planning or the evaluation process. This role for the farmer meets most of the reasons for locating plots on farms as listed by Lockeretz (1987).

In contrast, a second approach was essentially an extension of the farming systems research/extension philosophy and method (Hildebrand & Poey, 1985), where farmers were primary participants in the setting of a research agenda, search for relevant treatments, layout and implementation of the trial, and interpretation and use of results. The latter approach provides an environment in which the methodologies given by Taylor (1990) for on-farm research can be implemented: use of multidisciplinary research teams (including the farmer), whole-farm analysis of results where appropriate, design of long-term plots and treatments, and synthetic as well as analytical approaches to use of data. In the USA, the former approach has been favored by researchers from land grant universities, while the latter has been part of the agenda of farmer groups and other nonprofit organizations. The proponents of each approach find it difficult to communicate at times with others who do not share their definitions of what constitutes research, since each group has a relatively clear mind set of what is meant by "on-farm research," while in fact these definitions are quite divergent.

When a university-trained scientist uses the term "research," there is an assumption of an explicit and testable hypothesis, replicated treatments in a randomized pattern in a standard design, homogeneity of variances among treatments, control of experimental conditions, and relative uniformity of the experimental area—or some blocking pattern to handle variation in the field. These are the normal assumptions connected with the analysis of variance, and although they are not always strictly adhered to, we often make the assumption that they are being met. Saying that "standard statistical procedures were followed" implies all of the above even if the researcher (or farmer) did not really understand the statistical thinking very well.

Many of these criteria are not recognized nor understood by most farmers. They prefer trials that are fairly close to the home farm or under similar conditions or both, that have plots large enough to use commercial equipment, that show visible differences among treatments, that can reduce costs or increase profits, or that solve a constraint that was already perceived on their farm or in the area. In the real world we encounter comparisons from 1 yr to the next, from one field to another, from one farm to a neighbor's, or among strips in a field that have different treatments (e.g., cultivar or hybrids) with no replication. Although these comparisons do not meet the criteria recognized by the scientist to qualify as credible or valid research, the results are no less meaningful to many farmers. We do find that careful explanation in an extension meeting of some of the criteria used by researchers, for example replication and randomization of treatments, leads to a fairly quick understanding of the need for these methods and the importance for repeatability of the experience.

Are these two definitions of "research" mutually exclusive, or is there some middle ground where farmer creativity, land, and resources can be utilized for credible on-farm research? Over the past several years, there has been substantial work on large plots with few treatments, replication and randomization, and standard statistical analysis. Long strip designs used to compare two or three treatments were described by Thompson (1990), and are currently being used by a number of the members of the Practical Farmers of Iowa, among other groups. Rzewnicki et al. (1988) summarized these trials from Iowa as well as some from farms and from experiment stations in Nebraska. With plots that ranged from 65 to 400 m (200 to 1200 ft) long by four to eight rows wide and three to six replications per treatment, they found coefficients of variation (CV) from less than 1.0 to about 10%; the CV's were frequently less than 5%. Practical researchers who are familiar with the variation in most field experiments find these levels very acceptable.

How is it possible that such large plots have low CV's? Although we are only now testing these hypotheses by comparing large and small plots from the same field (Shapiro et al., 1989, 1990), it appears that a long and narrow plot goes across a range of variability in the field. A plot located adjacent with the same dimensions crosses the same gradient, and at any one point there is relatively less difference between the strips than there is across the gradient in each long plot. Thus the potential exists for planting contrasting treatments side by side, allowing use of full-sized commercial equipment and having a highly visible comparison, while still meeting the requirements of replication and randomization. This would appear to be one option for an individual farmer to collect credible data for one site in 1 yr, and use standard statistical techniques such as ANOVA, t-test, or paired comparisons to evaluate the trials. In one set of comparisons, the Clay County Corn Growers in Nebraska planted maize hybrids in unreplicated strip plots in four areas in the county, with similar conditions and the same hybrids in each test. Analyzed with farms as replications, there were CV's from 3 to 4% over the 5 yr of the tests (Rzewnicki et al., 1988). This opens the possibilities for individuals or groups of farmers to work in a cooperative research network and to develop a credible set of comparisons for use by them and by others. Each farmer becomes a part of the research and extension network, since these plots are used for field tours and the data for extension meetings before the next planting season.

Results from these large replicated or unreplicated trials in Nebraska represent one approach that can be taken by farmers in a highly mechanized, large farm situation. It is a challenge to the practical researcher or applied extension person to explain the basic characteristics of the trials, and to work directly with farmers in developing the research agenda.

INTERPRETATION AND EXTRAPOLATION OF RESULTS

The zone across which such results can be applied depends on how many sites were used for the trials, the soil and climatic characteristics of the sites, whether similar results could be expected from other sites in the region, and how credible or repeatable the results are from the experiments. Several dimensions of this question have been discussed above. In statistical terms, the potential for application of results across a range of environments depends on the significance of the specific technology by environment interaction. An example is the testing of hybrids across locations, and measuring the genotype by location interaction. When this is low, it is relatively easy to recommend one or a few hybrids across a wide area; when the interaction is large, there is a high degree of site specificity and need for unique choices for different locations.

It is important to consider the effects of replications, years, and locations in contributing to the value of results. Increasing number of locations and environments had little effect beyond about eight on the magnitude of the standard error of a genotype mean (Saeed et al., 1984). Increasing number of years from one to two substantially reduced the standard error of the mean, while adding an additional year had minimal effect. Likewise, increasing number of replications has little effect on the standard error. The influence of additional locations or environments is much greater than either adding years or replications to an experiment in order to reduce the variance of a mean, thus increasing the potential for detecting statistical differences among treatments in the experiment. Although it is less expensive to add replications in a single location, this is relatively ineffective in increasing the potential to detect differences. This is consistent with the above discussion on need for a large number of locations or environments for testing, and the relatively smaller need for replication in one site. The concept of single replications and a large number of locations is a cornerstone of current commercial hybrid testing strategies (Bradley et al., 1988). The efficiency of this procedure in a testing program has recently been described (Dofing & Francis, 1990). Replication at one site does improve the precision of measurement at that site in that year. But with multilocational (multiple environment) on-farm testing, the relevant variance is that among locations or environments. Therefore, multiple replications at one location contribute little to the potential extrapolation from that site to others, or to other years.

The challenge for the individual farmer is to decide what information really applies to his or her site, given the abundance of results from trials that are available from industry, university, or private sources. The better a farmer is able to characterize the farm and the individual fields, and the better the description of the conditions under which data were collected in other sites, the easier it will be to decide which data or recommendations are relevant. This is a practical way of defining recommendation domains, a topic already explored. The best place to look for relevant data is within the same recommendation domain as that where the field is located. It should be apparent that these domains are not defined only by geographical location, by soil type, or by any single factor. Likewise, it is possible that a single farm may encompass several domains. It is important to understand the concept, and to use this information to best access the most appropriate data before making production technology decisions.

PARTICIPATORY MODEL FOR RESEARCH AND EXTENSION

On-farm research trials and demonstrations for extension purposes have long been a staple component of comprehensive investigation and development programs in agriculture. Some of the reasons have been described above. There are even more compelling reasons today why research with individual farmers and groups of producers makes sense (Francis et al., 1990). There are limited research and extension budgets, with an increasing focus of Federal funds in the USA on basic work at the expense of applied research. This is a trend that is being followed by national research programs around the world, as scientists become better prepared for basic investigations and the glamor of genetic engineering and high technology solutions pervades the scientific community. In contrast to the range of ecological situations where farmers produce crops, the research establishments have relatively few experiment stations. Much of the research performed on these stations is reductionist in nature, with limited regard for the incorporation of new innovations in the total farming system. For these reasons, there is comparative advantage to conducting at least some of the research in a wider array of sites with collaborating farmers.

Another compelling reason for working directly with individual farmers and groups relates to distance from the controlled research site to the farm where results will be applied. This "distance" may take several forms (Francis et al., 1990). Geographic space in miles or kilometers from one site to another is the most commonly used measure of distance. Farmers are willing to travel certain distances to visit other sites, depending on culture and infrastructure (Rzewnicki, 1991). More important, perhaps, is the "ecological distance" from one site to another. For example, a low lying area with poor drainage and heavy soils may be a very short physical distance from a well-drained, lighter soil on a hillside, yet the soil conditions, appropriate cultural practices, and crops or varieties that are appropriate may be quite distinct. Finally, there may be "conceptual" or "psychological distances" between researcher and farmer, based on differences in education or experience, and these need to be bridged in order to effect a working partnership and a fully participatory system of research and extension. On-farm activities among people who have mutual respect for each others' talents and potentials to contribute can help to overcome these social distances.

The potentials of a participatory network of farmers and researchers can perhaps best be illustrated through use of an example. The unique contributions of the farmer in the total research process is highlighted. A number of additional examples, especially in farmer contributions to ideas for weed management, were recently summarized by Doll and Francis (1992).

Maize Yield Response to Nitrogen in Crop Rotations

In order to study the effect of N applications on maize yields in continuous maize and sorghum (*Sorghum bicolor*) compared to rotation of these cereals in Nebraska, a network of about 30 farmers was established to work with a project of the University of Nebraska. There has been great concern about the energy costs of this input in maize production, as well as potential for nitrate contamination of groundwater supplies that are frequently used for human and animal consumption. Supported in part through a grant from the Nebraska Energy Office, a university technician established contact with a number of farmers, many of whom were members of the Nebraska Sustainable Agriculture Society. All were interested in more efficient use of N, and in finding ways to quantify the effects of a cereal–legume rotation on response to this important nutrient.

In cooperation with farmers, fields and experimental sites were chosen, soil samples were taken, and laboratory test results discussed. Together the team determined realistic yield goals and developed N budgets considering all sources of this major nutrient. Each farmer thus derived a conservative but optimum level of N for the coming season. In most fields this N rate and a one-half rate were included, and in some fields a zero rate as well. On many fields these treatments were applied in replicated strips across the entire field. For the years 1988, 1989, and 1990 there was a predicted yield response to rates of 90 to 165 kg/ha (80 to 150 pounds N per acre), although the actual economic optima were lower in most fields (Franzluebbers, 1991). Following soybean, sweet clover (*Melilotum* spp.), or alfalfa (*Medicago sativa*) there was no economic response to applied N by either maize or sorghum under rainfed conditions. In irrigated fields, there was no economic response of maize yields to N if the maize followed alfalfa. The conclusions from this 3-yr project, as interpreted by farmers and project personnel, was that N generally is over applied under many conditions in Nebraska.

Statistical analysis followed the standard procedures described above in Model [1]. Where replicated treatments were present, the analysis and comparison were conductd on each farm. In a number of cases there was a single replication per farm, and these results were pooled with the replicated sites using a single mean per treatment per location, and locations used as replications. Regression analysis was used to compare the response of cereal yields to applied N in rotation compared to continuous culture. Grain yield response in maize and sorghum to applied N was measured at 29 sites with continuous cropping and at 57 sites in rotation with legumes and small grains. Continuous maize responded to N up to about 90 kg N/ha, the maximum level in the trials. Maize and sorghum following small grains or legumes showed only a modest response in some cases, not statistically significant, and not economically sound because the cost of N plus application was not offset by the increase in yield. This type of analysis is useful for grouping results of like treatments across sites.

To reach more farmers with this information, eight meetings were scheduled jointly by the Sustainable Agriculture Society and Nebraska Extension in early 1990. The objectives of the trials and methods were described, tables or figures presented, and the meeting turned over to farmers to interpret the data and derive results. A lively discussion ensued about results from the trials and how to apply them to specific fields. The university staff present were valuable as resource people, helping to explain why or why not crops were responding in specific situations. But the farmers were deriving their own recommendations, and the extension specialists were able to empower the producers to make these decisions. During the next summer of 1990, many of the trial fields were used for field tours and discussion on site. Farmers were in charge of describing what happened. These are both examples of participatory extension practices.

Should the farmer put more confidence in results from his or her own field trial, or from the aggregate analysis across sites? It is usually appealing to have one's own data from a field on the farm, where the cultural practices are known and the results appear to uniquely fit that farm. Whether these are the best data to use to predict next year's results depends on the similarity of cultural practices, hybrids, and soil conditions across the range of sites, and how likely those sites represent the potential range of possible rainfall events that may occur over a number of years. Since rainfall is the most limiting factor in most Nebraska sites each year in rainfed crop culture, it is possible that the mean performance over several similar sites will be a better predictor of next year's situation than the results from the single farm. We are seeking data and a method to analyze this situation. The decision by an individual farmer at the moment is a judgment call, and the best that we can do is to provide tools to help improve that judgement.

RESEARCH-EXTENSION AGENDA:
LARGE AND SMALL FARMERS

The approach and examples presented in this chapter illustrate the potential of an emerging paradigm, or shift in patterns of research and extension activities. This applies to both large and small farmers. Efficient research of the type being discussed here depends on recognition and full representation of the research domain or inference space. Scientists need to recognize that on-station trials are useful for research and development, but limited for surfacing knowledge about production realities. Both large and small farmers need to recognize that they represent these realities, and in a major way can contribute to their identification and solution.

Once the inference space is well understood, it is possible to follow with carefully designed activities to solve the problems associated with the primary constraints to sustainable production using resources and information from both large and small farms. Data gathered in-situ is critical, and on-farm trials that include the widest possible variety of farms are essential. Failure to include this range of environments can result in compromised research (bad for the scientist) that may produce flawed information and incorrect production decisions (bad for farmers). The need for many environments and a wide range of participants should be clear.

Large-scale farmers often have the resources to contribute to the research process, and can easily grasp the relevance of self-generated information to their own operations. Using the principles described above, their efforts can be more productive if combined across locations, and at times, combined with data from small-scale farm trials. A wide coalition of farmers, researchers, and extension specialists can bring together the resources needed to broaden the range of environments over which results can be collected. If large farmers are incorporated in a technology evaluation network with small farmers, the length of time required for testing can be reduced because environments (locations) can substitute substantially for years. This can lead to greater efficiency in use of scarce resources, both from farmers and the government. It may well be possible for many of the larger farmers, in conjunction with the national research and extension organizations, to pick up costs associated with on-farm research that will benefit them, and at the same time enhance the development of technology for all farmers.

The role of the research and extension specialists in this system needs to be clearly recognized. The upgrading of farmer participation and input in the research process does not devalue the scientists' role, but rather expands their capacity to recognize and work with real world problems that limit production and the range of solutions that may be available to solve them. The role of research specialists can, in fact, become more focused on describing the "why" behind questions in agriculture, ecology, biology and sustainability. The role of extension specialists can be to catalyze the exchange of information among a number of credible and relevant sources. The research process can be a rigorous statistical exercise, and it is possible to determine where results can be applied in the appropriate recommendation domains.

Assumptions about roles of different players, cooperation among farmers and scientists, relevance of information from various sources, and ultimate objectives of the systems need to be recast. This is the paradigm shift described above. Many agricultural infrastructures are set up with good intentions, but fail to produce anticipated results due to inadequate communication, limited scientific literacy among specialists and farmers, and a strained relationship between those who develop theoretical knowledge and those who focus on practical application. This is a problem in both developing and developed countries. The farming systems paradigm, and especially the on-farm research approaches described here can offer enormous potential that will benefit national agricultural infrastructure as well as sustainable agricultural production systems.

REFERENCES

Bradley, J.P., K.H. Knittle, and A.F. Troyer. 1988. Statistical methods in seed corn product selection. J. Prod. Agric. 1:34–38.
Braidwood, R.J. 1967. Prehistoric men. Scott, Foresman and Co., Glenview, IL.
Byerlee, D., L. Harrington, and D.L. Winkelmann. 1982. Farming systems research: Issues in research strategy and technology design. J. Am. Econ. Assoc. 5:899–904.
Chambers, R. 1981. Rapid rural appraisal: Rationale and repertoire. Public Admin. Dev. 1:95–106.
Chambers, R., A. Pacey, and L.A. Thrupp. 1989. Farmer first: Farmer innovation and agricultural research. Intermediate Technol. Publ., London.
Cox, D.R. 1958. The planning of experiments. John Wiley & Sons, New York.

Deming, W.E. 1953. On the distinction between enumerative and analytic surveys. J. Am. Stat. Assoc. 48:244–255.

Deming, W.E. 1975. On probability as a basis for action. Am. Stat. 29:131–137.

Dofing, S.M., and C.A. Francis. 1990. Efficiency of one-replicate yield testing. J. Prod. Agric. 3:399–402.

Doll, J.D., and C.A. Francis. 1992. Participatory research and extension strategies for sustainable agricultural systems. Weed Technol. 6:473–482.

Eberhart, S.A., and W.A. Russell. 1966. Stability parameters for comparing varieties. Crop Sci. 6:36–40.

Finlay, K.W., and G.N. Wilkinson. 1963. The analysis of adaptation in a plant breeding programme. Aust. J. Agric. Res. 14:742–754.

Francis, C.A. (ed.). 1986. Multiple cropping systems. Macmillan Publ. Co., New York.

Francis, C.A., J. King, J. DeWitt, J. Bushnell, and L. Lucas. 1990. Participatory strategies for information exchange. Am. J. Alter. Agric. 5:153–160.

Franzluebbers, A. 1991. On-farm results of fertilizer response in crop rotations. M.S. thesis, Univ. of Nebraska, Lincoln.

Freeman, G.H. 1973. Statistical methods for the analysis of genotype-environment interaction. Heredity 31:339–354.

Gilbert, E.H., D.W. Norman, and F.E. Winch. 1980. Farming systems research: A critical appraisal. Michigan State Univ., Rural Dev. Paper no. 6, Dep. Agric. Econ., Michigan State Univ., East Lansing, MI.

Harrington, L.W., and R. Tripp. 1984. Recommendation domains: A framework for on-farm research. CIMMYT Econ. Program Working Pap. 02/84. CIMMYT, Mexico.

Harville, D.A. 1976. Extension of the Gauss-Markov theorem to include the estimation of random effects. Ann. Stat. 4:384–395.

Henderson, C.R. 1975. Best linear unbiased prediction under a selection model. Biometrics 31:423–447.

Hildebrand, P.E. 1981. Combining disciplines in rapid appraisal: The sondeo approach. Agric. Admin. 8:423–432.

Hildebrand, P.E. 1984. Modified stability analysis of farmer managed, on-farm trials. Agron. J. 76:271–274.

Hildebrand, P.E. 1988. Technology diffusion in farming systems research and extension. HortSci. 23:488–490.

Hildebrand, P.E. 1990. Modified stability analysis and on-farm research to breed specific adaptability for ecological diversity. p. 169–180. In M.S. Kang (ed.) Genotype-by-environment interaction and plant breeding. Louisiana State Univ., Baton Rouge.

Hildebrand, P.E., and F. Poey. 1985. On-farm agronomic trials in farming systems research and extension. Lynne Rienner Publ., Boulder, CO.

Hill, J. 1975. Genotype-environment interactions—a challenge for plant breeding. J. Agric. Sci. 85:477–493.

Lockeretz, W. 1987. Establishing the proper role for on-farm research. Am. J. Altern. Agric. 2:132–136.

McCullagh, P., and J.A. Neider. 1989. Generalized linear models. Chapman and Hall, New York.

McLean, R.A. 1989. An introduction to general linear models. p.9–28. In Applications of mixed models in agriculture and related disciplines. Southern Coop. Ser. Bull. 343. Louisiana Agric. Exp. Stn., Baton Rouge.

Mead, R. 1988. The design of experiments: Statistical principles for practical application. Cambridge Univ. Press, Cambridge, England.

Norman, D.W. 1980. Farming systems approach: Relevance for the small farmer. Rural Dev. Paper no. 5. Dep. Agric. Econ., Michigan State Univ., East Lansing, MI.

Plucknett, D.L., and N.J.H. Smith. 1986. Historical perspectives on multiple cropping. p. 20–39. In C.A. Francis (ed.) Multiple cropping systems. Macmillan Publ. Co., New York.

Rhoades, R. 1989. The role of farmers in the creation of agricultural technology. p. 3–9. In R. Chambers et al. (ed.) Farmer first: Farmer innovation and agricultural research. Intermed. Technol. Publ., London.

Rzewnicki, P.E., R. Thompson, G.W. Lesoing, R.W. Elmore, C.A. Francis, A.M. Parkhurst, and R.S. Moomaw. 1988. On-farm experiment designs and implications for locating research sites. Am. J. Altern. Agric. 3:168–173.

Saeed, M., C.A. Francis, and J.F. Rajewski. 1984. Maturity effects on genotype × environment interactions in grain sorghum. Agron. J. 76:55–58.

Shapiro, C.A., W.L. Kranz, and A.M. Parkhurst. 1989. Comparison of harvest techniques for corn field demonstration. Am. J. Altern. Agric. 4:59–64.

Shapiro, C.A., A.M. Parkhurst, and W.L. Kranz. 1990. Update on the statistical implications of experimental units for on-farm trials. p. 30. *In* Agronomy abstracts. ASA, Madison, WI.

Singh, B.K. 1990. Sustaining crop phosphorus nutrition of highly leached oxisols of the Amazon Basin of Brazil through use of organic amendments. Ph.D. diss. Univ. of Florida, Gainesville (Diss. Abstr. 91-16049).

Snedecor, G.W., and W.G. Cochran. 1980. Statistical methods. 7th ed. Iowa State Univ. Press, Ames.

Southern Regional Bulletin. 1989. Applications of mixed models in agricultural and related disciplines. South. Coop. Ser. Bull. 343. Louisiana State Agric. Exp. Stn., Baton Rouge, LA.

Steel, R.D.G., and J.A. Torrie. 1980. Principles and procedures in statistics: A biometrical approach. 2nd ed. McGraw-Hill, New York.

Stroup, W.W. 1989a. Predictable functions and prediction space in mixed linear models. p. 39–48. *In* Applications of mixed models in agriculture and related disciplines. Southern Coop. Ser. Bull. 343. Louisiana Agric. Exp. Stn., Baton Rouge.

Stroup, W.W. 1989b. Applications of mixed models. p. 1–8. *In* Applications of mixed models in agriculture and related disciplines. Southern Coop. Ser. Bull. 343. Louisiana Agric. Exp. Stn., Baton Rouge.

Taylor, D.C. 1990. On-farm sustainable agricultural research: Lessons from the past, directions for the future. J. Sustain. Agric. 1:43–87.

Thompson, R. 1990. The Thompson-farm on-farm research. Rodale Inst., Emmaus, PA.

Tripp, R., and J. Woolley. 1989. The planning stage of on-farm research: Identifying factors for experimentation. CIMMYT Agric. Econ. Program Bull., Londres 40, Mexico.

Westcott, B. 1985. Some methods of analysing genotype-environment interaction. Heredity 56:243–253.

Whyte, W.F., and D. Boynton. 1983. Human systems for agriculture. Cornell Univ. Press, Ithaca, NY.

Wotowiec, P., S. Poats, and P.E. Hildebrand. 1988. Research, recommendation and diffusion domains: A farming systems approach to targeting. *In* S. Poats et al. (ed.) Gender issues in farming systems research and extension. Westview Press, Boulder, CO.

13 Designing Future Tropical Agricultural Systems: Challenges for Research and Extension

Charles A. Francis

Department of Agronomy
University of Nebraska
Lincoln, Nebraska

World human population currently is growing by about 90 million people each year (Urban & Rose, 1988), a global rate of increase of about 1.8% from our current population of 5.2 billion. In addition to global insecurity resulting from competition for scarce natural resources, the specter of insufficient food will haunt decision-makers in much of the developing world for the coming decades. Food needs and potential for increased production are directly related to global natural resource supply and the environment. As summarized by Brown (1990), "Food scarcity in developing countries is emerging as the most profound and immediate consequence of global environmental degradation, one already affecting the welfare of hundreds of millions." How has this situation developed?

During the past four decades there has been a rapid increase in world food production, yet in the past 5 yr this growth has appeared to stagnate, especially in African countries where food deficits are most acute. Brown (1990) outlines three historical trends that are converging to make it more difficult to increase global food production in the near future. The first is the scarcity of both new crop land that is suitable for intensified production and fresh water for expanded irrigation potential. The second is the apparent lack of new technologies that will promote a quantum leap in production efficiency, such as we experienced with hybrid maize (*Zea mays*) or the introduction of chemical fertilizers. The third is the negative effect of global environmental degradation on long-term potentials for food production. He states that, "Any one of these trends could slow the growth in food output. But the convergence of all three could alter the food prospect for the nineties in a way the world is not prepared for." Practical research in agricultural systems is needed to open new potentials for increased food production.

Challenges will be difficult in countries where productivity is already high, and potential responses from known technologies are expected to be low (Herdt, 1986, unpublished data). In contrast, the semiarid regions such as the West African Sahel have limited natural resources and few apparent new appropriate technologies available to provide massive increases in production (USDA-ERS, 1988, unpublished data). Further complicating the introduction of new technology is the cost and uncertainty of energy supplies, the appropriateness of technologies that are currently available, and the environmental impact of systems that rely on intensive inputs of chemical fertilizers and pesticides. There is growing concern about the unintended human impacts of many chemicals in the environment and in the food chain, and we see an accelerating search for alternatives. Research is needed to expand potentials in the Tropics for a sustainable food production into the future, and there is need for an extension program that will effectively demonstrate alternatives to farmers in each country.

The concept of sustainability has been implicit for decades in the planning of research, although there is increasing attention today to the changing conditions under which we will have to maintain land productivity and food production. There are biological, ecological, resource, and social dimensions to the challenge of sustainability, and it is important to consider the challenge in both a global and a long-term perspective.

Conway (1985) describes steady-state vs. fragile systems and their capacities to buffer shocks from outside the system. Two examples that reflect the buffering or sustainability in a biological sense are cultivars with maximum genetic diversity and high organic matter in soils that can buffer stress conditions. But sustainability is more than components. It requires careful study of tropical systems and their potentials.

Much research in biological systems has wide applicability, since principles can be studied and then validated in a range of climates and alternative cropping systems. There are many biological interactions or "integration efficiencies" that need to be better understood (Harwood, 1985), and their contributions to future tropical farming systems explored. Research priorities include crop improvement, soil fertility, pest biology and management, tillage and other measurement criteria. There also is need for research that is better focused on specific problems in key areas. Many technologies cannot be moved easily from one site to another. The role of extension will be to seek the range of adaptation of potential new practices, and this can be done efficiently in a participatory venture with farmers. Finally, it is important to be able to envision a desirable future, and to begin to make decisions today to cause that future to happen. There is a need to develop better access to information, and an international capacity to predict key factors and changes. These dimensions of developing future tropical cropping systems are discussed in the sections that follow. The chapter summarizes ideas from others in this book, and depends on the historical perspectives of Harwood (1990).

RESEARCH AND EXTENSION PROCESS

The process of research and extension has evolved through this century from (i) primarily empirical testing on the part of each farmer, to (ii) concentration of expertise in government agencies and universities, to (iii) involvement of a multiplicity of public and private groups. There is a growing concern about attracting greater farmer participation in the process, exemplified by broadly increased interest in farmer innovation and testing (Chambers et al., 1989). This concept builds on the principles of farming systems research and extension, described by Gilbert et al. (1980) and Norman (1980) and used over the past decade in a number of national research programs in developing countries.

There is a growing appreciation within formal research and extension organizations of the potential role of farmers in this process. According to Rhoades (1989), "Farmers have much of importance to say to scientists and . . . farmers' methods of practical research are complementary to those of scientists. . . . Farmers' knowledge, inventiveness and experimentation have long been undervalued, and farmers and scientists can and should be partners in the real and full sense of that word in the research and extension process." The reasons for on-farm research have been discussed by Lockeretz (1987) and Taylor (1990). A growing body of technical literature on the design and conduct of experiments under farm conditions and on the extrapolation of results from these trials to other areas is providing greater credibility among scientists for this type of research (e.g., Hildebrand & Poey, 1985; Rzewnicki et al., 1988).

IDENTIFICATION AND RELATIVE IMPORTANCE OF PRODUCTION CONSTRAINTS

Given the wide range of climates and types of soil that exist around the globe, there are substantially different constraints to productivity from one country or site to another. There are so many combinations of soil and climatic factors and interactions among these factors that it is difficult to establish the precise physical and climatological constraints for a given area without "on the ground" experience or surveys. The participation of farmers in this process can be crucial. There may be economic, political, or social reasons why productivity is limited in a given situation, and these reasons may not be apparent to a researcher or extension specialist who is familiar only with crops and soils. To design future tropical production systems, it is essential to understand what factors currently limit productivity in each site.

The challenge of putting priorities on these production constraints is not easily resolved. One method for establishing relative priorities was given by Parkhurst and Francis (1986). Lack of adequate rainfall frequently limits crop yields, yet the magnitude of this problem varies from year to year in terms of timing and severity of drought and even from one site to others near by. Biotic factors such as insect or weed infestation may vary greatly among regions or even among fields. Insect and weed populations depend on cropping history and previous problems with these pests, introduction of weed seed or adult insects from outside the area or field, and prevailing climatic conditions in a given season. Soil fertility needs depend on the crop to be grown, history of the field, organic matter, and many other factors. Constraints in the Tropics may differ from one season to the next or in a single year in the same field, e.g., flooding in the rainy season and specific insects in the dry season. Choice of a cropping sequence can influence the relative importance among production constraints. Thus, to decide which among these constraints is most important to research is not a trivial question. Yet it is essential to quantify the problems faced by farmers in their quest to improve productivity in order for the research program to make an impact.

Just as there is biological and economic buffering in diverse cropping systems, there is potential to build in flexible management decisions for the farmer to respond to production constraints. This buffering capacity is needed for the unforeseen and unexpected interactions that will occur in the process of listing constraints, seeking an array of solutions, predicting changes as a result of using the best solution, and analyzing responses to those solutions (Parkhurst & Francis, 1986). This iterative process of recognizing and reacting to perceived constraints is critical to successful research programs as well as farmer management.

ASSUMPTIONS ABOUT FUTURE RESOURCES AND CONSTRAINTS

In order to successfully set priorities for research and for agricultural development, it is essential not only to examine the constraints to production today, but also to anticipate the nature of limitations to productivity in the future. To design a research program based entirely on today's farming systems and constraints is to ignore rapid global increase in population, reduction in arable land per person, need for fuel, competition for other scarce natural resources, and future economic and social realities. For example, the world economic situation includes enormous deficits in foreign exchange and international debt in many countries, resource inequities between north and south, and individual and national decisions that focus on narrow gains for a few individuals and countries. This is the economic reality within which we need to design an agricultural strategy for increasing food production. Most land that can be farmed profitably and in a resource-efficient manner is already in production. Although there will be development of some new land resources, this venture is becoming increasingly expensive in terms of investment in infrastructure, maintaining fertility on fragile soils, or developing irrigation where water is limited. One key assumption is the need for intensification of production on existing lands. Some areas are highly fragile under current production systems, and it would be reasonable to restore these to natural vegetation or other less intensive uses.

Highly productive agricultural systems are strongly dependent on fossil fuels for making N fertilizer, processing other plant nutrients, producing pesticides, facilitating irrigation, mechanizing field operations, and processing and distributing food. This reliance on a diminishing resource brings into question the sustainability of current conventional agricultural systems. We must be concerned when a country's food security depends entirely on N derived through use of scarce fossil fuels.

Although agriculture uses a relatively small share of the world's total petroleum consumption, food systems must compete with transportation, personal needs for heating, cooling, and cooking, a wide array of industrial uses, and other human wants. Although highly efficient per unit of labor, high technology agriculture makes inefficient conversion of fossil fuel energy to food, compared to more extensive traditional systems. As stated by Dover and Talbot (1987), "Productivity without sustainability is mining" the resources on which those production systems depend. Since traditional sources of fossil fuels will become increasingly scarce and expensive, it is important to derive a research agenda that drastically improves the energy use efficiency of crop production. In a recent review, Goldemberg et al. (1987) describe alternative energy resources as well as approaches to energy use that can result in high-efficiency production systems. The assumption should be that fossil fuel energy will be increasingly scarce and expensive.

Will crops and animal species used for food be different one or two decades in the future? We depend on a narrow species range for food. In a short-term perspective of a few years, there do not appear to be major shifts in food species. It is likely that Southeast Asia will depend on rice (*Oryza sativa*), and the highlands of Latin America on maize (*Zea mays*) and potato (*Solanum tuberosum*). Yet if cropping patterns are observed over decades or centuries, there are significant and obvious changes in species. Maize was not known in East Africa until it was introduced by world voyagers some 500 yr ago, now it is the primary staple cereal crop in Kenya and Zimbabwe. Dry bean (*Phaseolus vulgaris*) from America is popular in Central Africia, and cowpea (*Vigna unguiculata*) and sorghum (*Sorghum bicolor*) from Africa are important crops in tropical America. There are likely to be changes in crop species in the long term, and planning in research needs to include a dimension of introducing and testing new species in current and potential future cropping systems.

Water for agriculture will become increasingly scarce and expensive (Postel, 1989). A rapid shift toward greater efficiency of use and recognition that there are multiple legitimate demands for this resource can help avoid shortages and reduce the environmental impact of lower water tables and stream flows. The average efficiency of water use worldwide is less than 40%, but there are known methods that can substantially increase this number (Postel, 1984). There is a need for national and local policies that will encourage greater water-use efficiency in agriculture (Rangeley, 1987). We must recognize that water will become increasingly limited for supporting crop production.

There is growing concern about the levels of pesticides used worldwide in crop production (e.g., McCracken & Conway, 1987). In part, this concern is for the energy cost of producing products, but there also is wide concern that residues from toxic chemicals will directly enter the human food system (Jeyaratnam, 1985). A greater concern should be the health and safety of farmers and farm workers who deal directly with these products (Bull, 1982). Bird et al. (1990) discuss the evolution of pest management from a presynthetic pesticide era through a synthetic pesticide era to the current integrated pest management era. They further envision a refinement of current systems to a future sustainable agriculture era in regard to pest management. This assumes that alternative strategies will be available through research. Unavailability of chemical products to many tropical farmers due to lack of infrastructure, non-effective application of products due to poor equipment or low levels of education, or lack of money to buy this technology make it imperative to search for alternatives.

Finally, the low level of economic interest in agriculture in many developing countries suggests the need for strategies that will (i) increase the level of awareness of the importance of food and investment in an environmentally sound agriculture (Brown, 1990) and (ii) stimulate the development of systems that are as dependent as possible on internal, nonpurchased resources (Francis, 1988). These are the realities of agriculture as we move into the 1990s, and the types of constraints that limit tropical food production. We need to take into account current systems and limitations, as well as those that will prevail at least a decade into the future when today's research results will become available to the farmer.

DETERMINATION OF SHORT- AND LONG-TERM PRIORITIES FOR RESEARCH

Given the urgency of food needs in the tropics, there is need for short-term solutions as well as long-term strategies to provide for human nutrition. These priorities will differ from one country or region to another, and there may be conflicts between the short- and long-term goals. In some cases it may be rational to sacrifice some short-term advantage in order to reach a greater long-term goal. Most important is the examination of the needs and the constraints to producing food, and to make explicit the priorities for development. A model for determination of research priorites in specific cropping systems was described by Parkhurst and Francis (1986). Future research efforts, including those funded by outside agencies, need to include longer-term commitments, and must include resource and environmental monitoring.

Adequate nutrients for successful crop production are lacking in many of the highly weathered soils of the Tropics (Sanchez & Benites, 1987). With an immediate and urgent need for cereal production, a short-term priority may be to determine the most economical yield response to added N or P, and a national decision to import these nutrients or to build a fertilizer plant to produce them at home. The longer-term solution may be design and promotion of cropping systems that integrate legumes into the sequence, provide a rotation of pasture with cattle or other ruminants with crop production, or some combination of these methods with low levels of applied fertilizers.

It is obvious that priorities will vary in different parts of a country where production constraints are different. Research for improved lowland paddy rice production in Southeast Asia may concentrate on insect and disease resistance to maintain high yields in an intensive, fertilized system. Research for nearby upland rice areas may focus on weed management and most efficient capture and use of rainfall. Short-term research on insecticides or fungicides to control serious pest problems may provide quick solutions and more stable production, while the plant breeders pursue a longer-term strategy of developing varieties that have genetic resistance or tolerance to the same problem. It is imperative that research planners not attempt to generalize about the most critical constraints to production, and that programs are designed with both short- and long-term goals for sustained productivity.

INTEGRATION EFFICIENCIES OF PRODUCTION SYSTEMS

In order to meet both short- and long-term goals in research for greater productivity, it is important to look carefully at the functioning of natural ecosystems. These systems hold lessons that can be used in our design of efficient cropping systems. For examples of the principles of agroecology, there are several recent texts and reference works available (Altieri, 1987; Altieri & Hecht, 1990; Carroll et al., 1990; Dover & Talbot, 1987; Gliessman, 1990). Ecological interactions in agroforestry and alley cropping systems are described by Atta-Krah and Kang in Chapter 6 of this publication. The cyclical nature of resource flow in natural systems and the interaction of new technologies with the biological activities of tropical cropping systems are important to sustainability.

CYCLICAL RESOURCE FLOW IN BIOLOGICAL SYSTEMS

Plant nutrients, water, and undesirable pests all exist in complex and interacting cycles in natural and human-controlled biological systems. In a global sense, water and nutrients are not lost from the total system, but are undergoing a continuous and dynamic movement from one site to another. As an example, cycling of nutrients in both natural and agricultural systems has been summarized by King (1990). In a natural system plants grow, die, and decay, adding nutrients through the residue to the soil, both to the organic and the inorganic fractions. One major loss comes from burning for clearing in slash and burn systems, or for weed control in others, and much N and S are lost from the field (Barker, 1990). Some nutrients enter from the atmosphere through fixation or with rainfall, and some are lost from the system through leaching, soil erosion, denitrification or volatilization. Nutrients are taken up by plants, and the cycle continues. In addition to these processes, in agricultural systems there are additions of nutrients through fertilizers and removal of nutrients with the harvest of crops or grazing of livestock. Removal of fallow growth also exports nutrients and reduces the potential for N fixation. The cycle can continue, and a sustainable production level maintained, if the nutrients exported from the system or cycle are replaced in some way. In order to take advantage of this inherent nutrient cycling, it is important to minimize losses from the system and to make most efficient use of what is added. Losses due to soil runoff with rain or wind erosion can be reduced by careful management. For example, vetiver (*Vetiveria zizanioides*) grass can be used to create contour strips to reduce erosion. Choice of crop and cropping system, management of residue, slope in each field, and conservation structures such as terraces can all affect potential for erosion and nutrient loss. Weil and Mughogho (Chapter 8 in this publication) describe nutrient cycling in systems that incorporate *Acacia albida*, a woody perennial. Reduced tillage or zero-till planting helps to maintain residue on the soil surface, and this residue will both break the force of rain drops and absorb moisture. An additional benefit is reducing water loss from the immediate crop environment where water is needed by current and future crops.

Other types of loss of nutrient loss are leaching and denitrification. Leaching losses depend on rainfall and temperature, soil type, crop, and the specific nutrient measured (King, 1990). Leaching can be minimized by limiting application rates of chemical nutrients or manure, and by using management methods that capture and hold nutrients in the organic phase. Denitrification is affected by soil organic matter, aeration of the soil, moisture level, soil chemical properties, and temperature (Alexander, 1961). Only a few of these factors can be affected by crop management. These are some of the complexities of nutrient cycling in soils, and the mechanisms by which nutrients are lost. Efficient nutrient application and use needs attention by the careful manager, in order to meet the needs of the growing crop, so as to not allow excessive nutrients to be placed in the system where they can easily be lost. Return of all or most crop residue to the soil can promote active recycling of nutrients in the system. Grazing crop residues can accomplish the same goal, although it is important to spread the manure around if possible to maximize its use by crops and minimize losses.

The example of nutrient cycling is used to illustrate biological cycles in the field, and how a tropical cropping system can be manipulated to increase the efficiency of cycling and reduce losses. Nutrient cycles interact with those of water, plants, insects, pathogens, weeds, and other components of the system, and these interactions are critical to total system efficiency.

INTERACTION OF TECHNOLOGY WITH INTEGRATION EFFICIENCIES

Development of a more productive, input-intensive agriculture has involved the introduction of a number of controls on the environment, including fertilizers, pesticides, irrigation water, and others. In a sense, the modern farmer/manager is able to both increase productivity and enhance stability of crop production by dominating the natural climate and soil environment on which agriculture formerly depended. In a number of ways, this has given us the potential to feed today's large human population. At the same time, this introduction of technology has changed some of the efficiencies of natural systems and has impacted potential sustainability.

Classical yield response curves to most applied nutrients, if those elements are limiting to production, are quadratic in form with a decreasing return to additional units of the nutrient at higher levels of application. An example of decreasing efficiency of nutrient use is given by Power and Doran (1984) for N use in the USA. From 1959 to 1979, N fertilizer applied on four major crops increased from 2.2 to 8.6 million Mg, an increase of 295% in total application. During the same period, average crop yields increased from 2274 to 3853 kg/ha, an increase of only 69% in yields. There was a rapid move away from rotation of cereals with legumes such as clover (*Trifolium pratense, T. repens*), lespedeza (*Lespedeza* spp.), hairy vetch (*Vicia villosa*), and alfalfa (*Medicago sativa*), and toward a continuous cereal pattern that now depended more heavily on imported N fertilizer. The integration efficiencies were lost due to changes in cropping pattern that eliminated or reduced the use of rotations and legumes on many farms. It is likely that elimination of deep-rooted legumes in the rotation reduced the potential for upward nutrient retrieval by these crops to bring the nutrients into the root zone of a succeeding cereal (Harwood, 1985). Such loss in efficiency must be avoided in design of future tropical cropping systems.

Insect, plant pathogen, nematode, and weed species interact intimately with crops in the field and are highly influenced by climate, cropping patterns, and cultural practices. Populations of target pests are strongly reduced, at least in the short term, by applied chemical pesticides. There is substantial evidence that diversity in species and vegetation such as that found in some multiple cropping systems significantly reduces insect pest problems (Altieri & Letourneau, 1982; Altieri & Liebman, 1986). In natural systems, the dispersion of host species and diversity of ecosystems can restrict the spread of undesirable plant pathogens (Browning, 1975), and the same is true in species-diverse cropping systems (Thresh, 1982). Crop diversification has been used to manage undesirable nematodes through decoy or trap crops (Altieri & Liebman, 1986). Use of continuous cropping of single species rather than crop rotation can interrupt some of the natural controls on unwanted species in the field and make control more expensive. Application of general contact chemical products that kill or suppress a wide range of insect or weed species may at the same time eliminate some biotic components of the system that are useful in controlling others. Also, in some small-farm situations, the weeds may be a source of forage for animals. Water use by weeds and by crops must be balanced, taking into account the goals of the family for the total system.

These are examples of how integration efficiencies in complex systems may be changed by the use of specific pest control technologies. There is no question of the need for intervention by the farmer in the use of new technologies to increase productivity. We are learning to better anticipate the effects, both intended and unintended, of these interventions. But we need to examine the entire system, including the farm family resources and goals. If we can understand the impacts of a range of alternative technologies, in terms of resource use, environmental effects, and long-term profitability, it will be easier to make rational choices about what technologies are most sustainable for the future.

RESEARCH PRIORITIES FOR A SUSTAINABLE AGRICULTURE

Given the highly variable nature of agriculture around the Tropics, and the great diversity that exists in both systems and constraints to production, it is difficult to list a set of absolute research priorities for improving sustainability of production. In general, research needs to focus on cropping practices and systems that will make more efficient use of land and other production resources. Agriculture must be profitable in both the short and the long term. Solutions to increased food production need to be politically viable and socially equitable, in order to improve access to available food by people across the economic spectrum. Specific priorities are listed in the areas of plant genetics and crop improvement, nutrient sources for crops, strategies for pest management, tillage and soil/water management, diversity in systems design, and economic and other criteria for evaluation of productivity. This list is not exhaustive, but rather it is exemplary of the types of research called for by authors in this book and by national programs in the tropics. These ideas are summarized from the other chapters and reflect recent publications and research by Miller (1988, unpublished data), and Francis (1990).

Innovative ideas about research and priorities for a sustainable agriculture must be explored within the context of the current conventional agricultural research community. Given the structure of research institutions, funding sources, peer review, and reward systems that follow a paradigm based on conventional agriculture (Macrae et al., 1989), it is likely that sustainable agricultural activities may not be encouraged by the establishment. Agricultural research generally takes place in a policy vacuum (Ruttan, 1982; Busch & Lacy, 1983). In the absence of well-defined national goals in most countries, individual scientists or directors of research generally establish their own research goals (Macrae et al., 1989). Without a clear public set of goals or policies, the research direction is thus subject to the bias of each individual scientist or administrator and to pressures from funding sources or other vocal parts of society. Such challenges to the research agenda vary greatly from one country to another, yet lack of clear goals is a general problem that exists in most national governments. Within this context, the following specific basic and applied research directions are described.

Plant Genetics and Crop Improvement

Much of the effort in the international agricultural research centers and in national research programs has been in crop improvement. Early successes of new rice and wheat (*Triticum vulgare*) varieties in the Tropics and Subtropics spurred efforts in this direction. Much of the breeding and testing has been done under experiment station and large farm conditions. Although there is broad adaptation in many of the new cultivars, there is little doubt that valuable germplasm has been discarded in the process that could be useful to low-resource farmers in the Tropics. Although there is a high expectation that cropping conditions can be improved through the introduction of fertilizers, pesticides, irrigation, and other inputs, it is unlikely that the infrastructure and capital needed to make this happen will soon be available in many parts of the Tropics. Plant breeding for sustainability solves this by adapting genetic combinations for the production environment, rather than altering the environment by adding expensive inputs. Increases in productivity through improved germplasm and new cultivars is a highly desirable option, especially through programs focused on breeding for less favorable cropping conditions. Important research priorities in this area include:

- Continue selection and testing of new cultivars for tolerance or resistance to major insect and pathogen problems in crops
- Initiate breeding programs designed to make crops more tolerant to periodic drought and low total water availability, and more efficient in water use
- Select cultivars tolerant to less favorable soil environments, including both nutrient deficiencies and toxic levels
- Breed for early maturing crops that can fit into sequences where there is limited growing season or moisture
- Initiate breeding programs for adaptation to complex intercropping and relay cropping systems in the tropics
- Select for nutrient-use efficiency, especially for N, P, and K where these elements are limiting
- Design breeding methods that can test new combinations in early generation, and can move testing into farm conditions as early as possible
- Take advantage of new breeding techniques made possible by "genetic engineering," such as wide crosses, efficient selection methods, gene transfer
- Integrate breeding efforts into a total agronomic research program that includes manipulation of cropping systems and cultural practices

Alternative Nutrient Sources And Nutrient Cycling

Sustainability of the supply of nutrients for plant growth is central to maintaining productivity. There has been a wealth of research and data collected on responses of principle crops to applied fertilizers. Less information is available on alternative sources of plant nutrients and on responses of crops to lower levels of fertilizers and alternative sources of nutrients in the Tropics. There is an emerging literature on rhizobotany and genetic control of root morphology. Some of the key areas in which research is needed include:

- Determine sources of N and other major nutrients in tropical soils and their fate in cropping patterns with different levels of technology
- Study legumes adapted to tropical climates, their potential contributions to nutrients for cereal grain crops, and include research on potentials of woody legume species
- Understand and quantify the role of deep-rooted crops and trees in promoting the cycling of nutrients from lower soil strata
- Study cycling processes of N and other nutrients in the organic and inorganic fractions as well as in the soil solution in tropical soils
- Experiment with methods of legume incorporation for maximum nutrient recovery in tropical soils and climatic conditions
- Test reduced tillage methods and management of crop residues, including the effects of termites, ants, and other insects on residues
- Investigate methods of integration of animal manures and crop residues with cropping patterns to maximize nutrient use
- Study dynamics of rhizosphere organisms in different cropping sequences with primary tropical crops
- Pursue basic studies and practical agronomy work to understand plant-microbe interactions to improve symbiotic fixation of N
- Determine plant–microbe interactions of mycorrhizae and other associations that improve plant nutrient uptake under low-input conditions
- Test modifications of the soil rhizosphere through introduction of different organisms to promote crop growth, reduce toxins, reduce pathogens
- Examine genetic differences in root morphology and function that contribute to efficient nutrient use and cycling
- Study complex multiple cropping patterns and the nutrient dynamics among cereals, root crops, legumes and agroforestry species components
- Measure impact of perennial polycultures of herbaceous and woody plants on microorganisms and rhizosphere dynamics
- Determine simple indicators of key nutrient deficiencies, and interpretations of soil tests for accurate fertilizer and other nutrient use
- Study complementary effects of interacting sources of plant nutrients, their relationship with rainfall and other climatic variables
- Determine the economic and environmental impact of alternative approaches to improving soil fertility in tropical soils

Understanding Pest Biology and Management Strategies

Emphasis in pest control will shift toward pest management through understanding the dynamics of weed, insect, plant pathogen, and nematode populations and their interactions with soil microorganisms and other flora and fauna. This extension of the integrated pest management concept to greater resource use efficiency and environmental concern is parallel to similar concerns and research interests in temperate regions. Some of the specific research activities that are needed include:

Studies of Weed Biology, Dynamics, and Cultural Control

- Study primary tillage and cultivation effects on tropical weed competition, populations, and species dynamics
- Determine effects of cultural practices such as planting dates, crop densities, spatial organization, fertilization, irrigation on weeds in cropping and agroforestry systems
- Investigate weed/insect interactions under a range of tropical cropping systems and climatic conditions
- Quantify the effects of various tropical intercropping systems on weed growth and population dynamics over time in both annual and perennial species
- Pursue basic research on tropical weed biology and ecology, including weed competition with other weeds as well as with crop species and cultivated trees
- Study genetic diversity in weed populations, allelopathic suppression of germination and growth, and relationships with micorrhizae
- Continue studies on biological control of tropical weeds, including surveys of natural predators and bioherbicides

Studies of Plant Pathogens and Nematodes

- Study regional dynamics of tropical plant pathogens and the seasonal effects and annual fluctuations of pathogen incidence
- Quantify the climatic effects on pathogen and nematode dispersion and on survival in alternative crops and systems
- Understand how balance between beneficial and harmful microorganisms is affected by tropical climate, cropping patterns, fertilization and other practices
- Develop alternative methods to reduce or replace chemical pesticide use, including biocontrol and manipulation of tropical cropping systems and practices
- Continue selection for genetic tolerance or resistance to plant pathogens and nematodes and the interaction of resistance with climate and system
- Determine the influence of tropical multiple cropping systems on plant pathogens and nematodes, and manipulation of systems for control
- Develop quantitative methods to evaluate economic levels of infection and how these can be used to make economically rational decisions for control
- Study the interactions of plant pathogens and nematodes with other soil microorganisms, flora and fauna, and how these are influenced by tropical cropping systems

Studies of Insect Population Dynamics and Management

- Improve understanding of ecology of tropical cropping and tillage systems and how they influence insect pests
- Study scouting methods, economic levels of infestation and damage, and how these can be used to determine when control methods are needed in the field
- Investigate regional pest dynamics, insect dispersion and reproduction, insect predictions, and how management should be organized on a regional basis
- Evaluate insect/plant pathogen/weed/crop/soil microorganism interactions and how they can be managed for reducing economic damage to tropical crops
- Study balances among beneficial and harmful insect species and how these are affected by tropical crop management and system design
- Investigate a range of cultural and biological control methods, including system design, agronomic practices and biological agents
- Continue emphasis on genetic selection for insect tolerance or resistance in tropical crop species, a proven and highly successful venture in crop improvement
- Study basic biology of insect reproduction and insect interactions with host species and crop plants, and how management can influence these interactions
- Determine the effects of tropical multiple cropping and agroforestry systems on insect populations and damage to crops, and how these can be modified to improve control

Methods developed for integrated pest management should be extended to more regions and types of cropping systems in the Tropics. This approach takes into account the insect, weed, and pathogen populations from one season and one year to the next, and includes an evaluation of the expected crop loss from infestation or infection. The method also includes a wide consideration of options for pest management, including changing crops or cultural practices, choosing genetically resistant varieties, including biological control as part of the system, and using limited applications of carefully targeted pesticides as a last option.

Tillage, Soil and Water Management

Research priorities in tropical soil and water management include methods for efficient use of these resources by current farmers and conservation of soils for future generations. To maintain or improve productivity, it is essential to develop management strategies to keep soil in place and enhance its nutrient status and physical properties for crop production. Since the vast majority of tropical cropping areas are rainfed, it is imperative to capture and store as much rain as possible in the soil profile where it can be used by current and subsequent crops. Research ideas have been described by Lal (Chapter 16 in this publication), by Brown and Thomas (1990), and by Sanchez and Benites (1987). Some of the priorities are:

- Study alternative primary tillage operations to assess their energy use and efficiency, potential to maintain residue on the surface, and control weeds
- Investigate reduced and zero-tillage alternatives and their adaptability to tropical cropping systems, climates, and soils
- Evaluate the effects of permanent crop cover, using multiple cropping and agroforestry systems that are diverse in time and space, to avoid need for tillage
- Study alternative methods of water trapping or harvesting and storage within the soil profile for use by subsequent crops
- Test low-cost structures for water storage and distribution, and the efficiency of limited water use for irrigation
- Select crops and cultivars that are well adapted to reduced tillage systems in the tropics
- Select crops and combinations of species that are efficient in their use of limited water resources, and that can survive periodic drought conditions
- Study intensive tropical crop rotations, intercrops and agroforestry patterns that can reduce primary tillage costs and increase efficiency of water use.

Species Diversity And Alternative Systems Design

Extensive research has been done on the agronomic components of cropping systems, including nutrient response patterns, tillage and planting options, row spacings and plant densities, weed and other pest management, and economics of crop production. Until recently, research in the Tropics has concentrated on monoculture systems, those assumed to represent the future of a "modern agriculture." Relatively little attention has been given to species diversity, intercropping, agroforestry, and other more complex options in system design (Francis, 1986; MacDicken & Vergara, 1990). Over the past decade, there has been an acceleration of interest in complex cropping systems, and the information from this work is gradually entering the literature. Among the alternatives under study are intensive intercrop systems, annual and perennial mixtures, alley cropping, and other systems generally grouped under the term agroforestry. The current emphasis on agroecology indicates an awareness by research specialists that ecological interactions are important to the design of cropping systems, and that much relevant experience can be gained from prior ecology studies that will be useful to agriculture. Agroforestry systems hold much promise for tropical regions (Barker, 1990). Here are some of the research areas that will receive priority:

- Develop an easily accessible information base and retrieval system that will allow efficient use of past research results on tropical cropping and agroforestry systems

- Continue study of basic and applied aspects of rotation effects and the biological impacts of previous crop on current crop productivity
- Study cover crops, grain and forage legumes, and other methods of introducing diversity into cropping systems over time
- Test current and modified farmer intercropping practices with new varieties, alternative cultural practices, and pest control options
- Study the transition from chemical intensive to low-chemical input systems and the biological and economic impacts of such a transition
- Evaluate the relative contribution to productivity, stability, and sustainability of alternative production inputs, given a limited resource base
- Examine the biological, economic, and social implications of improved systems of species combinations as contrasted to monoculture systems
- Design and test tropical perennial/annual systems that appear to have promise as agroforestry alternatives

Economic And Other Criteria For System Evaluation

Methods for the economic evaluation of agricultural sustainability are currently being developed, since there is need for analysis over a longer than usual time frame and it is essential to factor in the future costs and availability of resources. This goes beyond a conventional analysis of cropping systems for net return, stability of income, and risk. A number of criteria also need to be considered in the elaboration of an evaluation framework for sustainable agricultural productivity in the Tropics. Economists would argue that these new dimensions can be taken into account in the application of conventional analyses, if assumptions can be made on the current and future costs of various inputs. Others would argue that some criteria are more difficult to quantify, including social and political impact of alternative systems, health and safety aspects of agriculture, gross sustainable product (as calculated in Indonesia), ecologically sustainable productivity, and contributions of alternative systems to quality of life. For these measures, there is need for new methods, for additional evaluation criteria, for new assumptions about what is desirable for the long term, and for introduction of human values into the measurement process. Among the research priorities are:

- Evaluate economics of alternative cropping patterns under a range of input cost and crop price scenarios, including potential future energy costs
- Quantify the long-term cost of permanent soil loss and the short-term costs of regenerating fertility in badly eroded areas
- Evaluate risk and uncertainty in alternative tropical cropping patterns, and the impact of crop/animal product diversity on sustainable productivity
- Assess current and projected future costs of nonrenewable resources needed for success in improving productivity in tropical systems

- Incorporate health and safety criteria in the evaluation of alternative types of input use and ways of increasing productivity of systems
- Calculate the "gross sustainable product," after developing the criteria needed to assess long-term costs of both inputs and outputs from national production systems
- Determine ways to measure the ecologically sustainable product based on balance of use of renewable and nonrenewable resources
- Seek methods of incorporating social objectives of individual countries into the economic evaluation of system success, and develop models for use

This list of research priorities is comprehensive, but not precisely tailored for any specific country or region. These are among the potential objectives that should be considered in the development of a set of unique priorities for each national development program and for regions or states within countries. Constraints, resources, and opportunities are unique in each set of circumstances, and the research program needs to be carefully organized to fit those conditions. More important than this list is the *process* that leads a director and research group through the identification of human needs, constraints to productivity, assumptions about future availability of production inputs and infrastructure, and eventually the selection of specific research questions and procedures to follow to create a more sustainable food production and supply sector in the country.

INVOLVING FARMERS AND INSTITUTIONS IN PARTICIPATORY EXTENSION

There is an emerging awareness of the importance of involving farmers in the research and extension process. Rhoades (1989) summarized the need to recognize and then tap into indigenous knowledge. From the integration of farmer knowledge with science, a number of options will emerge in the design of new tropical cropping and crop/animal systems. Since research resources are scarce, and since extension often does not reach those people most in need of services, it is important to consider how participatory methods could be used in transfer or sharing of information. Methods of implementing on-farm research activities as well as a continuous sharing of information through extension are described by Hildebrand and Poey (1985).

Hayward (1987, unpublished data) notes that extension must be viewed in its relation to other components of development, including research, education, marketing, and international trade. Priorities must reflect stage of development of a region or country, availability of appropriate technologies, and infrastructure to make them available to farmers. Among the priorities identified by Hayward of the World Bank include setting the right priorities and recognizing that these will differ among countries. The "training and visit" (T & V) model is seen not as a recipe, but as an approach that can be adapted to different circumstances. Extension is not widely recognized as a viable agent of change within the development process in some tropical countries, and this needs to be corrected. Extension has a role in knowledge development as well as dissemination, and the advantage many extension workers bring to the job is an ability to see whole systems rather than only the components seen clearly by some researchers. It is critical to develop, train, and sustain extension people in their jobs, upgrading their skills and providing adequate incentives. One challenge through much of the Tropics is to make the extension process and service cost-effective, primarily through working with groups and seeking the methods that will provide the greatest multiplier effect. The World Bank has recognized the key importance of extension, and has provided substantial loan monies to help the systems develop.

The concept of extension has become far more than putting a group of messengers in action to carry production practices from the research sites to farmers. In most countries the farmer is increasingly faced with a number of sources of information, and one of the big challenges is to sort out these often conflicting ideas to decide what is right for each farm and field. Some of the key focus areas for extension as summarized by Francis (1990) include:

Focus on Systems Rather than Components

There are likely to be continued and incremental improvements in productivity due to new varieties, fertilizer formulations, or weed control practices, yet the substantial advances in the future will come from understanding complex systems and the interactions that make biological agroecosystems function.

Focus on Efficient Use of Resources

Scarcity, cost of fossil fuels, and lack of access to this energy source by many farmers in the Tropics requires that we develop systems that make most efficient use of available internal resources on the farm. Those resources and inputs that are purchased from outside need to be used in the most efficient possible manner. Without a specific product to sell, extension can provide objective advice on input use.

Focus on Information as a Production Input

One of the ultimate renewable resources is information, a resource that is not used up and in fact grows more valuable with use. Information is creatable, expandable, diffusive, transportable, substitutable, and sharable; this unique input can be used effectively to increase and sustain productivity. There is need to collect, organize, and share information across crop, discipline, and system boundaries, as well as across national boundaries.

Focus on Participatory Systems for Information Development

One benefit of on-farm research when the farmer is a full partner in this process is the spin-off in potential diffusion of that information to neighbors and friends. When farmers are involved with choice of problems to confront, with sorting out alternative solutions, designing treatments, implementing the tests in the field, taking data and interpreting results, they are full owners of the process as well as the idea. These same farmers can be valuable extension people in their immediate communities or spheres of influence.

Focus on Process Rather than Product

There is a prevalent concept that researchers develop a formula, menu, or series of products in a package that will solve all production constraints. Farming systems research/extension methods summarize a process through which farmers, with the help of specialists, design solutions for their own constraints. They are empowered to own both the process and the solution. Extension programs in the Tropics need to adopt educational programs that will foment this type of empowerment.

Focus on Diversity of Enterprises

Although many farmers in the Tropics have not moved to specialization to the same degree as their counterparts in temperate countries, there is still a danger in focus on one or a few crops or products. Biological diversity in a cropping or crop/animal system mimics to some degree the diversity and thus stability that is found in natural systems. Greater diversity in products for sale can enhance the economic sustainability and reduce the overall risk in an operation. Farmers need to be flexible enough to react to market opportunities, multiple potential uses of their crops, and ways to add value to their products on farm.

Focus on Community in Addition to Farm and Family

Without community, there is little quality of life for the rural family. It is important to design systems on the farm as well as micropolicy that will encourage the health and vitality of rural communities. More than an infrastructure to support the farming or ranching enterprises, this focus includes attention to people's needs for education, recreation, social gatherings, for religious practices, and for building community and quality of rural life.

Priorities in extension programs in tropical countries need to consider the challenges found in many developing regions: improving diet, nutrition and health; conserving and managing natural resources and the environment; unique and appropriate alternatives to subsistence-level farming; importance of risk in production decisions; land tenure issues, especially for small farmers; availability of production inputs and credit; development of markets and infrastructure; impact of national agricultural policy; family and community economic well being; and incentives for young people to enter agriculture.

Sustainability will be enhanced in tropical countries if these issues are addressed. Farmers, social scientists, and rural planners generally agree that owner/operator systems in agriculture provide incentives for economic progress, productivity, and stewardship. Carefully designed national policies, as well as practical and relevant extension programs can help these systems develop.

ENVISIONING THE FUTURE IN AGRICULTURE

The conventional approach to viewing the future is to track past trends up to present and extrapolate those lines in the future. This assumes that technology will stay the same or will continue to be developed and become available at approximately the same rate. The method further assumes that resources will continue to be available, and incentives provided by governments. This approach to predicting the future is essentially a reactive model based on current trends. To subscribe to this approach is essentially to ignore the potential of humans to reflect on their activities, to observe and interpret changes in the global ecosystem, and to use reason to evaluate alternative strategies for the future.

A more exciting and empowering process for the future is to project ahead a number of years or decades and envision a desirable environment or situation. This can be done for a single field or farm, for a community, for a nation, or for the world. One could describe a field that has a certain level of organic matter, annual/perennial crop mixtures with multiple uses, multiple cropping pattern that uses available resources around the year, and system that depends entirely or almost completely on internal, renewable resources. For the farm, one could describe the mix of crops and livestock, a secure food supply, greater nutritive value of foods, the value-added products that would provide steady income and lower possible risk, and diversity in biology and economic activities. For the community one could imagine clean and well-organized streets, market places, ample parks and green areas, library and school facilities, a viable business sector to meet consumer needs, professional services such as a medical doctor and facilities, dentist, and others, ample adult educational opportunities, religious organizations, and all the elements that contribute to quality of life. These would be available to rural community dwellers as well as those who live on nearby farms. A similar envisioning could be done for a region, a country, or for the entire planet.

The next and critical step is to begin making decisions today to cause that desirable future to happen. This may or may not make rational economic sense in the short term, but if the goals are well in mind and broadly accepted by all the family or the community, the short-term sacrifices will result in a long-term positive result. This process empowers individuals, families, communities, and groups of human communities to take charge of their future, rather than preparing them to react to whatever outside forces cause to happen. In order for neighboring farm families or groups of communities to work toward a common vision, they must rationalize differences in goals when there is conflict over use of physical space or resources. Such a visioning process is exciting, dynamic, and requires ownership by all the members of the family, community, or nation. The process could be called "visioning a desirable future," or creating an environment in which we and our descendents would like to be. It is useful to remember that the future begins tomorrow, and that we can begin to make decisions today to cause that desirable future to happen. This is one approach to envisioning a sustainable agriculture in the Tropics and wherever farmers and ranchers live and crops are grown.

ACKNOWLEDGMENTS

Comments by Drs. Tom Barker, Ray Weil, Ken Cassman, and Max Clegg are gratefully acknowledged, as well as secretarial support by JoAnn Collins and Pat Martinez.

REFERENCES

Alexander, M. 1961. Introduction to soil microbiology. John Wiley & Sons, New York.

Altieri, M.A. 1987. Agroecology: The scientific basis of alternative agriculture. Westview Press, Boulder, CO.

Altieri, M.A., and S.B. Hecht. 1990. Agroecology and small farm development. CRC Press, Boca Raton, FL.

Altieri, M.A., and D.K. Letourneau. 1982. Vegetation management and biological control in agroecosystems. Crop Prot. 1:405-430.

Altieri, M.A., and M. Liebman. 1986. Insect, weed, and plant disease management in multiple cropping systems. p. 183-218. In C.A. Francis (ed.) Multiple cropping systems. Macmillan Publ. Co., New York.

Atta-Krah, A.N., and B.T. Kang. 1993. Alley farming as a potential agricultural production system for the humid and subhumid tropics. p. 67-76. In J. Ragland and R. Lal (ed.) Technologies for sustainable agriculture in the tropics. ASA Spec. Publ. ASA, CSSA, and SSSA, Madison, WI.

Barker, T.C. 1990. Agroforestry in the tropical highlands. p. 195-227. In R.G. MacDicken and N.T. Vergara (ed.) Agroforestry: Classification and management. John Wiley & Sons, New York.

Bird, G.W., T. Edens, F. Drummond, and E. Groden. 1990. Design of pest management systems for sustainable agriculture. p. 55-100. In C.A. Francis et al. (ed.) Sustainable agriculture in temperate zones. John Wiley & Sons, New York.

Brown, H.C.P., and V.G. Thomas. 1990. Ecological considerations for the future of food security in Africa. p. 353-377. In C.A. Edwards et al. (ed.) Sustainable agricultural systems. Soil and Water Conserv. Soc., Ankeny, IA.

Brown, L.R. 1990. The illusion of progress. p. 3-16. In L.R. Brown et al. (ed.) State of the world, 1990. Worldwatch Inst., Washington, DC, and W.W. Norton and Co., New York.

Browning, J.A. 1975. Relevance of knowledge about natural ecosystems to development of pest management programs for agroecosystems. Proc. Am. Phytopathol. Soc. 1:191-194.

Bull, D. 1982. A growing problem: Pesticides and the third world poor. Oxfam Public Affairs Unit, Oxford, England.

Busch, L., and W.B. Lacy. 1983. Science, agriculture, and the politics of research. Westview Press, Boulder, CO.

Carroll, C.R., J.H. Vandermeer, and P.M. Rosset. 1990. Agroecology. McGraw Hill Publ. Co., New York.

Chambers, R., A. Pacey, and L.A. Thrupp. 1989. Farmer first: Farmer innovation and agricultural research. Intermed. Technol. Publ., London.

Conway, G. 1985. Agroecosystem analysis. Agric. Admin. 20:31-35.

Dover, M., and L.M. Talbot. 1987. To feed the Earth: Agro-Ecology for sustainable development. World Resour. Inst., Washington, DC.

Francis, C.A. (ed.). 1986. Multiple cropping systems. Macmillan Publ. Co., New York.

Francis, C.A. 1988. Internal resources for sustainable agriculture. Gatekeeper Ser. no. SA8. Int. Inst. for Environ. and Dev., London.

Francis, C.A. 1990. Future dimensions of sustainable agriculture. p. 439–466. In C.A. Francis et al. (ed.) Sustainable agriculture in temperate zones. John Wiley & Sons, New York.

Gilbert, E.H., D.W. Norman, and F.E. Winch. 1980. Farming systems research: A critical appraisal. Michigan State Univ. Rural Dev. Paper no. 6. Dep. Agric. Econ., Michigan State Univ., East Lansing, MI.

Gliessman, S.R. 1990. Agroecology: Researching the ecological basis for sustainable agriculture. Springer-Verlag, New York.

Goldemberg, J., T.B. Johansson, A.K.N. Reddy, and R.H. Williams. 1987. Energy for a sustainable world. World Resources Inst., Washington, DC.

Harwood, R.R. 1985. The integration efficiencies of cropping systems. In T.C. Edens et al. (ed.) Sustainable agriculture and integrated farming systems. Conf. Proc., Michigan State Univ. Press, East Lansing, MI.

Harwood, R.R. 1990. A history of sustainable agriculture. p. 64–76. In C.A. Edwards et al. (ed.) Sustainable agricultural systems. Michigan State Univ., Lansing, MI.

Hildebrand, P.E., and F. Poey. 1985. On-farm agronomic trials in farming systems research and extension. Lynn Rienner Publ., Boulder, CO.

Jeyeratnam, J. 1985. Acute pesticide poisoning: A Third World problem. World Health For. 6:39.

King, L.D. 1990. Sustainable soil fertility practices. p. 144–177. In C.A. Francis et al. (ed.) Sustainable agriculture in temperate zones. John Wiley & Sons, New York.

Lal, R. 1993. Technological base for agricultural sustainability in sub-Saharan Africa. p. 257–263. In J. Ragland and R. Lal (ed.) Technologies for sustainable agriculture in the tropics. ASA Spec. Publ. ASA, CSSA, and SSSA, Madison, WI.

Lockeretz, W. 1987. Establishing the proper role for on-farm research. Am. J. Altern. Agric. 2:132–136.

MacDicken, K.G., and N.T. Vergara. 1990. Agroforestry: Classification and management. John Wiley & Sons, New York.

Macrae, R.J., S.B. Hill, J. Henning, and G.R. Mehuys. 1989. Agricultural science and sustainable agriculture: A review of the existing scientific barriers to sustainable food production and potential solutions. Biol. Agric. Hort. 6:173–219.

McCracken, J.A., and G.R. Conway. 1987. Pesticide hazards in the Third World: New evidence from the Philippines. Gatekeeper Ser. no. SA1, Int. Inst. Environ. and Dev., London.

Norman, D.W. 1980. Farming systems approach: Relevance for the small farmer. Michigan State Univ. Rural Dev. Pap. no. 5. Dep. Agric. Econ., Michigan State Univ., East Lansing, MI.

Parkhurst, A.M., and C.A. Francis. 1986. Research methods for multiple cropping. p. 285–316. In C.A. Francis (ed.) Multiple cropping systems. Macmillan Publ. Co., New York.

Postel, S. 1984. Water: Rethinking management in an age of scarcity. Worldwatch Pap. no. 62. Worldwatch Inst., Washington, DC.

Postel, S. 1989. Water for agriculture: Facing the limits. Worldwatch Pap. no. 93. Worldwatch Inst., Washington, DC.

Power, J.F., and J.W. Doran. 1984. Nitrogen use in organic farming. p. 585–598. In R. Hauck (ed.) Nitrogen in crop production. ASA, Madison, WI.

Rangeley, W.R. 1987. Irrigation and drainage in the world. p. 29–35. In W.R. Jordan (ed.) Water and water policy in world food supplies. Texas A & M Univ. Press, College Station, TX.

Rhoades, R. 1989. The role of farmers in the creation of agricultural technology. p. 3–9. In R. Chambers et al. (ed.) Farmer first: Farmer innovation and agricultural research, Intermediate Technol. Publ., London.

Ruttan, V.W. 1982. Agricultural research policy. Univ. of Minnesota Press, St. Paul, MN.

Rzewnicki, P.E., R. Thompson, G.W. Lesoing, R.W. Elmore, C.A. Francis, A.M. Parkhurst, and R.S. Moomaw. 1988. On-farm experiment designs and implications for locating research sites. Am. J. Altern. Agric. 3:168–173.

Sanchez, P.A., and J.R. Benites. 1987. Low-input cropping for acid soils of the humid tropics. Science (Washington, DC) 238:1521–1527.

Taylor, D.C. 1990. On-farm sustainable agriculture research: Lessons from the past, directions for the future. J. Sustain. Agric. 1:43–87.

Thresh, J.M. 1982. Cropping practices and virus spread. Annu. Rev. Phytopathol. 20:193–218.

Urban, F., and P. Rose. 1988. World population by country and region, 1950–86, and projections to 2050. Econ. Res. Serv., USDA, Washington, DC.

Weil, R.R., and S.K. Mughogho. 1993. Nutrient cycling by apple ring acacia in agroforestry systems. p. 97–108. *In* J. Ragland and R. Lal (ed.) Technologies for sustainable agriculture in the Tropics. ASA Spec. Publ. ASA, SSSA, and SSSA, Madison, WI.

14 Strategies for Developing a Viable and Sustainable Agricultural Sector in sub-Saharan Africa: Some Issues and Options

Abbas M. Kesseba

International Fund for Agricultural Development
Rome, Italy

This chapter seeks to outline some of the major elements which govern the performance of the region's agricultural sector and to discuss the factors and issues which must be addressed in formulating strategies for a viable and sustainable agricultural sector in Sub-Saharan Africa (SSA). As argued in sections two and three of the chapter, sustainability is an old concern, but the search for clarity in defining its operational content continues to present a challenge. Nowhere is this challenge greater than in SSA, the nature of which is described in the fourth section. The fifth section identifies the policy, technological and institutional issues that must be addressed in shifting agriculture onto a sustainable path in SSA, while section six provides some initial ideas on leverage points for this. Before concluding on an optimistic note in section eight, section seven poses questions for further thought.

SUSTAINABLE AGRICULTURE: OLD CONCERNS, NEW PROBLEMS

Even before man began to practice sedentary agriculture, the need for sustainable harvesting systems to meet human food requirements was constantly addressed. Through the ages, humans shifted from hunting and gathering to sedentary forms of agricultural production systems which also have continued to be shaped by the need to maintain the productivity of the resource base. The traditional practice of shifting cultivation exemplifies a strategy adopted to pursue sustainable agriculture. Such practices were adopted to maintain the fertility of the resource base. As land became a limiting factor, these systems started to break down under population pressure, and their sustainability was questioned. The search for alternatives has continued right up to this day.

Technological innovations which led to the use of efficient and effective devices such as the Shadouf (the counter-poise lift), the Sakia (the scoop-wheel) and basin irrigation—all used in ancient Egypt, or the construction of bench terracing, elsewhere in the ancient rice (*Oryza sativa* L.)-growing systems on the hillslopes of Southeast Asia, are examples of traditional water control, and soil and moisture conservation measures. Such traditional technology was sustainable in the prevailing spatial and temporal contexts.

During the twentieth century, demographic pressure further elicited increasingly sophisticated responses which offered the possibilities of intensification of resource use. These were accompanied by the development and transfer of modern technology which came in packages of inputs including seeds of high-yielding varieties, chemical fertilizers, biopesticides, better controlled water supply, etc. As the introduction of such technology led to increased demand for farm labor, associated with different sets of management practices accompanying the packages, a significant development in farm mechanization also followed. This is reflected in the development and use of mechanical equipment from seed drills to tractors and combine harvestors.

Extensive literature has been written on the impact of these developments on the economic and biological sustainability of the farming system. It has generated diverse views about what is the best strategy to address the need to constantly improve productivity from limited arable land resources, with due consideration to the conservation of those resources and, of course, the economics of the alternative strategies in question. To date, the debate remains unresolved in spite of significant strides taken in the recent past to identify and articulate economically viable and environmentally sustainable strategies. Unfortunately, as expected, few options have been developed which adequately address agricultural production systems in the more adverse physical and climatic conditions. However, it is in the light of such conditions that sustainability has now become the overriding issue.

Part of the problem lies in the fact that sustainability has been notoriously difficult to define coherently, especially in terms of its operational content. Moreover, it is often not viewed in a dynamic context, as it should be. The generation of new technology facilitates the achievement of ever higher levels of productivity. In most instances, the goal achieved is that of transcending a given "conventional" yield threshold. The new level of productivity could well be considered the ceiling of biological productive capacity at that given point in time. It may even be responsible for "lifting" a production system and placing it into a sustainable path.

However, demographic change continues to present new demands requiring a constant response from production systems. Achieving higher equilibria, in turn, requires a continuous stream of technology and each new kind of technology provides the opportunity to reach a new optimum. In the context of population expansion, therefore, an agricultural system is sustainable only in the context of its current carrying capacity which is transient—in the short run it is defined by a given technology. In the medium to long run, without successive changes in productivity through appropriate technological change, the system would be rendered unsustainable.

Technology change has not kept pace with the rapidly deteriorating productive capacity, especially in the context of fragile marginal agroecosystems. Whenever these areas have sought to increase food production through intensification of production systems, it has invariably led to environmental damage; meanwhile economically attractive conservation-based technology is hard to develop. This has led to a consistently falling trend in per capita food output in these areas—a matter of grave international concern, especially since the phenomenon is now becoming common to better endowed high potential areas as well.

While the term "sustainable agricultural development" is vague, various attempts have been made to define more clearly the concepts and elements which could serve to translate it into operational terms. There is some consensus for viewing sustainable development as a process directed towards inducing defined economic change which can result in an ecologically sustained pattern of resource use and is implemented in a participatory manner.

The Consultative Group on International Agricultural Research (CGIAR) adopted in 1988 a definition of sustainable agriculture: "Sustainable agriculture should involve the successful management of resources for agriculture to satisfy changing human needs while maintaining or enhancing the quality of the environment and conserving natural resources."

For a more profound understanding of this definition it is necessary to consider the meaning of the individual words used: (i) *successful* implies that the production system should generate adequate income and otherwise be economically viable and socially acceptable; (ii) *management* includes policy decisions that can affect agriculture at all levels, from national government to the individual producer; (iii) *resources* includes inputs and manufactured goods coming from outside the agricultural sector (e.g., agricultural chemicals, machinery, etc.); and (iv) *maintaining the quality of the environment* suggests that changes in the environment or the availability of natural resources should not threaten the capacity to meet changing needs and that production needs should be met without unnecessarily damaging natural ecosystems.

It is argued that it is primarily population growth and poverty that threaten the possibilities for sustainable agricultural development. The increasing population densities lead to increased pressure on natural resources and force poor farmers to expand to more marginal areas resulting in a need to intensify agricultural production. The traditional production systems are in harmony with the environment prior to exposure to various forces of pressure. Such systems usually do not provide adequate possibilities for intensification. Thus, new production systems that can build on the resilience of the traditional systems that can be characterized as "open production systems" utilizing external inputs and are able to produce surplus, need to be developed. Only through such systems will it be possible to relieve the pressure on natural resources and ensure sustainable agricultural development.

The widespread concern for sustainability of agricultural production systems is not confined to SSA. Processes of environmental degradation, manifested in deforestation, desertification, soil erosion and deterioration in water availability and quality, affect substantial areas in all other parts of the developing world. These processes are a threat to sustainable agriculture; however, they constitute effect rather than cause.

The sustainability of agricultural production systems is determined by a complex web of forces which have a profound influence on the management and use of natural resources. In the case of SSA the complexity is extraordinarily great.

THE SEARCH FOR GREATER CLARITY

The U.N. Conference on the Human Environment of 1972 was one of the first occasions on which production systems were discussed within the context of threatened environmental sustainability. While increased emphasis was placed on the environment as a result of that Conference, the widespread famines which followed, inspired the World Food Conference of 1974. The conference was convened to address agricultural production systems in the context of the need for reversing the alarming food security situation throughout the developing world. It also articulated the need to ameliorate the plight of the poorest of the rural poor. The conference led to the establishment of the International Fund For Agricultural Development, (IFAD), with a specific mandate to address the particular concerns of the poorest target groups in increasing agricultural production, reducing undernutrition and alleviating their poverty, a substantial percentage of whom live in SSA.

Since then, several international initiatives have further enriched discussions on agricultural development, on the one hand, and the environment on the other—for instance, in the context of rural production systems, the World Conference on Agrarian Reform and Rural Development (WCARRD, 1979)—and, in the context of the natural resource base, the formulation of the International Conservation Strategy, (International Union for the Conservation of Nature, 1980). However, the environment and sustainable agricultural issues were discussed virtually in isolation, while in reality they are inextricably linked. Linking these divergent strands of concern, the Report of the World Commission on Environment and Development (WCED) of 1987 represented an important attempt to clarify some of our perceptions about the relationship between environment and development and thus provide some basis for establishing a framework for addressing agricultural development and environment in an integrated manner. This was further developed by the IFAD-initiated International Consultation on Environment, Sustainable Development and the Role of Small Farmers, in 1988.

In building awareness on the relationship between the environment and development, the above-mentioned initiatives paved the way for the U.N. Conference on Environment and Development (UNCED, 1992), where a comprehensive global strategy for sustainable development, known as Agenda 21, found expression. The linking of the environment with development also is evident in the Agenda 21 program areas on Sustainable Agriculture and Rural Development (SARD), which emerged from the Food and Agriculture Organization (FAO)/Netherlands den Bosch Conference on Agriculture and the Environment (1991). However, while the SARD programe areas are useful to the extent that they list some of the major requirements for action in the agricultural sector, each program remains a compartmentalized element of a broad strategy that does not provide a conceptual basis for establishing priorities, clarifying trade-offs and evaluating alternatives in the quest for sustainability.

The intellectual evolution towards a perception concerning what it takes to pursue sustainable agriculture is, thus, still ongoing. Considerable progress has been made towards achieving the convergence between environment and development at the conceptual level. This is encouraging and mirrors IFAD's own experience in the field, at the operational level.

By virtue of an orientation towards resource-poor conditions which the Fund has adopted as a result of its efforts to reach its target groups, its projects have inevitably had to address the natural resource constraints and opportunities facing them, in pursuit of sustainable rural poverty alleviation objectives. This concern is embodied in Agenda 21 emerging from UNCED.

In pursuit of its specific mandate, IFAD has made its own modest contribution to the understanding of the self-reinforcing mechanisms which govern the rural poverty—environmental degradation nexus. It recognizes that the processes which cause environmental degradation and rural poverty, and contribute to the complex linkages between the two, have not been studied rigorously enough. These processes are often site-specific and are governed by structural, technological and cultural causes which do not readily lend themselves to solutions adapted from elsewhere.

Thus, the considerable efforts of the international development community to define the parameters of the sustainable development process which culminated in the UNCED initiative, have been promising in terms of guiding a gradual shift to global sustainability. However, much more still needs to be done to identify the key leverage points for assisting poor rural populations to effect a transition to economically viable and ecologically sustainable agriculture. Nowhere is this challenge greater than in SSA.

Developmental efforts in SSA continue to focus on intensification of agricultural production without due attention to sustainability factors. At the same time, insufficient attention to conservation-based principles and practices has undermined the sustainability of agricultural systems in SSA. The IFAD's strategy for investment in the region is increasingly placed within the context of the dynamics of the rural poverty and resource degradation processes. There also is clear recognition of the fact that, over time, growing populations need to go beyond mere maintenance of the resource base, through increasing resource productivity.

THE SETTING

The Policy Context for Agriculture

The SSA is among the poorest developing regions of the world. Twenty-nine out of thirty-four of the world's poorest countries are in this region. Throughout the 1980s, it witnessed falling per capita incomes, with increasing hunger and malnutrition, while climate and soil constraints, interacting with inappropriate resource use, were accelerating environmental degradation and the depletion of the resource base. The cumulative impact of these factors reinforced by unregulated demographic pressures had a debilitating effect on the overall productivity of agriculture in the region. A population of about 600 million in 1989 is projected to rise to about 900 million by the year 2000, if the average annual growth rate of 3.3%—the highest in the world, continues to prevail.

The population has grown beyond the capacity of the existing rural services (the infrastructure, financial services, health, input supply, marketing, etc.) for poor small farmers and their households. This strain on services undermines developmental efforts to alleviate rural poverty and takes its toll on the sustainability of agriculture. However, population dynamics should not be made the scapegoat for all of the region's woes, although they certainly share in the responsibility for poor food security performance.

A number of other factors, notable in the realm of the policy environment are equally responsible for many of the constraints facing SSA agriculture. Unfortunately, there is no single consistent strategy mounted in response to the agricultural crises affecting the region.

Several initiatives such as the Lagos Plan of Action for the Economic Development of Africa 1980–2000, the Harare Declaration of 1984 and "Africa's Priority Programme for Economic Recovery 1986–90" of the Organization of African Unity (OAU), displayed concern over the agricultural crisis in Africa. These documents also identified many of the causes of this crisis, among them the general neglect of the food crops subsector in most SSA countries—continuing from the colonial period well into post-independence Africa.

At the same time, international development institutions also have taken a comprehensive stock of the constraints facing the agricultural sector in the region and all recognize the need to break the trend of poor agricultural performance. However, given the multiplicity of donors and development aid agencies working in SSA, the prescriptions offered often differ considerably. This has led to uncoordinated efforts which counteract in such a way as to pose barriers to the implementation of individual programs.

The Lagos Plan adopted by African heads of state and by the OAU, had explicit objectives for "bringing about self-sufficiency in food and diminishing dependence on exports." Another position held by the World Bank and the International Monetary Fund, reflects a broad objective of "enabling the African Countries to reduce their Balance of Payments deficits," recommending a more outward-oriented trade strategy. While the merits of each position are debatable under various economic conditions, the fact that there is a fundamental dichotomy in the strategic economic policy agenda, has led to self-defeating initiatives.

For instance, the pursuit of policies aimed at food self sufficiency on the part of SSA countries has been, at most, a qualified success. Fifteen out of thirty-five countries of the region had reached food self sufficiency by 1985 to 1986 and were, in fact, temporary net exporters (FAO, 1978-1988). However, this achievement is questioned since the countries included, ironically, many with the lowest income levels and a high incidence of rural poverty. High degrees of self sufficiency in these countries must therefore be seen in the context of low effective demand and not in terms of success achieved through production expansion.

The principle need in these countries, therefore, is to achieve improved household food security, which entails the provision of sufficient and stable food supply to all stata of society including the rural poor, especially women and children. It is often argued that this can be achieved not only through improved domestic agricultural production (for which the individual country may not have a suitable resource base) *but also* from the production of exportables (for which an individual country may possess a comparative advantage), which will pay for food imports and improve food availability. However, the latter approach needs to be pursued catiously through an appropriate mix of macroeconomic and sectoral policy reforms, since individual and collective inter- and intrasectoral impact can be highly detrimental to some segments of the population, and even to the economy as a whole, depending on the prevailing circumstances in each country; thus it may not lead to the desired responses to specific policy instruments.

The position held by the Bretton Woods institutions is that the promotion of "efficient" allocation of resources led by undistorted market signals through a comprehensive set of macroeconomic policy changes would stimulate domestic economic growth, including that of the agricultural sector. This position relies on a strategy of specific economic adjustment at the core of which is fiscal discipline, real exchange rate devaluation and trade liberalization. In particular, it demands reduction in the domestic budget and the external deficit through fiscal and monetary measures, elimination of distortions in the factor and product markets (the rationalization of subsidies and removal of exchange and other biases against exportables), reallocation of resources from the production on nontradables to that of tradables, and improved efficiency of public sector institutions. These prescriptions are packaged in structural adjustment policies (SAPs) which have been widely applied in a number of SSA countries, albeit with mixed results.

Figure 14-1 shows the relative export prices for five major SSA export commodities over the period 1961 to 1987. It clearly shows the decline of world market prices for coffee, (*Coffea arabica* L.), cocoa (*Theobromo cacao*), copper, cotton (*Gossypium hirsutum* L.) and petroleum, being the main export commodities of the continent. This has led to a serious decline in SSA countries' foreign exchange earnings, increasing the balance of payment deficit and undermining their capacity to respond to structural adjustment

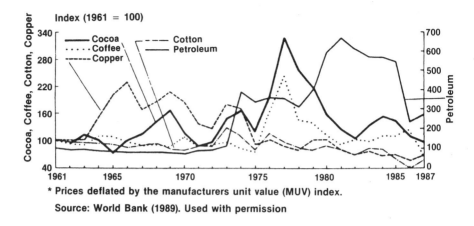

* Prices deflated by the manufacturers unit value (MUV) index.

Source: World Bank (1989). Used with permission

Fig. 14-1. Prices for five major SSA export commodities, 1961 to 1987.

policies within the time frame available to them. Thus the international terms of trade have negatively affected the African economies and are likely to continue inflicting negative impacts unless the Uruguay Round of negotiations on the terms of trade can successfully reach an agreement regarding agricultural commodities.

Many of the economies of countries in the region have specific domestic structural constraints and "rigidities" (e.g., restricting resource mobility) which impede successful adjustment initiatives, especially in the short run. Distortionary domestic policies of the past also have created circumstances which are not conducive to demand management. Attempts to check aggregate demand (through curtailment in public expenditure or increased interest rates) can have a detrimental impact on output (contractionary effect), unless other distortions, that limit resource movement for instance, are removed first.

Seeking to reorder sectoral policies without due regard to existing imbalance between domestic absorption and domestic supply, (stabilization policy) has led to a new set of distortions. Sectoral reallocations (between tradeables and nontradeables) through the introduction of strategies to influence production and exchange (privatization) without due regard to the above-mentioned imbalances, has further skewed the impact of the adjustment process in favor of the "haves" with significant trade-offs in terms of imposing high costs on the "have-nots." Given the general higher incidence of resource-poverty among female-headed households (a prominent feature in southern Africa) and their lack of access to production factors, they have less opportunity to respond to price and other incentives under SAPs; hence the impact of these policies is even less favorable to poor women.

The process of structural economic adjustment may take its toll on the domestic economy (especially in the short-term) in terms of distorting real wages, through adverse distributional effects which may arise due to higher food prices, and threatened household food security, especially among the poor segments of the population. Many of ths risk-averse governments in SSA would therefore be legitimately wary of pursuing such a policy in the future, unless forced by circumstances which drive them towards borrowing foreign exchange. A number of countries in SSA which embraced SAPs (and the attractive external assistance which came with these policies) have de facto abandoned some of the policy components.

Stringent fiscal discipline under structural adjustment regimes has led to the switching of budgetary allocations to economic sectors other than agriculture. The resultant low resource allocation for agricultural research, rural infrastructure, extension, etc., has now been reinforced by ill-conceived domestic policies which have been traditionally biased against food and agricultural commodity markets.

The pursuit of adjusting the external balance under SAPs also has not been entirely trouble-free. A collective export-led strategy has been pursued by all the countries in the region (at great cost and investment on the part of individual countries). However, it has been amply demonstrated that hostile world market responses (protection measures and market disruption leading to falling international commodity prices) can undermine the envisaged advantages and, in fact, ultimately worsen the economic situation in the region.

Supply and Demand for Food

Future demand for food and related agricultural products in SSA countries is uncertain. Forecasts for the year 2000 consistently point to food gaps of 15 to as much as 50 Mg [or maize (*Zea mays* L.) equivalent], depending on population projections and assumptions about food production growth. If current trends continue, the recurrent famines witnessed during the 1980s may well continue throughout the 1990s and beyond. The prospects of global climate change have added yet another dimension of uncertainty.

Food shortages in SSA can take two forms. They may be chronic and represent a continual trend of production shortfalls, or they may be transitory; for example, sudden food crises due to crop failure or an economic or civil crisis. Household food security of poor rural populations may require interventions not only to address food availability (production/supply) but, as argued above, also access to food (demand). Increased food production and locational availability is indeed a necessary first condition. But access of individual households to food often suffers from seasonal fluctuations in income, affecting purchasing power. Furthermore, intrahousehold disparities in access to food make some members more vulnerable to food shortage, i.e., women and children.

At the macrolevel, the need to improve food self reliance in SSA would require a considerable increase in the current production growth rate of 2% per annum. Given the current shortfall in food availability, notwithstanding the drought conditions currently prevalent throughout southern Africa, with no decrease in fertility rates, domestic food production in the region would have to increase at a growth rate of over 5% per annum in order to avoid food imports by the year 2000.

The limit to area expansion for increasing food production has been reached in many areas. In most parts of the region, production increases can be achieved mainly through diversification and intensification of natural resource use. The potential of the natural resource base is limited, soils are inherently poor and their fertility is declining. Forage resources are either over- or underutilized. Traditional systems which relied on extensive management to allow natural resource recovery no longer suffice. In the absence of resource conservation measures, the areas are poorly adapted to continuous arable farming. The use of higher levels of inputs such as chemical fertilizers or pesticides, increases environmental risk in the absence of appropriate husbandry. The use of higher levels of inputs such as chemical fertilizers or pesticides, increases environmental risk in the absence of appropriate husbandry. At the same time, increasing productivity through use of external inputs also raises questions as to technical appropriateness, institutional feasibility, economic viability and social acceptance. Moreover, changes in domestic policy (agricultural price policy, agrarian reform, etc.) and rationalization of international trade arrangements (tariffs and other protectionist measures) are important prerequisites to sustainable agriculture in the region.

The three main factors which will influence SSA demand for food in the coming decades will thus be population growth (currently increasing by 3.1% each year, the highest in the world), income, and urbanization. Population growth is related to the need for food, but effective demand depends on income and the purchasing power it affords. Meanwhile, urbanization will result in a shift in demand from coarse food grains, roots and tubers to the finer and more costly cereals such as wheat (*Triticum aestivum* L.) and rice and to livestock products. It will also affect the source of food supply, more of which would have to come from production for markets as opposed to small-scale, subsistence agriculture.

Therefore, while local food production per capita is declining, increasing urbanization is changing consumption patterns. Carefully designed strategies are necessary to control this trend and to ensure adequate recycling of financial resources to the rural areas, to facilitate investment and employment generation, and to alleviate poverty. The dimensions of rural poverty in SSA are vast and complex. Apart from inadequate incomes, household food insecurity and the consequent hunger and malnutrition, contributing factors include lack of access to social services, adequate production resources, rural financial services, useful technology, off-farm employment opportunities and economic incentives.

The Need for Sustainability

The need for sustainability of agricultural production cuts across the board. It is an issue which affects high-potential areas and marginal areas alike. Although it can be argued that it is in the latter areas, where the poverty of resources is more pronounced, that the main pressure points threatened by unsustainability exist. While it is recognized that elements which undermine sustainable agriculture in the high potential areas also need to be addressed, the issues and strategies considered by this chapter are purposely focused more specifically on the poverty of the marginal areas. This is because there are vast high-potential areas in the region, where the carrying capacity is still far from being exhausted. Moreover, the resolution of many of the issues at stake would be equally relevant to the entire fabric of agroecological and socioeconomic conditions found in SSA.

THE ISSUES

The goal of sustainable agriculture in the region can be achieved, only through action which is based on strategies responding to the broad range of relevant issues needing attention. These issues seem to fall naturally into three clusters: (i) policy issues, (ii) technology issues, and (iii) institutional issues.

Policy Issues

Low Investment in Agriculture

An issue common to most developing countries is the low priority accorded to agriculture. The meager allocation of budgetary resources to institutions serving the agricultural sector is a reflection of a historical practice. In SSA, the basis for changing this practice could lie in bringing to the attention of policymakers the fact that the agricultural sector receives much less in public expenditure than its contribution to gross domestic product (GDP) in percentage terms.

Inadequate investment in services for the rural sector accounts for lack of effective research and extension, poor infrastructure and inacessible markets. The smallholder has thus little incentive to produce more than is necessary to meet household subsistence needs, and thus policymakers view smallholder agricultural production as meant for local consumption alone. Since commercial agriculture is often the main source of surplus production catering for urban and export markets that generate revenue, the government's domestic resource allocation process works to the detriment of agricultural smallholders, and the consequent further marginalization of smallholder subsistence agriculture lowers its ability to achieve sustainability.

Among other factors that have a bearing on recent low investment in the agricultural sector in SSA is the impact of stringent structural adjustment regimes. Under pressure to curtail budgets, many governments in the region have had to reduce the allocation of national budgetary resources to the Technology System (Research and Extension); this reduction has been particularly damaging to small-scale farmers, especially in low-potential areas, including the generally neglected marginal areas, where most of the poor reside. The financial sustainability of such institutions which serve the agricultural sector becomes threatened. It has been predicted that, even with

significant advances in agricultural research, demographic pressures will offset
any growth in productivity achieved through the introduction of new technology. Therefore, more resources should be channeled to research and extension in order to exploit the potential of the less favorable areas while
satisfying the demand and need on the part of the low-production and less
sustainable areas for essential goods.

However, recent trends are pointing in the opposite direction. While the
demand for generating technology for sustainable agriculture increases, the
budgets of National Agricultural Research Systems (NARS) in SSA are declining in real terms, resulting in an exodus of scientific personnel (as high as
18% in some cases) and operational budgets which are inadequate to pursue
meaningful long-term research on such technology. Inadequate budgetary
allocations also influence the quality of practical and scientific education for
extension agents and researchers in developing countries, thus reducing the
ability of NARS to design, test and adapt useful but complex sustainable
agricultural technology.

Short-term Gains versus Long-term Stability

Policymakers in most SSA countries have generally favored strategies
for maximizing short-term gains (for example to demonstrate political
achievement). These are often incompatible with the need for optimizing output on a long-term basis (i.e., intergenerational equity); thus they undermine
the sustainability of the resource base.

Short-time horizons also guide decision-making at the farmer level, especially under the risky, highly variable production conditions of rainfed
areas. Farmers have felt compelled to maximize output for the short-term
satisfaction of current basic needs, even if this amounts to mining the resource
base. The issue for use of sustainable agricultural practices is that of recon-
ciling conservation of the resource base for the future with the more immediate survival needs of the present.

The need at national level for mobilization of foreign exchange earnings to settle debts and pay for imports may drive policy makers in developing countries to pursue short-sighted policies of encouraging cash crop
production for export without an adequate assessment of demand-related
constraints in the world market. Such production requires inputs as well as
labor. Resource-poor farmers attempting to participate must resort to mining the available natural resources, unless they are adequately equipped with
conservation-based production technology and access to inputs which can
replenish fertility on a continuos basis. The related issue of incentives for
resource management under adverse conditions is a crucial one.

Policy Incentives, Land Tenure and Farmer–Resource Base
Interaction

Besides dealing with the problem of population growth, policymakers
also must address other pressure points affecting sustainable rainfed agricul-

ture in developing countries. Among the major policy issues which need to be addressed are the use of fiscal instruments and agrarian reform.

Economic Policy Instruments. Domestic agricultural policy must promote sustainable land use. In SSA countries national policies often have reflected a lack of understanding of the consequences of policy change on land use at farm level. Incentives such as subsidies for credit and inputs can create market distortions leading to unsustainable farming practices. However, their arbitrary removal also can lead to resource degradation, unless farmgate prices adjust upward to permit the continued application of external inputs.

Thus while sustainable agriculture is dependent on the economic decisions of the farmer, farm management with this objective may become difficult when national economic policies are aimed at balancing trade and reducing budgetary deficits. It is then left to the producer to choose the most appropriate technological options available, and to decide on levels of inputs external to the farm, based on perceptions of short-term risk and economic incentives. An overall consistent and coherent macroeconomic policy addressing all subsectors of the rural economy would be of considerable importance under the circumstances, but this is virtually nonexistent in SSA countries. The conflicting effects of policy measures aimed at particular variables, without due consideration to linkages with other variables, frequently defeats the originally intended purpose.

The macroeconomic policies of SSA governments are often inherently biased against the agricultural producer. For instance, where agriculture does succeed in producing a surplus, it is often undermined in countries which, for political reasons, provide cheap food for the urban population through artificially low producer prices for agricultural products. Government policy to promote efficient resource use entails resorting to the use of complicated procedures of pricing controls, fiscal and monetary policy, trade policy and institutional and agrarian reform which may serve to regulate resource use through the creation of incentives/disincentives.

As discussed earlier, adjustment measures to improve the allocative efficiency to rejuvenate economic performance using the various instruments mentioned above, can have a mixed impact. Especially in the short run, the adjustment process can have trade-offs which are detrimental to the sector which traditionally faces unfavorable terms of trade—in the case of SSA invariably, it is the agricultural sector which has been undermined. The impact on the farmer and his or her response, in terms of unsustainable resource use, are some of the undesirable results of well-meaning, albeit, deficient policy measures.

Agrarian Reform. Traditional social and legal status in the region is responsible for unequal access to resources, this, in turn, can have negative effects on agricultural sustainability. But land tenure legislation to correct this problem is not popular with policy makers in these countries. It is well known that lack of land tenure increases the risk of resource degradation. Landless or tenant farmers are willing to accept such a risk since their decision is based on discounting future benefits from a resource to which they may not have permanent access. Furthermore, if there is no security of land tenure, motivation to invest in the conservation of such resources is naturally low.

Government legislation on land policy, especially common property rights, must take into account traditional mechanisms and controls on individual access to communal resources. In many SSA countries, sheer demographic pressure has defied the enforcement of restrictive legislation and has transformed common property into de facto open access resources.

Legislation on land titling, usufruct and private property rights, must therefore be geared towards location-specific conditions and must strategically address the target groups involved. This is important, since well-intentioned, private property rights legislation also can undermine the (traditional) access of vulnerable groups (including the landless and poor rural women) to formerly common natural resources unless the traditional land rights system is fully taken into account. This affects their inputs to the production system, equity and income security, with severe repercussions in the absence of off-farm employment opportunities.

Technology Issues

The choice of strategies of address sustainable agricultural production depends on the technological options available. There are limited options in terms of expansion of area under cultivation, sustainable increases in agricultural production will come from intensification of resource use. This implies the need for more efficient utilization of all available inputs and for improvements in farm management practices.

Strategies for developing a viable and sustainable agricultural sector in SSA must be based on technology which not only increases productivity but also ensures the sustainability of the resource base. Sustainable agricultural technology includes genetic improvement of crops and livestock (older and new biotechnology); proper husbandry (including maintenance of adequate ground cover, plant and animal nutrition), resource conservation and development (including soil and water conservation, reforestation, pasture development, improved land and water use), mechanization, integrated management of pests and diseases; and postharvest technology (storage, processing, marketing, especially in the context of poor rural women who traditionally process food and often use large amounts of fuel wood through inefficient processing techniques).

Some of the required technology is not yet available, and virtually all of what is available requires adaptation to local conditions. It is ironical that we have failed to tailor our technological knowledge to obtain a positive response from farmers. The more marginal the resource conditions, the more intractable the problem.

There is vast potential for research to address technology development for the marginal rainfed areas of SSA. Techology must fit the conditions under which it is to be used. However, policy assistance (price policies) may well be required to secure uptake and to encourage the necssary adaptations to production constraints due to the agroecology of SSA, the socioeconomic conditions, local rural infrastructure, farmers' resource endowments and postproduction constraints.

Resource allocation within the agricultural research system has been skewed in favor of the better-endowed areas and farmers, with generation of technology mainly focused on high-potential areas, where the economic and physical constraints to production are less pronounced or removable through investment in resource development, such as irrigation. If sustainable agricultural technology generation is to be achieved for the less well-endowed regions of SSA, the research approach and methodology must be overhauled. To an extent this has already commenced, although the need for a holistic research approach which goes beyond the farming system to the broader rural systems, is not yet fully recognized. The wider rural system affords a focus that includes aspects of off-farm employment and income-generating activities, i.e., livestock enterprises, agroprocessing, etc.

Much traditional technology has existed in harmony with the fragile natural resource base in SSA and is, in many instances, built on sound principles of resource management which take into account elements of risk and uncertainty. But many examples of such technology have broken down due to external pressures (population in particular). New and improved technology is needed, which also could retain many of the elements of sustainability inherent in traditional farming systems. Farmers have themselves shown great ingenuity in developing improved practices to meet local needs, for example mixed farming, intercropping, multistorey cropping etc., which have served to minimize risks. These innovative practices have evolved within the framework of the traditional farming system and could be adapted based on modern technological advances.

The emphasis on increasing yield, as an imminent response to the pressure on limited resources, led to a shift from resource-based to science-based agricultural systems. Improved productivity depends on agricultural technology to ensure better control of the cropping environment, the selection of appropriate crops and varieties adapted to the agroecological conditions, soil fertility management incorporating nutrient recycling and fixation (alley farming, crop rotation with legumes, etc.), prudent application of chemical fertilizers, better weed and pest control (biological control/integrated pest management), better dryland water management and hybridization for crop varieties which can respond to the need for more stable production systems.

Research on crop and animal protection is a major area needing focus in SSA, since pests and diseases such as the cassava mealy-bug (*Planococcus citri* R.), green spider mite (*Tetranychus* spp.), bacterial blight (*Xanthomones cumpestris* pv. manihotes) etc. can cause heavy preharvest crop losses while animal diseases [e.g., pest tick, (Argasidae and Ixodidae families), trypanosomiasis (transmitted by protozoa of the genus *Trypanoseae*), foot-and-mouth disease (caused by entero virus of the Picornavirus group, genus *aphthovirus*) and rinderpest (caused by Paramyxovirus genus morbillivirus) have impaired livestock productivity in the region considerably. Particularly affected are smallholders who cannot afford to lose their meager produce year after year. There is great scope for research on Integrated Pest Management (IPM) technology which is both cost-effective and environmentally sustainable. However, biological control of pests requires painstaking research which is both time- and resource intensive. More attention is needed in this area as new agricultural pests cause problems which cannot be solved with available biological control. In many instances, the solutions are already well advanced and the issue is application of what is known. Further research is required to refine the solutions for greater applicability.

In the livestock subsector, further gains in animal productivity could be achieved through genetic improvement, animal nutrition and the control of pests and infectious disease. Biotechnology in animal production has developed far beyond the artificial insemination of dairy cattle. Embryo transfer has become practical in developed countries. The DNA and cell-fusion technology have led to direct gene insertion. Future application of biotechnology in animals will include genetic engineering to control diseases, digestive physiology for improving animal nutrition, stress immunity and energy conservation.

Biotechnology could thus offer many opportunities for research and have wide application in solving many problems relating to crop and animal production. Advance biotechnology is being increasingly used in plant breeding programs and should receive strong support because of the promise it holds for the development of varieties of crops possessing characteristics of high yield, salt tolerance, drought tolerance, etc. However, there is a trade-off between the high cost of biotechnological research and the expected payoffs, which may not be forthcoming in the immediate future, given the long-term nature of such research.

The development and adoption of improved technology based on the semidwarf, high-yielding varieties of wheat and rice, and the complementary agrochemical and irrigation inputs, led to the "Green Revolution" on the Indian subcontinent and destroyed the myth that smallholders were constrained by traditional customs/cultural practices which inhibited their acceptance of high-input management strategies. It also gave great credibility to agricultural research.

In SSA, however, many physical and institutional constraints prevented replication of the Asian experience. Apart from low inherent soil fertility in the region, soil moisture is highly variable and irrigation is limited, resulting in low adoption of green revolution technology. Irrigation development is limited due to high costs and poor utilization in the short- to medium term. Whereas the green revolution depended on crop response to high inputs, the challenge in much of the region is to improve crop and livestock performance under marginal, low-input technology conditions. Of course, there are certain exceptions, for instance, the case of hybrid maize in the better-endowed areas.

Institutional Issues

The inadequacy of investments in the physical and institutional infrastructure necessary for technology testing and dissemination prevents fulfillment of the productivity potential inherent in available technology. Institutional issues, interacting with policy variables and the environment, are partly responsible. To date, very few national services in SSA have an explicit, clear-cut extension strategy. Without specific targeting, extension services have tended to favor the better-off farmers.

The region's technology transfer system is often not suited to the needs of the small farmer, who tends to minimize risk by diversifying production. This results in complex crop-farming systems and integrated crop-livestock systems. Extension agents in most SSA countries are usually equipped with recommendations for single commodities in monocropping systems, which are of limited value to the diversified and complex production system of the risk-averse smallholder.

There is, thus, a critical need for ensuring *feedback/information flow* from farmers to researchers. However, conventional extension systems (including the T&V system) have failed because of the inherent top-down nature of the research–extension–farmer nexus. How does one actually ensure a two-way dialogue "between" the farmer and researcher under the circumstances prevalent in most technology systems of developing countries of the region? Are there effective mechanisms which are appropriate to the technology triangle, research–extension–farmer, in which the farmers can influence the research and extension system by molding it in such a way as to cater to their technology needs? These are some crucial questions which must be addressed by an overall sustainable agriculture strategy.

Extension workers in most of SSA are not a very reliable medium of technology transfer. Lack of proper training and approach/methodology results in extensionists not playing an adequate role as "messenger." Information flow is sometimes very selective, depending on the sensitivity of extension agents. Further research is required to fit extension delivery to the farmers' needs. The growing complexity in sustainable agricultural technology and in the farming problems they address requires skill-oriented (vocational) training of farmers, including women. The institutional structure within SSA countries is inadequate to cater to this need.

Nongovernmental organizations (NGOs) can play an important role in this regard as well as in testing/experimenting new alternative approaches to sustainable agricultural technology. They also are better placed to identify community needs and to assist in strengthening grassroot institutions. Effective linkages should be established between institutions responsible for technology research and NGOs.

Another important institutional issue concerns the infrastructure required to address the technology needs of women farmers in SSA. This is especially important in countries such as Lesotho, Botswana and countries of West Africa, where the number of de facto women-headed households is increasing as a result of male migration. Given their multifarious responsibilities for production, processing, preparation, storage and marketing of traditional rainfed crops, women farmers are a vital source of information and feedback for the research and extension system. Unfortunately extension services are often oblivious to this disadvantaged, though highly important target group.

Some of the causes of women's limited access to agricultural extension in the region are: (i) inadequate outreach due to institutional rigidities and difficulties in employing women extension agents, (ii) lack of sensitivity to cultural taboos, etc., which discourages extension contact with women farmers, and (iii) lack of adequate motivation to reach women farmers as there are no special rewards for this.

WHERE DO WE GO FROM HERE?

Strategic Elements to Consider

As indicated earlier, a single, coherent policy framework for sustainable agriculture throughout the region is inappropriate, policies applicable to one country may not necessarily be appropriate to another. The same is true of intracountry regions and areas. However, certain strategy elements can individually, or in combination with others, provide location-specific solutions.

A phased strategy has been suggested in situations where strong/radical policy measures (including land/legal reforms) are either undesirable or unpopular among policymakers, or where appropriate technology aimed at avoiding unsustainable practices (monocropping, shortening of fallow periods, etc.) is still not available.

A combined strategy can involve a two-phased approach—a short-term strategy of low-risk, low external input-based systems, combined with a longer-term strategy to increase off-farm income and improve yields in high-potential areas, which represent potential relief zones for migration from pressured marginal areas. Complementary to the above, there is a need to reconcile short-term economic objectives with long-term objectives of ecologically sustainable agriculture. Attention must be accorded to a policy-driven, process-oriented development approach in which government legislation may guide land-use patterns through ecologically sensitive production incentives, especially where technology with built-in incentives for resource management is not available.

Apart from rationalizing land-tenure policies to encourage conservation-based production practices, the question of providing additional incentives (fiscal strategies addressing pricing, subsidies/taxes as part of an overall incentive framework) is paramount; however, price incentives alone may not be sufficient, given the time lag between costs incurred for conservation activities and the benefits accruing in the long term.

Transitional short-term incentive measures such as food-for-work and forms of emergency transfer may not contribute to sustainability either, by definition, given their piecemeal, temporary nature. However, the crucial role played by such measures cannot be ruled out in situations where the incentive has, in fact, helped to overcome the constraints towards reaching and maintaining a state of sustainable conservation-based agriculture during the "transition" period.

In many instances, there is a strong incentive and rationale for policy measures to relieve pressure on natural resources by increasing off-farm income and employment opportunities. The whole issue of incentive systems is extremely crucial from the point of view of inducing sustainability. Instruments of "external" inducement to achieve resource management on-farm should be directed at production technology where built-in economic benefits in the short-term, resource-management parameters are not visibly discernible. Moreover, direct or indirect incentive measures must be applied within the context of the institutional capacity and delivery/recovery mechanisms; in the case of direct incentives, transfer payments such as cash in the form of wages, credit, etc., and kind in the form of food, agricultural inputs, etc., or as in the case with indirect legal, social or economic measures such as tax benefits for conservation-oriented land use (forestation, fodder crop production, etc.) or agricultural price adjustment measures which discourage production of commodities associated with natural resource degradation (timber production in rain forest areas), or through agrarian reform (secure land tenure arrangements). The issue of institutional structures required to facilitate policy implementation is discussed further in the final subsection.

Political agendas in SSA countries reflect low national commitments to research on staple food crops, traditional food crops, and marketable agricultural products for smallholders. Multilateral development institutions such as IFAD support research to generate sustainable technology for resource-poor farmers, but the funds available for such research are well below requirements. The international development community is making efforts to induce governments to increase resource allocation for agricultural research. Thus, the development of sustainable agricultural technologies for SSA conditions can receive its fair share of financial support. Governments also are being encouraged to develop the indigenous research and extension institutional capacity to generate and deliver useful technology on a self-reliant and sustainable basis.

The specific roles of poor women farmers must be reocgnized and supported. There should be explicit policy directives to improve the provision of financial services to women. Policies to involve rural women in participatory development programs must recognize their multiple and sometimes competing roles within their households and on the farm.

There is a dearth of tools of analysis which can adequately reflect the costs of environmental degradation and benefits of rehabilitation, in order to formulate suitable policies. Analytical tools are required to make policymakers aware of the "real" or economic costs of soil erosion, forest destruction, etc., so that development plans adequately respond through policy measures which encourage the integration of arable agriculture, forestry and other related subsectors. More reliable data also is needed for a better evaluation of natural resources. If accepted for economic analysis, this information can influence decisions on policy variables (prices, taxation, subsidies and other fiscal instruments) as well as assist in investment analysis (choice of projects - appraisal methodologies) for sound allocation of capital funds among sustainable rural development activities).

Technological Options

The above-outlined policy elements of a sustainable agricultural strategy for SSA must be closely linked to a complementary technology suitable for the diverse physical and socioeconomic parameters, and must address resource-poor conditions.

High-Potential Areas versus Low-Potential Areas

Research for generating sustainable agricultural technology faces an uphill task in the light of declining soil fertility which is characteristic of many SSA countries. Mere investment in agricultural research for yield maintenance will not be enough to satisfy a growing demand for agricultural output. Should reliance for increasing productivity be restricted to irrigated high-potential areas? Or could rainfed agriculture in the more marginal areas also contribute effectively toward sustaining per capita food production levels? Some hard choices will have to be made. Neglecting low-potential areas is neither socially desirable nor politically feasible. Since there are limited funds available for agricultural research, priorities need to be established.

In areas of low potential, technology development must follow optimal resource-use strategies. Land and water are perhaps the most crucial elements for sustainable agriculture in the rainfed drylands of the region. Water management takes the form of planned use of watersheds and needs technology for water harvesting to permit limited irrigation for soil moisture conservation, etc. Good land management may call for some investment of capital in the marginal drylands that are not suitable for crop production but may involve tree planting to create a pasture development resource (Sylvo-pastoral system), or an income from tree farming for a variety of purposes. Such alternative land-use systems are less risky, and often more productive and remunerative, than imposing crop agriculture on unsuitable sites.

In the hilly areas of countries such as Rwanda and Lesotho, technology for conserving fragile surface soils needs further refinement. For instance, labor-intensive conventional bunding systems are not always relevant to small-farming systems and do not enlist the participation of the farmers themselves. It is important to assess the relative efficiency of traditional technology, using boundary waterways, field drains, waste weirs etc., at alternatives to the conventional system of graded or contour bunding.

Pastoral societies show a marked resistance to transferring capital holdings in the form of livestock to cash. Further socioeconomic research is required to improve our understanding of the various motivations and incentives which can be offered to persuade livestock owners in fragile rainfed conditions to participate in sustainable production activities. Increasing forage production by the intensive use of underground water combined with efficient management can transform stock production in arid grazing lands. Forage banks and cut forage stall feeding have been developed as viable strategies although they have not always succeeded as a result of socioeconomic problems.

Dryland cropping strategies in SSA must, of necessity, rely on increased inputs (crop nutrients and water) if higher crop yields are to be maintained. Would crop residues and animal manure management suffice or are additional mineral fertilizers essential in order to achieve higher productivity in rainfed dry lands? The role of the International Fertilizer Development Centre (IFDC), and the newly established African Centre for Fertilizer Development would be important in conducting further research in the area of soil fertility management; for example on P adequacy, in collaboration with other relevant International Agricultural Research Centers (IARCs) in Africa.

Agroforestry-based technology has shown increasing promise, especially in enhancing the N content of soils in semiarid lands. Alley farming integrates livestock into the crop-based farming system. Further adaptive research is needed on the competition between trees and associated crops for moisture, and on the labor requirements—especially where the opportunity cost of labor is high. This research should be encouraged by development aid agencies, in order that conservation technology, truly useful under local conditions, is generated and disseminated.

Striking an Energy–Feed–Soil Fertility Balance

Making adequate energy and feed available on farms, without compromising soil fertility, is a challenge—especially under resource-poor conditions. A strategy is needed for the availability and optimal utilization of renewable resources, i.e., woody species for fuel wood, feed resources and replenishing soil fertility, especially in marginal areas. A fuel wood deficit is a common constraint in semiarid Savannas. The development of agroforestry and sylvo-pastoral systems, which can be an effective strategy, needs to be promoted in pastoral and agrarian societies where there is no tradition of multipurpose tree management.

Livestock kept by sedentary populations must be fed throughout the year. Feed storage supplemented by feed purchase during periods of scarcity of natural grazing often will be beyond the reach of poor farmers. A technology must be developed to allow the extensive grazing of indigenous renewable vegetation under efficient management, including the regulation of grazing. It has been confirmed (FAO, 1990) that controlled grazing, although not yet implemented on a wide scale, could slow down and even reverse desertification in the Sahelian region. In more humid zones, the establishment of feed gardens is a feasible option for tethered animals.

Input Constraints and Low-input Technology Options

Poor shareholders in better-endowed rainfed areas may resort to "low-external input" agriculture. Where farmers cannot afford to purchase, for example, agrochemical fertilizers, pesticides and mechanized equipment, these can be substituted, to an extent, by farm-grown/natural biofertilizers, organic/green manures, and biological control agents. Judicious use of external and internal inputs to "high-input" technology could be extended to the resource-poor farmer. Where labor constraints make mechanization an attractive option, low-cost farm machinery must be developed, taking into account the size of holdings, etc.

In a number of areas, the woman is the primary producer—either traditionally, or by virtue of the acquired status of woman-headed households due to male migration/absence. Technology development must therefore not focus merely on the physical and biological aspects of production but must also consider labor and socioeconomic constraints, and give specific attention to women's requirements.

Research Agenda

The above technology options must be refined, adapted and disseminated. Furthermore, the newly emerging areas of frontier research must address the technology requirements of the poor farmers in the region.

While there is a need to focus on research areas which would contribute towards increasing the sustainability of farming systems, in the regional context, emphasis should be on resource-poor conditions in which traditional extensive farming and livestock systems cannot be sustained. Research for increasing sustainability includes the development and adaptation of new technology which makes optimal use of land and water resources and which can help combat the pests and diseases undermining the yield potential of the agricultural production system. The development of new technology must build on the traditional practices and risk aversion strategies of SSA farmers under difficult agroecological conditions.

There are significant possibilities offered by advances in biotechnology and genetic engineering which, if properly harnessed, can optimize utilization of the natural resources available to resource-poor farmers in SSA. One of the most promising and revolutionary aspects of biotechnological research in the future lies in the ability to manipulate recombinant DNA and enable transfer of elite genes from one species to another. It is necessary to find mechanisms for reorienting some of this research to those commodities which are relevant to the consumption and production patterns of resource-poor farming systems in the region. The application of biotechnology would be most relevant in the context of the resource poor, since they would receive the highest payoff in terms of the production expansion possibilities that could be generated.

Future research also must consider socioeconomic factors such as household food security (food consumption, nutrition), health, and labor productivity which are inseparably bound, in the agricultural production systems of the region. Research also must address potential returns under various types of subeconomies, on small-farm production of marketable food and agricultural output, preferably high-value products that can bring a good return on investment and labor. Tree products such as fruits and gums combine this potential with greater ecological sustainability. Livestock and fish husbandry yield products that are subject to firm demand in urban and export markets.

Risk aversion on the part of smallholder farmers in Africa leads them to embrace a strategy of diversification of production systems. This should provide the basis for steering research toward more diversified crops and sources of income for poor smallholders, including rural women. It would involve a focus on high-value crops and livestock (including fisheries and forestry) as well as rural microenterprises for the target population, with special emphasis on the gender issues which impinge on the economic advancement of poor rural women. Appropriate postharvest technology for handling, storage, processing and marketing also are keys to poverty alleviation and household food security (nutrition) for the rural (and peri-urban) families who can grow perishable produce (vegetables, fruit, and milk) on small plots of land. Research is required in the context of these activities to help rationalize the focus on poor farmers producing at near-subsistence level in marginal areas and to alleviate the risk of saddling IFAD target groups with crops that constitute permanent elements of their "poverty trap."

In the context of resource management, it would be important to develop methodologies for appropriate valuation of natural resources as well as their extractive or nonextractive uses. It also would be critical to develop tools of analysis which can help discern farmers' perceptions, influencing their choice of one particular technology over another. Important parameters requiring further investigation include farmers' implicit discount rate (time preference) and degree of risk aversion, i.e., modeling the parameters governing farmers' decision-making.

Research Methodology

The issue of appropriate research management and methodology is crucial to the development of sustainable rainfed agricultural technology. The main emphasis should be on the generation of technology which takes into account the actual and perceived needs of the beneficiaries. Such emphasis is afforded by on-farm research. On-farm research consists of a rigorous procedure comprising diagnostic surveys, adequate identification of beneficiary needs, the search for appropriate technology, testing of technology via on-farm trials and the generation of recommendations for adoption. A strong foundation for participatory research is thus established, based on ecological and socioeconomic concepts and principles.

The IFAD-sponsored seminar on the "Generation and Transfer of Technology for Small Farmers," which took place in Seoul, Korea, in 1988, recommended the On-farm Client-Oriented Research (OFCOR) approach developed by International Services for National Agricultural Research (ISNAR), which is increasingly acknowledged as a suitable way to organize and manage research to address farming systems' requirements for technology. The OF-COR encompasses an array of earlier methodological approaches (farming systems research, cropping systems research, farmer-back-to-farmer research).

The functions performed by OFCOR within the research system are: (i) the application of problem-solving approaches focused on farmers as primary research clients, (ii) identification of major farming systems and client groups to define technology needs, (iii) application of an interdisciplinary perspective to diagnose problems and set priorities, (iv) adaptation and evaluation of technology in farmer-managed trials, (v) promotion of farmer participation as collaborators and evaluators of alternative technology, and (vi) taking into account feed-back from farmers to ensure relevance of research to client needs and promotion of linkages with extension for more effective technology transfer.

It is important, however, that the on-farm perspectives consider off-farm environment and the whole range of productive opportunities to meet the farmers' needs. Since OFCOR represents only a partial solution to the problem of attaining sustainable and equitable agriculture, a more broad-based Rural Systems Research (RSR) is required in order to genuinely address the range of issues which are exogenous to the farm, but affect the livelihood of the farm household and impinge on the sustainability of agriculture. This implies that every resource (ownership and access), activities (on- and off-farm), product (subsistence oriented or market oriented) and (real) income (flows of cash, food and net addition to stocks (reserves and assets) all must be taken adequately into account by research in order to foster sustainable livelihoods and agricultural development in SSA.

The IFAD is among the few multilateral financial institutions which are increasingly emphasizing on-farm research methodology in Fund-supported agricultural research. At the same time, IFAD's lending operations are drawing upon the results of Fund-financed research and complementing it with information on rural systems in the target areas and with site-specific in-depth studies to develop projects confined to addressing the particularities of those areas. The IFAD's Special Programme for sub-Saharan Countries Affected by Drought and Desertification (The Special Programme) has successfully tested and disseminated a number of sustainable agricultural technology packages through projects for poor farmers in SSA countries, including packages for soil and water conservation, agroforestry, small-scale irrigation and water harvesting and pasture management, while at the same time addressing off-farm employment generation to integrate a variety of activities to improve the sustainability of livelihoods of the rural communities in SSA, all predominantly dependent on agriculture. Feedback from these efforts can assist in the adequate conduct of RSR. Where projects do not exist, specific NGOs/Private Voluntary Organizations (PVOs) can assist in supplying the organizational and technical inputs required to provide information feedback by obtaining and testing technology from national research stations, by acting as clearing houses for information on local institutional conditions and providing the crucial information on the potential adoptability of technology. The conduct of RSR can be facilitated greatly by such involvement.

Institutional Orientation, Strengthening and Sustainability

A number of international development institutions are devoting substantial effort to the search for new directions in agricultural extension focused on poor resource rural areas. The main idea behind the newly emerging research and extension is that of a participatory technology system termed "client-demand system." The client, being the farmer, is the focal point and the farmer's interaction with the technology system ensures that the particular requirements of resource-poor farming are incorporated into the technology generation/dissemination process.

Any comprehensive strategy for sustainable agriculture in SSA must, therefore, incorporate the elements and operational thrust of this extension/technology transfer approach. How relevant is such a system to the region's agriculture? Are there specific conditions on the continent which may inhibit a client-demand extension system? Is it mandatory for a client-demand system to require group formation/association of farmers, in order that their needs be better articulated? These questions only can be answered if a concerted effort is made on the part of the development community to apply this framework through concrete development projects designed on the basis of a "client-oriented participatory philosophy" and to test the efficacy of the system under the actual socioeconomic and agroecological conditions prevalent in the region.

The role of NGOs can be particularly useful in establishing self-sustaining participatory institutions capable of addressing the various elements of the client-demand system by promoting the required interaction between the technology system and beneficiaries.

In addition to the foregoing strategies, some of the complementary measures which could lead to greater efficiency in the sector include: (i) increasing the role of the private sector, especially in marketing and the use of farm inputs and agricultural products; (ii) improving rural financial services to poor smallholders, including poor rural women heads of household; (iii) developing rural infrastructure; and (iv) increasing investment in vocational training (for farmers) and higher education so as to increase scientific and technical knowledge on the basis that "there is no substitute for indigenous human resource and institutional capacity-building to promote self-reliant growth."

The above measures only can be made possible by strengthening and establishing viable indigenous institutions which both directly and indirectly play a crucial supportive role in SSA agriculture through the provision of rural financial services, input supply channels, marketing cooperatives or in the conduct of agricultural research and extension of agricultural technology. The provision of these services on a sustainable basis requires both macro- and micropolicy reforms in most countries of the region.

Few SSA countries are endowed with adequate infrastructure of public and private institutions serving the agricultural sector. The cost of reaching the agricultural sector in rural areas is high, while budgetary resources are limited because of competing claims for resources, especially under tight budgetary conditions imposed by structural adjustment program. Alternative institutional arrangements should be explored which can address the current constraints.

The most attractive institutional alternatives has been the mobilization of cooperative movements as instruments of harnessing the required services which public institutions fail to deliver. Evidence has shown that marketing, credit and production cooperatives, if properly developed, have an exemplary record in this respect. Powerful rural groups have demonstrated that strong participatory structures and pressure groups have successfully lobbied for and received a substantial redirection in rural services which have responded to the needs of the agricultural sector.

In the African context, for instance, the Kikuyus of Kenya have demonstrated their ability to gain access to technical know-how, inputs and capital. In Tanzania, the Ujamaa movement revealed the potential for mutual cooperation and self-help by using traditional organizations as the vehicles of modernization and change. In West Africa, traditional, sustainable forms of cooperation have succeeded, especially when focused on short-term activities, with the participation of farmers in the construction of storage facilities, dissemination of information on producer and consumer prices, in the provision of facilities for advanced credit on a portion of the value of produce, etc.

A number of issues need to be addressed with regard to institutions providing rural financial services. One important issue is that of establishing sustainable financial services which can cater not only for the richer farming community but also for poor smallholders who constitute a major target group. Under conditions of low risk-bearing capacity, indebtedness, etc., which are common problems in the region, prima facie, credit delivery may not represent a very viable operation. However, a number of credit delivery efforts focused on the resource-poor have in fact succeeded in several SSA countries.

There are several examples of how credit cooperatives for onlending or retailing credit have led to cost effectiveness through simple lending procedures, more than has been possible through the high transaction costs associated with formal financial institutions such as banks, which are more appropriate in wholesaling credit to institutions providing rural financial services. The role of nonformal credit institutions can thus be explored in the context of providing cost-effective loan disbursement and recovery. Group lending also permits savings in credit administration—group responsibility for screening applications and ensuring repayment has succeeded in many parts of Africa, including remote areas of Wolamo in Ethiopia or Quthing district in Lesotho. Where such arrangements have not worked efficiently, other innovative arrangements need to be explored. Of significance is the need to sensitize development banks and commercial financial institutions and offer them a sufficient incentive to indulge in a modest degree of risk-taking by providing rural on-lending services to resource-poor farmers with limited (or no) collateral. This is crucial in situations where capital markets are imperfect or even nonexistent. In the context of smallholder credit-oriented projects, a contingency or risk/guarantee fund established by development institutions could go a long way towards providing the necessary insurance in the form of loan refinancing, under circumstances which might otherwise lead to high default rates.

In many instances, the absence of formal rural financial services can be compensated for by participatory mechanisms such as farmers' cooperatives, these can play a crucial role in credit and input delivery systems as well as in credit recovery. Grassroot participation ensures greater institutional sustainability in the absence of direct assistance in terms of sustained government administrative or financial support. However, in order to be effective, participatory mechanisms under nonfree market conditions do require some national commitment and state intervention, for instance in pricing, distribution and regulation of marketing practices i.e., in creating a conducive environment for the successful functioning of cooperative institutions. Sustainable participatory institutions must be created in the context of the site-specific sociopolitical and economic conditions within which self-help and grassroot participation is to be encouraged, using economies of scale and social control over resource allocation and mobilization. Successful cooperative movements cannot be replicated in the absence of these prerequisities.

The basic premise remains that sustainability is inextricably linked with effective beneficiary participation. Participation must be achieved at four levels—in decision making, program implementation, equitable distribution of benefits and evaluation. These distinct elements of participation for sustainability become relevant or realistic when the question of equity and access to resources (social justice) is resolved. However, equity in access to resources may not necessarily lead to their sustainable use unless other institutional issues of resource management are incorporated in the farming system and an adequate incentive structure created and supported.

The sustainability of certain public sector institutions such as national agricultural research centers, training institutions, etc., which are not financially autonomous and depend on exogenous and government budgetary support, is an issue which needs to be resolved in the macropolicy context. Governments under structural adjustments programs are seldom able to make adequate fiscal budgetary allocations to such institutions, nor finance the recurrent costs associated with them. In such circumstances, structural adjustment policies should institute counterstrategies to cushion such adverse structure impediments to the financial sustainability of institutions serving the agricultural sector and formulate fiscal disciplinary measures more sensitive or conducive to the well-being of the sector which, in the final analysis, is the engine of the export-led strategy promoted by structural adjustment programs. This is because in the SSA context it is the agricultural sector which is capable of generating exportable surpluses and already provides the major share of the region's total exports.

Moreover, the sustainability of institutions serving other needs of the SSA agricultural sector raises further issues. Institutions which support agricultural services commercially need to be strengthened in two areas—increased efficiency in delivery with improvement in geographical outreach and better orientation to address those target groups which require the services most, including the subsistence subsector.

CRITICAL QUESTIONS FOR FURTHER THOUGHT

While the key elements of a sustainable agriculture strategy have been touched upon, the overall issue remains how to harness the region's collective human and physical assets for sustainable agricultural growth. In considering possible strategies for sustainable agriculture, the following questions must be kept centrally in mind:

1. Does the region have particularly unique structural problems which have not been fully grasped by the international development community? If so, how can these be addressed effectively so as to achieve sustainability of institutions serving the agricultural sector of SSA countries?
2. Have institutional issues been ignored by the development process in the region? Would the strengthening of delivery systems be a sufficient condition for sustainable agriculture?
3. Is there a dearth of technology for sustainable agriculture which is particularly suited to the socioeconomic and marginal agro-ecological conditions of the region, where demographic pressures lead to problems of overcrowding? Are there sufficient alternative employment-generating opportunities to offset this?
4. Are traditional organizations and social systems responding fast enough to the changing socioeconomic circumstances in SSA?
5. If some sustainable agricultural technology already exists, does it lie on the shelves of research institutions or have there been adequate efforts to adapt and disseminate it to the intended beneficiaries?
6. Have there been any deliberate attempts on the part of policy-makers to institute agricultural/economic policy and legislative (agrarian) reforms in order that the benefits of science and technology may be reaped equitably?
7. Has the impact of external factors, (structural adjustment, distortions in world market prices, etc.) on the agricultural sector of SSA been correctly assessed?

Many of these concerns have not been addressed conclusively within the context of this chapter. It is clear, however, that the issues of public policy and planning for sustainable development, such as those relating to human resource development, institutional capacity, demography and the issues of technology must be discussed concurrently, in order to formulate an overall strategy for sustainable agricultural development in SSA.

In light of the inherently difficult physical, economic, legal and social parameters in SSA, resolving the issues already discussed is no mean task. However, it must be said that the continent also is endowed with vast human and natural resources, and is the beneficiary of a substantial amount of official bilateral and multilateral development assistance. Sustainable agriculture in SSA *can* be pursued if a coherent strategy is formulated and is operational within the framework of the particular conditions prevailing in each SSA country.

In spite of the problems associated with the political, technological and institutional deficiencies outlined in this chapter, the region's agricultural resources can be used to advantage, especially since this sector will continue to be the primary source of economic growth and food security in the foreseeable future.

The clear message derived from the issues raised above, is that the major challenge facing African agriculture is how to cope with the needs of a growing population. While this is seen as a "problem" by many, it could, in fact, serve as an "opportunity," if harnessed in an effective manner.

The basic element of such a strategy would be to view the agriculture sector of SSA as a vehicle for sustainable agricultural growth. This would come from transforming the sector into a major employer of its growing labor force and expanding its productive capacity to compete on world commodity markets (and in the process earn the foreign exchange needed to spur economic growth in other sectors). Its vicinity to European markets and the seasonal patterns which give parts of the continent a comparative advantage in the production of many crops, would have to be exploited to the full, if such a strategy is to succeed. However, such a strategy must be placed within the context of resource management concerns and intergenerational equity.

Many SSA countries are making a commendable effort to introduce structural and political changes in their countries to promote growth: the case of Botswana, for instance, achieving the world's highest economic growthrate in the 1980s. This fact reveals the region's intrinsic capacity to transform its economy from one reflecting chronic food insecurity to one of sustainable growth. The task is formidable, but is certainly within the realm of the possible. The international development community must rally to assist the continent to transform its agriculture, instead of succumbing to the pressures of those who speak of "aid fatigue."

Finally, concerted regional and international action would be required to bring about the required transformation. Environmental degradation cuts across political boundaries and regional cooperation would need to be fostered to address watersheds, river basins and other resources shared by SSA countries. International cooperation must be emphasized, not merely to address issues of food trade but also to source collaborative research and technology transfer (including TCDC and twinning arrangements with developed country institutions, etc.), thus ensuring that the much hoped-for *Second*

Green Revolution, which could be triggered by the achievements of new bio-technology research, will become a reality.

CONCLUSIONS

This chapter has not attempted to provide any answers to the complex problems facing the sustainability and productivity of agriculture in SSA. Instead, an attempt has been made to present a catalogue of issues, options and questions relevant to the development of strategies for a viable and sustainable agriculture in SSA. There are two good reasons for this. First, it allows a better appreciation of the broader picture in all its complexity, before the individual pieces of the puzzle are examined individually. Second, and more compellingly, it can be argued that no universally applicable set of solutions exists. The transition to sustainable agriculture is a major challenge everywhere, but in no region is the challenge greater than in SSA.

While differences do exist on the best way forward, there appears to be a growing consensus on at least a few issues facing agriculture in SSA. There is agreement among many that, in the face of rapidly growing populations, "sustainable" agriculture cannot mean "steady state" agriculture. For agriculture to be sustainable its productivity must increase over time. The transition to sustainable agriculture must be seen as an ongoing effort. This implies that dynamic strategies need to be evolved that allow agriculture to respond to changing needs and priorities at the local, national and international levels.

Any strategy that emerges is likely to be based on the recognition that the agricultural sector, including the untapped potential of the smallholders and resource-poor farmers, is a promising vehicle for sustainable development in SSA. Another basic element is the recognition that a dynamic agricultural sector that is sustainable will require a rational management of the country's capital stocks—not only man-made capital, but also human and natural capital. A third element, will, by necessity, lie in the maximum application of the principles of comparative and absolute advantage.

It also is clear that, in the overall context of urgently required structural adjustments in the world economy, any durable strategies will need to developed on a case-by-case basis reflecting the particular conditions prevailing in each country of the region. Ready-made prescriptions based on a priori judgements have proven themselves to be deficient far too often.

However, the evolution of tailor-made strategies, will be no mean task, given the difficult structural and transitory challenges facing countries in the region. The international community can play a useful role as a facilitator for this evolution, by acting on several fronts and on different levels to broaden the set of technological, institutional and policy options that are available to governments and farmers in their quest for greater sustainability and productivity in the agricultural sector. In addition, governments can be assisted in ensuring the internal and external consistency of any strategy for sustainable agriculture that finds expression at the national level and avoiding the pitfalls of a piecemeal approach to economic development which does not pay due attention to the interrelationships both between sectors within an economy and between economies.

The challenge now lies in identifying the key leverage points for effecting a transition to sustainable agriculture by examining issues on a case-by-case basis. It is hoped that this chapter, in attempting to specify some of the broader issues and options which might govern the evolution of strategies for a viable and sustainable agricultural sector in SSA, will catalyze more focused efforts to do identify and act upon these leverage points. These individual efforts by NARS, and others at the local level, should proceed in line with an understanding of the complex interlinkages that exist.

REFERENCES

Ahmad, Y.J., S. El Serafy, and E. Lutz (ed.). 1989. Environmental accounting for sustainable development. World Bank, Washington, DC.

Atta-Krah, A.N., and P.A. Francis. 1987. The role of on-farm trials in the evaluation of composite technology: The case of alley farming in southern Nigeria. Agric. Syste. 23:133–152.

Baxter, M. 1989. New developments in agricultural extension. p. In N. Roberts (ed.) Agricultural extension in Africa. World Bank, Washington DC.

Bhagwati, J.N. 1968. Export promoting trade strategy: Issues and evidence. Dev. Policy Issues Ser. Rep. VPERS7. World Bank, Washington, DC.

Binswanger, H., and P. Pingali. 1988. Technological priorities for farming in sub-Saharan Africa. World Bank Res. Observ. 3:81–98.

Binswanger, H.P., and J. von Braun. 1991. Technological change and commercialization in agriculture: The effect on the poor. World Bank Res. Observ. 6:

Carr, S.J. 1989. Technology for small-scale farmers in sub-Saharan Africa: Experience with food crop production in five major ecological zones. The World Bank Tech. Pap. 109. World Bank, Washington DC.

Cheru, F. 1992. Structural adjustment, primary resource trade and sustainable development in sub-Saharan Africa. World Dev. 20:497–512.

Cobbina, J., and A.N. Atta-Krah. 1991. Methodological and analytical issues in on-farm alley farming research. In H.J.W. Mutsaers and P. Walker (ed.) On-farm research in theory and practice. Int. Inst. Tropical Agric., Ibadan, Nigeria.

Conway, G.R., and E.B. Barbier. 1988. Sustainable agriculture for development. Int. Inst. Education and Dev. 1986.

Consultative Group for International Agricultural Research. 1986. Impact study. CGIAR, Washington, DC.

Consultative Group on International Agricultural Research/Technical Advisory Committee. 1988. Sustainable agricultural production: Implications for international agricultural research. Tech. Adv. Comm. Secretariat, FAO, Rome.

Delano, R.B. 1983. The farmer as an innovative survivor. p. 187–194. In J.W. Rosenblum (ed.) Agriculture in the 21st century. J. Wiley and Sons, New York.

Eicher, C.K. 1989. Sustainable institutions for African agriculture development. Working Pap. no. 19. Int. Serv. for Natl. Agric. Res., The Hague, Netherlands.

Farrington, J. and A. Bebbington. Institutionalisation of farming systems development. Are there lessons from NGO-Government links? Paper for FAO Expert Consultation on the Institutionalization of Farming Systems Dev., Rome. 15–17 Oct. 1991.

Food and Agriculture Organization. Production yearbook. 1978 to 1988. FAO, Rome.

Food and Agriculture Organization. 1981. Agriculture: Toward 2000. FAO, Rome.

Food and Agriculture Organization. 1978. Report on the agro-ecological zones project. Vol. I. FAO, Rome.

Food and Agriculture Organization. 1982. Follow-up WCARRD: The role of women in agricultural production. FAO Comm. on Agric., Rome.

Food and Agriculture Organization and the Ministry of Agriculture. Nature management and fisheries of the Netherlands: Criteria, instruments and tools for sustainable agricultural and rural development.

Gaiha. R. 1991. Structural adjustment and household welfare in rural areas: A microeconomic perspective. FAO Econ. and Social Dev. Pap. no. 100. Rome, FAO.

Idachaba, F. 1987. Policy issues for sustainability. p. 18–53. In T.S. Davis and L.A. Schirmer (ed.) Proc. of the 7th Agricultural Sector Symp. The World Bank, Washington DC.

International Fund for Agricultural Development. 1989. Comparative review of agricultural research and extension in IFAD projects. Draft Rep. IFAD, Rome.

International Fund for Agricultural Development. 1988. Environment, sustainable development and the role of small farmers: Issues and options. Sustainable development and the role of small farmers. Proc. Consultation on Environ IFAD, Rome.

International Union for Conservation of Nature and Natural Resources. 1986. Review of the protected areas system in the Afro-tropical Realm. Prepared in collaboration with the U.N. Environment Programme. Int. Union Conserv. Nature & Natural Resources,

Kesseba, A.M., J.R. Pitblado, and A.P. Uriyo. 1972. Trends of soil classification in Tanzania: The experimental use of the 7th approximation. Soil Sci. 23:

Kesseba, A.M. 1970. On the future of soil survey and land use planning for agricultural development in Tanzania. In Proc. of the 1st Conf. on land Use in Tanzania. University Press, 1970.

Kesseba, A.M. (ed.) 1990. Technology systems for small farmers, issues and options. Westview Press, London.

Kesseba, A.M., and S. Mathur. 1990. Technology systems for small farmers, issues and options. In A.M. Kesseba (ed.) IFAD's role in the generation of sustainable agricultural technologies. Westview Press, London.

Lele, U. 1989. Agricultural growth, domestic policies, the external environment and assistance to Africa. Lessons of a quarter century. Managing Agric. Dev. in Africa, Disc. Pap. I. World Bank, Washington, DC.

Mathur, S. 1987. Domestic development policies and growth patterns. p. In R. El-Ghonemy (ed.) Dynamics of rural poverty. FAO, Rome.

Mellor, J.W. 1988. Food policy, food aid, and structural adjustment programs: The context of agricultural development. Food Policy 13:10–17.

Mellor, J.W., C.L. Delgado, and M.J. Blackie. 1987. Priorities for accelerating food production growth in Sub-Saharan Africa. p. 353–376. In John W. Mellor et al. (ed.) accelerating food production in Sub-Saharan Africa. Johns Hopkins Univ. Press, Baltimore, MD.

Mellor, J.W., C.L. Delgado, and M.J. Blackie (ed.). 1987. Accelerating food production in sub-Saharan Africa. Johns Hopkins Univ. Press, Baltimore, MD.

Merrill-Sands, D., and D. Kaimowitz. The technology triangle: Linking farmers, technology transfer agents, and agricultural researchers. ISNAR, The Hague.

Odhaimbo, T.R. 1988. The innovative environment for increased food production in Africa. p. 38–51. In B.W.J. LeMay (ed.) Science, ethics, and food. Proc. of a Colloq. Organiced by the Smithsonian Inst., Washington, DC. The Smithsonian Inst. Press,

Okigbo, B.N. 1988. Sustainable agriculture in tropical Africa. p. 323–352. In C.A. Edwards (ed.) Sustainable agricultural systems. Soil Water Conserv. Soc., Ankeny, IA.

Oram, P. 1988. International agricultural research needs in sub-Saharan Africa: Current problems and future imperatives. CGIAR, Washington DC.

Pearce, R. 1990. Traditional food crops in sub-Saharan Africa: Potential and constraints. Food Policy 15:374–382.

Sobhan, I. 1977. Agricultural price policy and supply response: A review of evidence and an interpretation of policy. World Bank, Washington, DC.

Spencer, D.S.C. 1991. Collecting meaningful data on labour use and farm size for economic analysis associated with on-farm trials in sub-Saharan Africa. In H.J.W. Mutsaers and P. Walker (ed.) On-farm research in theory and practice. IITA, Ibadan, Nigeria.

Streeten, P. 1987. Structural adjustment: A survey of the issues and options. World Dev. 15:1470–1482.

Swaminathan, M.A. 1988. Sustainable development systems for small farmers: Issues and options. In Consultation on Environment. IFAD, Rome.

U.N. Department of International Economic and Social Affairs. 1980. Patterns of urban and rural population growth. U.N. Dep. of Int. Econ. and Social Affairs, New York.

U.N. Development Programme and the World Bank. 1989. African economic and financial data. U.N. Dev. Programme and the World Bank, Washington, DC.

van Keulen, H., and H. Breman. Agricultural development in the West African Sahelian region. A cure against land hunger? Centre for Agrobiol. Research, The Netherlands.

World Commission on Environment and Development. 1987. Our common future. Oxford Univ. Press, Oxford, England.

World Bank. 1989. Sub-Saharan Africa: From crisis to sustainable growth. World Bank, Washington DC.

World Resources and Institute, International Institute for Environment and Development, and Environment Programme. 1988. World resources 1988–89. Basic Books, New York.

World Resources Institute. 1988. World resources 1988–1989. World Resourc. Inst., Washington DC.

15 Sustainable Agriculture in sub-Saharan Africa: Perception, Scope, Complexity, and Challenges

Bede N. Okigbo

*United Nations University Programme on Natural Resources in Africa
Nairobi, Kenya*

Sub-Saharan Africa (SSA) is a region that is facing acute food crisis and needs not only sustainability in agricultural development but also in food security. This paper is devoted to an analysis of the comprehensive review of sustainable agriculture concepts, nature, peculiarities and strategies for its development in SSA by A.M. Kessaba (Chapter 14 in this publication). This paper addresses those issues that have either not been stressed enough or missed, and highlights peculiar issues and perspectives on sustainable agriculture and environment in SSA that are often overlooked in order to more effectively clarify the situation in this part of the world.

SUSTAINABLE AGRICULTURE, PERCEPTION, SCOPE, COMPLEXITY AND CHALLENGE

This section focuses on events that indicate the increasing global concern about environment, resources and their degradation under the pressures of rapid population growth, increasing intensity of farming and other development activities aimed at producing enough food, fiber and shelter and satisfying multifarious needs and demands including man's materialistic cravings. It reviews a number of global conferences, discussions and actions taken to conserve the resource base and assist the poor to gain access to food and basic needs. Included under measures taken in the establishment of agencies such as International Fund for Agricultural Development (IFAD) that has a stipulated and specific mandate "to address the particular concerns of the poorest target groups in SSA and elsewhere." The paper, however, does not effectively address several elements (concepts, perceptions, scope, complexity and challenges) referred to in the subhead of this section. Pertinent points deemed necessary to stress are briefly indicated below:

1. The emergence of sustainability as a *buzzword* and *paradigm* for development is the culmination of environmental concerns which started to gather momentum from the early 1960s, spurred green parties and political activism worldwide, and culminated in the report of the World Commission on Environment and Development (1988) which was accepted by the General Assembly of the U.N., governments, development agencies and donors as the policy guideline for development.

2. I beg to differ in the observation that processes that cause environmental degradation and rural poverty and the complex linkages between the two have not been studied enough. We now have a good idea not only about the causes which are different in various parts of the world, but it also is necessary to emphasize that not only poverty in the developing countries of SSA is associated with environmental degradation. Progress and affluence in the developed countries also are associated with degradation of the resource base. Some of the industrial and agricultural technologies introduced from the developed countries, and also their transported waste, contribute to environmental degradation in SSA. Moreover, I strongly feel that the problem lies in the marked inequalities that exist in the world in human resource development, institutional capacities and especially appropriate technology development capabilities that are necessary for generating and finding location-specific solutions on a more even keel in all science-based development programs in agriculture and other sectors.

3. *Sustainability* and *sustainable agricultural development* cannot be more clearly defined in more concrete and practical terms since the former is an abstract noun and the latter is an adjective qualifying a process of concern. Yet we are aware that where sustainability or sustainable agricultural development is achieved, the end process is a desirable one both in the present and in the future. If the definition of sustainable agriculture or sustainability in agricultural development is regarded as nebulous, we are at least more conversant with the characteristics that sustainable agriculture should possess and/or exhibit. If this is not so, it would be difficult to determine strategies, plan and execute programs for development of sustainable agriculture.

There is no doubt of the nature of sustainable agriculture as: a *bioeconomic business activity, science and art in which resources (e.g., climate and soils) are manipulated, managed or otherwise taken account of and inputs (e.g., fertilizer, manure, machinery and energy) are orchestrated in such a manner within the context of available technologies and prevailing socioeconomic conditions as to create favorable environmental conditions for the crop(s) and/or animals to attain levels of productivity that satisfy increasing needs and demand for food, fiber, feed and other objectives without degrading the environment or preferably enhancing the resource base.* Even if the definition remains unclear, obvious elements that characterize sustainable agriculture, conceptually should include:

1. Ecologically sound management or manipulation of resources such as climate and soil so as to maintain levels of productivity that minimize risk of failure and satisfy increasing demand on a continuing basis.

2. Orchestration and use of different inputs in quantities, quality, sequences and timing that are efficient and entail reduced reliance on costly external inputs such as agrochemicals, and more reliance on internal inputs such as can be attained by recycling of materials and use of biological processes, for example, N fixation for maintenance of adequate levels of productivity with minimum adverse effects or even rejuvenation of the resource base.

3. The use and marriage of appropriate elements of traditional, conventional and emerging technologies that are ecologically sound, economically viable, accessible and user friendly under different socioeconomic circumstances. These technologies are continuously adapted or finely tuned to specific location changing circumstances.

4. Science and technology-based agricultural development requires that all the above activities are backstopped by effectively linked, well-coordinated and efficiently functioning education/training, political/administrative, research/extension, marketing/pricing, credit and agribusiness institutions and infrastructure that ensure sound policies, priorities, strategies, plans, program formulation, continuity and commitment in resource allocation, sound management and program execution in overall agricultural development activities. In other words, there must be an ability to generate, adapt, continuously evaluate, monitor and work in partnership with farmers in testing, modifying and adapting new technology. Deficiencies in one or more of the above aspects have resulted in low agricultural and food productivity, low incomes, lack of food security and poverty in SSA.

All these indicate the complexity of sustainable agricultural development. The overall scope and complexity is comprised of the fact that sustainable agricultural development is an integral component of general economic development and calls for a holistic and systems approach since all sectoral development strategies, policies, plans and program activities must be compatible in order to synergistically interact in such a way as to conserve, minimize degradation or better still enhance the resource base. If this does not occur, the lack of sustainability and resource-based conservation in one sector such as industry could threaten agricultural productivity and sustainability. Thus sustainability in development is a multifaceted phenomenon involving a wide range of interdisciplinary activities in all sectors and at different levels (local, national, regional and global). For a more detailed discussion on the complexity and challenge of development of sustainable agriculture in SSA, see Okigbo (1991).

The Setting

This section paints a fairly comprehensive picture of the situation in SSA albeit a very gloomy and somewhat one-sided picture. While it is not incorrect to emphasize the extent of poverty in SSA by noting that 28 out of 34 of the world's poorest countries are in this region, it must be pointed out that with many small countries of less than 5 million people, the largest number of the poorest people in the world are not in SSA. It also is necessary to observe that poverty in SSA is not mainly a problem of low agricultural productivity, population explosion and inadequate policies for food production. It is the result of interaction among several crises that are agricultural, demographic, economic, ecological and political in nature. While it is observed that the Lagos Plan of Action, Harare Declaration and Africa's Priority Program for Economic Recovery display concern about the agricultural crisis and identified the causes of the crisis, they also contain well-articulated policy instruments, guidelines and strategies for finding solutions to the crisis. Unfortunately, these guidelines have not been reflected in programs of action.

It also should be emphasized in this section of the paper, that some progress has been and is being made. For example, significant progress has been made in the development of technologies for boosting agricultural production, human resource development and strengthening of national agricultural research systems (NARS) at the International Agricultural Research Centers (IARCs). Four of these, namely International Institute of Tropical Agriculture (IITA), International Livestock Center for Africa (ILCA), International Laboratory for Research in Animal Diseases (ILRAD), and West African Rice Development Association (WARDA), are located in the region and International Crops Research Institute for the Semi-Arid Tropics (ICRISAT) has a Sahelian Center and active regional programs, while the International Service for National Agricultural Research (ISNAR) also has effective institutional building or strengthening activities in partnership with NARS. An international research center (ICIPE) is making valuable contributions in insect pest management. In IITA, for example, an IFAD-supported biological control program has made valuable contributions in the combating of the cassava mealybug (*Phenococcus Manihoti*, Mat-Ferr) and green spider mite (*Mononychellus tanajoa*, Bondar)—two devastating pests of cassava (*Manihot esculenta*, Crantz), a major staple food crop especially for the poor in SSA. Moreover, large areas of high potential in the *Middle Belt* of West Africa rendered of minimum agricultural use and of low density human settlement by parasitic diseases such as onchocerciasis, trypanosomiasis and bilharzia have now been freed from some of these diseases and are available for settlement and agricultural production. I do not mean that the picture painted of the environmental setting in Chapter 14 of this publication is wrong. Rather, I am emphasizing the need to strike a balance and avoid painting a one-sided gloomy one.

The Issues

This represents a fairly comprehensive discussion of the issues that are relevant. However, comments on some of these are necessary. Under "Policy Issues" related to "Low Investment in Agriculture," it is not just enough to observe that "recent low investment in the agricultural sector in SSA is the impact of stringent structural adjustment regimes." The full sequence of events in the vicious cycle should be pointed out. The agricultural crisis forced many countries in SSA to increasingly rely on food imports to satisfy demand, this coupled with drought and fuel crisis of the early 1970s resulted in heavy debt burdens with the situation exacerbated by decline in commodity prices for the last two decades. The overall result is that the international debts of African countries rose from $108 billion (U.S. $) in 1980 to almost $220 billion (U.S. $) in 1987 (U.N., 1990). About 26% of Africa's total export earnings is spent on debt servicing while formal debt service obligations amount to about 33% of export earnings (U.N., 1990). With overvalued currencies and structural adjustment programs as components of IMF conditions for economic recovery, many countries are under serious economic stress. Mention must be made of the heavy military expenditures of African countries. Between 1975 and 1979, at 1980 price/exchange rates, this amounted to about $9 billion (U.S. $) or more than 50 times the annual budget of the CGIAR. At least 20 countries spend more than 10% of their budgets on the military and at least 19 of them spend more than 5% of their gross domestic product on the military and less than 1% on agricultural research (Kidron & Segal, 1987).

As regards "Short-Term vs. Long-Term Stability," it should be noted that at the farmers' level, even in developed countries, soil or environmental degradation takes a long time to become visibly manifest. Therefore, farmers tend to ignore it and invest on short-term beneficial measures. It is only through education and training that farmers can learn more about preventive conservation measures as a part of routine farm practice. Even then, governments will have to find ways of assisting farmers in investing for the enhancement of the resource base for future use.

The section on "Policy Incentives, Land Tenure and Farmer Resource Base Interaction" discusses many pertinent issues, some of which are controversial and either need caution in tackling them or could result in serious hardship and even civil strife if recommended measures are implemented now. For example, insistence on removal of subsidies has many implications on sustainability and of adverse effects on the farmer's adoption of conservation measures and use of certain inputs. For instance, with devaluation of currencies in structural adjustment programs, prices and earnings from export products decline even though volumes of exports rise. But input costs escalate. Under such situations, how can farmers in SSA be denied of input subsidies while developed country farmers are subsidized and their products are protected in the world market by many policy measures? How can equity be achieved if subsidies are withdrawn from farmers while urban masses are favored by policies which subsidize food in one way or another?

The problem of agrarian reform in SSA is often emphasized as contributing to a lack of sustainability in agriculture because it is argued that farmers do not carry out long-term improvement where they do not have title to the land. Land tenure problems in SSA are explosive. Land reform in Africa has only been successful in a few instances where land redistribution has been carried out where a few landlords have a monopoly on the land or where as a result of European settlement, large areas of land have been taken from natives or indigenous peoples. Otherwise, traditional communal tenure has built-in stability in minimizing landlessness. Moreover, with the exception of fragmentation that may result from adherence to it, in many areas of high population density, land allocated to families becomes more or less privately owned land and farmers are not deterred from long-term improvements if they can afford them and recognize the necessity. The only kind of agrarian reform that may be necessary with the full understanding of the farmer is land consolidation where holdings are highly fragmented. Communal tenure, except in areas of pastoral nomadism, is giving way gradually to permanent individual ownership. Sustainability may be hampered more by agrarian reforms now than in the future.

Technology Issues

Again this is a comprehensive review but comments are necessary here and there.

1. As regards mechanization, it is not just enough to ensure appropriate levels of mechanization. It also is necessary to ensure an appropriate type of mechanization. In addition to this, appropriate technologies involving mixes of biological, cultural and other techniques may be used to eliminate mechanization that may involve more energy use or have adverse effects on the environment. For example, adequate residue management may eliminate extent of use of mechanical and chemical weed control measures.

2. The problem of research favoring better endowed areas and farmers is the result of copying strategies and technologies that have been used in the agriculture of temperate developed countries. Due consideration should be given to the location specificity of agriculture, agricultural systems, subsystems and their component technologies. Horizontal transfer of technologies often have met with failure.

3. Two or more decades of farming systems research (FSR) aimed at ensuring that: (i) a holistic approach is adopted, (ii) technologies are tailored to farmers' needs and circumstances, (iii) there is more of *bottom-up* than *top-down* approach and farmer participation at all stages of technology design, testing, adoption, monitoring and so on, and (iv) adoption of on-farm adaptive research (OFAR) methodology referred to in Chapter 14 of this publication have not made much progress in SSA. This is due largely to weakness of NARS, decline in the level of funding, facilities and resource allocation to agricultural research and deficiencies in overall institutional capabilities. After 20 yr of FSR, SSA still does not have enough socioeconomists in number and quality. In dealings with gender issues in on-farm research, SSA still relies on expatriate women socioeconomists and on more men than women.

4. It is rather too simplistic and not quite correct to claim that traditional technologies have broken down. If they have broken down, it is then contradictory to observe that "farmers have shown great ingenuity in developing improved practices to meet local needs" and that "innovative practices have evolved within the framework of traditional farming systems and

could be adapted, based on modern technological advances.'' What has become obvious is that in developing sustainable agricultural production systems in SSA, due consideration should be given to the location specificity of agricultural production, failures of horizontal transfer of farming systems and component technologies from one part of the world to another and the need to ensure economic viability, ecological soundness and cultural acceptability for widespread rapid adoption. This makes it imperative to adopt integrative approaches. First, selected elements of traditional technologies and systems have to be integrated with selected appropriate and compatible elements of conventional or "modern" technologies and also emerging ones. Second, there is need to adopt integrated pest and disease management systems as indicated in Chapter 14 of this publication. Third, there is need to use integrated watershed management in land use and commodity production diversification. Fourth, there also is need to integrate efficient reduced inorganic fertilizer use with the use of organic manures and biological processes such as N fixation and mycorrhizal phosphate nutrition. Last, there is need for integration of commodities in diversification of production, for example, integration of arable crops with perennials, integration of crops and animals, integration of crops, livestock and aquaculture where environmental conditions and circumstances permit.

5. As regards technological priorities in sustainable agriculture in SSA, the paper is silent on postharvest technologies. This is a serious omission because many food imports to SSA are done in order to satisfy demand for convenience foods and not mainly to satisfy demands due to rising income and population growth. Moreover, food imports also are used to satisfy demand due to increased mobility, changing roles of men and women and the need to ensure more uniform supplies throughout the year.

6. The Green Revolution had its second-generation problems in Asia and elsewhere most of which have been solved. The problem in SSA is how to develop human resource and institutional capabilities to address the wide range of commodities and ecological zones in a situation where some of these commodities only started to be addressed by research in the last decade or so.

7. As regards gender issues, it is not realistic to group problems of women in areas of southern Africa that suffer from migration of men to the mines and related jobs in the Union of South Africa with those of women in West Africa. In West Africa, considerable differences occur among ecological zones, cultural and other groups.

Institutional Issues

This is perhaps the most serious problem. Although several salient elements of these were discussed, the paper is silent on several other issues. These include: (i) educational and training institutions with curricula that are not relevant to the needs of countries of SSA, and in agriculture they are not vocationally adequate to produce self-employed farmers and innovative researchers, and very few women are being trained at all levels and disciplines, (ii) national research/extension systems which manifest many deficiencies in funding, facilites, coordination and management with universities often isolated from the system in which only they are uniquely mandated to cover almost all of the full range of basic, strategic, applied, adaptive, maintenance and developmental research, (iii) political and administrative institutions which are unstable, corrupt, poorly governed, managed, and manifest many deficiencies, and (iv) poor communication services with inadequate communication between policy makers and researchers, between researchers and the public, between researchers and extension and so on. Moreover, there are very poor institutional linkages, for example, among education/training, research, extension and the farmer.

Where Do We Go from Here and Strategic Elements to Consider

Under this section, I have only few observations to add to those I have made above. These are briefly indicated below:

1. The problem of land tenure in SSA is peculiar and explosive and should not be treated as urgent as is emphasized. It requires more time for education, caution and planning.

2. Development of sustainable agriculture necessitates more emphasis to be given to adult education, reorientation of farmers and researchers and measures to improve training in indigenous knowledge systems and combating rapid globalization of culture. This is very important because during the last 100 yr, much effort has been put into wrongly educating scientists that many traditional practices are primitive and in making farmers feel that they are using unscientific and inappropriate technologies. Now that we are adopting participatory on-farm adaptive research approaches in which we require them to interact with researchers fully and freely, they need to be given the confidence to contribute in dialogue and interaction with researchers. On the other hand, researchers need to know more about traditional farming systems and be attitudinally reoriented to be humbly ready to learn from farmers.

3. Development of sustainable agriculture requires changes and improvement in management of research. There is a need for: (i) more efficient management of interdisciplinary research teams, (ii) more basic research on environment and monitoring of changes in environmental variables, (iii) more long-term monitoring of the treatment effects in research, and (iv) long-term experimentation on such production practices as crop rotations.

Technological Options and Other Issues

Here, only three issues need to be considered. The first is the need to give higher priority to technologies for utilization of millions of hectares of hydromorphic or valley bottom areas of high potential. The second is to ensure use of appropriate technology mixes that involve efficient resource management. In this regard, sustainability does not call for low input technologies only as such, but for what is ecologically sound, economically viable and culturally acceptable. The third issue to be emphasized is that sustainability in development and in agricultural production in SSA can only be achieved by Africans in Africa. Cooperation with scientists in other parts of the world and in various institutions with the assistance of donors also is imperative for success and rapidity in achieving breakthroughs.

INTERNATIONAL FUND FOR AGRICULTURAL DEVELOPMENT, DEVELOPMENT INSTITUTIONS, SUSTAINABLE AGRICULTURE IN AFRIDA AND CONCLUDING REMARKS

The final section of the paper emphasizes IFAD's approach to research and extension which is based on a participatory technology system designated as "a client-demand system" with the client identified as the smallholder. After raising several questions to be answered about using such a system in SSA, it is observed that the only way of obtaining answers to these questions is to test the system and philosophy in a real-life development program. The paper ends with a listing of complementary measures that include many of the requirements and accelerators of agricultural development that should accompany the strategies outlined in Chapter 14 in this publication. In addition, the need for international and regional action, cooperation among countries in combating environmental degradation and various aspects of collaboration in research and technology transfer were stressed as necessary if "a hoped-for green revolution" is to be achieved on SSA through applications of "new biotechnology research."

CONCLUSIONS

While agreeing with many of the issues, strategies and requirements suggested for achieving sustainable agriculture in SSA, I deem it necessary to conclude with the following observations:

1. The IFAD's so-called "client-demand system" and "client-oriented participatory philosophy" does not differ significantly from recent developments in on-farm participatory and adaptive approaches in farming systems research developed by IARC's in partnership with some NARS. Moreover, considerable progress has been made by the U.N. and other development agencies and even research institutions outside Africa in various ways that could significantly contribute to the development of sustainable agriculture in SSA. The major constraint appears to be deficiencies in human resource development, institutional capacity building, efficient resource management and creation of an overall enabling environment that ensures that SSA has the capabilities for generating, borrowing, adapting, to otherwise acquiring technologies in order to achieve breakthroughs in sustainable agriculture in a more self-reliant manner within a minimum time frame.

2. The poverty, limitations in resources, weakness and deficiencies of NARS and unfavorable economic environment in SSA has made Africa a battlefield for development agencies that are more or less competing for influence and dominance. Some of these efforts are counterproductive and others involve policies and paternalistic practices that nurture greater dependence rather than self-reliance of countries of SSA. Unless there is greater coordination of effort, provision of adequate, balanced and sustained assistance in human resource development and institutional capacity building on a sustained basis for 10 or more years, achievement of sustainable agriculture and general development would be impossible.

3. Development of sustainable agriculture in SSA calls for a research and development agenda of a much wider scope and depth than conventional agricultural production research. It calls for sustained allocation of adequate levels of resources, commitment and efficiency in resource management rather than for conventional agricultural research and development which SSA was unable to achieve prior to the advent of the current sustainability paradigm in development when the economic environment was more favorable. African universities have a very crucial role to play in this regard.

4. Special agenda for sustainable agriculture and overall sustainable development exists for developed and developing countries including SSA if the guidelines stipulated by the World Commission on Environment and Development (1988) report is adhered to. The report calls for changes in attitudes, quality and nature of economic growth, policies, strategies, planning and execution of development programs. In addition, it calls for more effective collaboration and peaceful coexistence in which the developed countries have a special role to play in bringing about more equity in the sharing of natural resources. Unless the various factors which contribute to the current unfavorable global economic environment for SSA and the heavy debt burdens are eliminated, it would be impossible for the region to develop and use capabilities that can equip it to take advantage of the significant progress being made in research and development at the IARC's and elsewhere worldwide in the development of technologies that constitute components of science and technology-based sustainable agriculture. The most limiting factors in SSA are time, followed to some extnet by institutional deficiencies.

5. There are already enough well-articulated policy guidelines and strategies for sustainable agriculture in the region. The recent wind of change towards participatory democracy in the region, end of the Cold War and cravings for peace leading to the end of civil wars and civil strife in some countries auger well for the development of sustainable agriculture in which African policy makers and researchers have vital roles to play.

REFERENCES

Kesseba, A.M. 1993. Strategies for developing a viable and sustainable agricultural sector in sub-Saharan Africa: Some issues and options. p. 211–243. In J. Ragland and R. Lal (ed.) Technologies for sustainable agriculture in the Tropics. ASA Spec. Publ. ASA, CSSA, and SSSA, Madison, WI.

Kidron, M., and R. Segal. 1987. The new state of the world atlas. Rev. Simons and Schuster, New York.

Okigbo, B.N. 1991. Development of sustainable agricultural production systems in Africa: Roles of international agricultural research centers and national agricultural research systems. p. 1-14. *In* Distinguished African Scientist lecture series no. 1. Ibadan, Nigeria.

United Nations. 1990. Report of the expert group on Africa's commodity problems: Towards a solution. U.N. Gen. Assembly Doc. A/45/581 of 5.10.1990.

World Commission on Environment and Development. 1988. Our common future. Oxford Univ. Press, Oxford.

16   Technological Base for Agricultural Sustainability in sub-Saharan Africa

Rattan Lal

Department of Agronomy
The Ohio State University
Columbus, Ohio

The problem of agrarian stagnation in sub-Saharan Africa (SSA) has defied national and international relief efforts for more than a quarter of a century. Rapid increase in human wants and needs has stressed natural resources beyond limits, resulting in a widespread problem of soil and environmental degradation, e.g., accelerated erosion by wind and water, desertification, rapid deforestation, expansion of agriculture to marginal lands, siltation of waterways and reservoirs, and pollution of rivers and lakes. The human dimension of the tragedy is equally discouraging. Despite massive emergency aid, starvation and malnutrition persist.

Persistence of the agrarian stagnation raises several questions: (i) Are natural resources (soil, water, climate) of SSA adequate to meet ever-increasing human wants and needs? (ii) Are technologies available to manage soil and water resources to substantially enhance productivity? (iii) Are viable technologies being adopted? If the answer to all these questions were affirmative, we would have no problem. A negative answer to any of these questions requires the development and implementation of effective strategies to overcome the limitations. Otherwise, the negative trends presently dominating SSA agriculture will only worsen.

ADEQUACY OF NATURAL RESOURCES

The SSA covers a large area of 2231 million hectares. However, only about 6% of the total area is presently under arable land use (Table 16-1). The SSA also has a wide range of soils, landscape, vegetation and climate. Agriculturally significant annual rainfall ranges from 500 mm in the Sahel and arid regions to more than 5000 mm in the wet zone of Mount Cameroon. The growing season length ranges from less than 75 d in arid regions

Table 16-1. Land use in Africa (recalculated from FAO, 1990).

Land use	Africa	Northern Africa	South Africa	Sub-Saharan Africa
			10^6 ha	
Total area	3030.7	677.8	122.1	2230.8
Land area	2964.1	676.2	122.1	2165.8
Arable land	168.1	23.1	12.4	132.6
Permanent crops	18.8	3.4	0.8	14.6
Permanent pasture	890.8	107.5	81.4	701.9
Forest and woodland	683.3	18.8	4.5	660.0
Other land	1203.1	523.4	23.0	656.7

to year-round cropping in the humid zones. Vegetation is equally diverse, ranging from xerophytic scrub to tropical rainforest.

All major soil orders occur in SSA. Soils of the humid region are typically old and highly weathered Oxisols and Ultisols. Soils of the subhumid and semiarid regions are Alfisols, and coarse-textured Psamments. Alfisols, Aridisols and Plinthitic soils predominate in arid and semi-arid regions. The heavy-textured Vertisols containing predominantly high-activity clays, also occur extensively in semiarid climates and are agriculturally important soils. It is important to understand that interspersed throughout the region are some highly productive soils, e.g., alluvial soils along flood plains and inland valleys and soils of volcanic origin, such as those in the highlands of Rwanda, Burundi, Zaire, and Cameroon (see Chapter 17 in this publication).

Although not generally recognized, Africa also is endowed with a fair amount of water resources. Presently irrigated land area is only about 5 million hectares (Table 16-2). However, a reasonable level of irrigation potential exists that awaits development and exploitation. The land, along alluvial flood plans and inland valleys, is generally more suitable for small-scale irrigation than for large-scale grandiose schemes.

Africa also has the human capital needed to develop physical resources. Recent decades have seen the development of manpower with all the skills needed for the broad range of human needs. Technical manpower is especially strong in populous countries, e.g., Nigeria, Ghana, Kenya, and Zimbabwe. In fact, unemployment and underemployment of trained personnel has contributed to mass exodus to Europe, North America and the Middle East.

Table 16-2. Currently irrigated land area in Africa (FAO, 1986).

Irrigation	Land area
	10^6 ha
Modern	
Very large scheme (> 10 000 ha)	1.74
Large and medium scheme (500–10 000)	0.90
Traditional	
Small scale	2.38
Total	5.02

Table 16–3. Population carrying capacity of Africa for different scenarios (FAO, 1984).†

Carrying capacity at different input levels	People	Ratio to population of 1975
	millions	
Low input	1 120	3
Intermeciate input	4 608	12
High input	12 920	24

† Actual population in 1975 was 380 million.

Scientifically used, soil resources of Africa are adequate to support an acceptable standard of living for the current and future populations of SSA. A study conducted by FAO (1984) showed that SSA can support 1120 million people at low levels of input, 4608 million at intermediate levels of input, and 12 920 million at high levels of external input (Table 16–3). The actual population of SSA at the time was about 400 million. Similar surveys conducted by De Vries and De Wit (1983), and Higgins (1984) showed that soil and water resources of Africa are adequate to support several times the present population.

Despite high potential and vast resources, it is ironic that extent of soil and water degradation in Africa is equally alarming. Natural resources are severely degraded because of mismanagement, exploitation for short-term gains, and widespread practice of low input subsistence farming (Lal, 1988, 1990). Resource-based continuous cropping even at low level of productivity can lead to an average nutrient loss of 10 kg N, 1.8 kg P and 7.1 kg K $ha^{-1} yr^{-1}$. The rate of nutrient loss is about twice as much in Eastern Africa, and is likely to increase because of the increase in demographic pressure and intense cropping.

EFFECTIVENESS OF AVAILABLE TECHNOLOGIES
FOR MANAGING SOIL AND WATER RESOURCES

Africa has an impressive history of high-quality research data. Some of this research was initiated in early 1930s. History of research in soil and water management and crop improvement was summarized by Lal (1992).

The research base has led to the development of a wide range of technological options with a potential for the sustainable use of soil and water resources. Examples of relevant soil and water management technologies for different ecoregions are outlined elsewhere in this volume (see Chapter 19 in this publication). Some technical innovations have witnessed cyclic patterns of re-discovery once about every 20 yr. For example, tied ridges of soil and water conservation in semiarid regions were recommended in the 1950s and 1960s (Prentice, 1946; Dagg & McCartney, 1968). Agroforestry, and more specifically the use of *Leucaena* spp., also was recommended in this same period (Pereira & Beckley, 1953; Jurion & Henry, 1969), and later in 1980 by International Institute of Tropical Agriculture (IITA). Mulch farming always has been a favorite technology, not only for soil and water conservation, but also for soil fertility enhancement and nutrient cycling. Most of these station-proven technologies are technically sound.

TECHNOLOGY ADOPTION IN SUB-SAHARAN AFRICA

An important question that has repeatedly been asked is whether technically viable and station-proven technologies are being adopted. The answer to this question is no. Most technological innovations have proven successful in on-station experimentation and in research-managed on-farm trials. However, farmers of SSA have not abandoned the age-old traditional systems based on hoe, machete, and the match box. The absence or poor adoption of improved and apparently high-yielding technologies deserves the attention of sociologists, anthropologists, policy makers, and extension specialists. One of the principal reasons for the low rate of adoption is the topdown approach of research, without the participation of the farmer in prioritizing critical issues, defining research methods, and in validating and adopting the technology by fine tuning it to local conditions. Researchers often perceive a research problem according to their assessment of the farmer's constraints to enhancing production. Researchers design methodology for on-station or on-farm experimentation, develop a hypothesis, collect and analyze data and publish results without interaction with farmer. It is not surprising, therefore, that the so called "improved technology" is often rejected by the farmers of SSA. Agricultural sustainability is extricably linked with the recognition of the farmer being the premier research client and with the farmer's effective participation.

Has response by donor agencies been timely, adequate and effective in providing financial assistance to overcome the crisis and alleviate sufferings? An answer to this question is vividly presented by Lele (1991). It is argued that over the three decades ending in 1990, billions of dollars have been transferred from developed countries to Africa. It seems, however, that most of this aid has been rather ineffective in stimulating growth, breaking the vicious cycle, and alleviating poverty and human suffering. The problem lies both with national policies and donor perception. Furthermore, donors need to coordinate their assistance with regard to long-term development strategies and institutional building.

Hudson (1989, unpublished data) conducted a survey of FAO-funded programs and listed reasons for their success or failure. Overall, the success rate was about 25% for projects initiated in 1970s and 56% for those initiated in 1980s. Similar conclusions of low success rate (12–40%) were arrived at by a survey conducted by the World Bank (1984; 1985, p. 38–43; 1986). Hudson listed the inappropriate technology among the principal reasons for project failure. He concluded that technology should be appropriate and tested locally; offer short-term, on-site benefits, and large increments (50–100%); require affordable inputs, especially labor; not include foregone benefits, e.g., giving up land; not include any increased risk; and be in tune with existing social factors, e.g., the separate roles of men and women in agriculture.

FARMER PARTICIPATION IN TECHNOLOGY ADOPTION

In this context, International Services for National Agricultural Research (ISNAR) has put forward a useful strategy. The concept termed "on-farm client-oriented research" (OFCOR) appears to be an important strategy for involving farmers in the decision-making process at all stages of research program, technology development, and validation (see Chapter 14 in this publication). The principal goal is to adopt a problem-solving approach focused on the farmer as the primary client.

Another facet of the OFCOR approach is to create off-farm employment opportunities to reduce the proportion of the population directly dependent on agriculture through development of agrobased industries. The suggestion of rural systems research (RSR) also is a useful strategy to generate off-farm opportunities to support farm enterprise, e.g., input-based industry, cash and credit system, processing and storage, marketing and transport, etc. Sustainable agriculture cannot be developed without the creation of off-farm supportive industries.

"The client-oriented participatory philosophy" must take roots in a system aimed at revitalizing agriculture in SSA. The farmer must be rewarded for his/her contribution and be assured of the expected returns.

RESEARCH AND DEVELOPMENT PRIORITIES

Development priority should concentrate on intensification of agriculture in regions of high productive potential. These are large areas with soils of high inherent fertility and assured water supply, e.g., flood plains, hydromorphic valleys and inland swamps, soils of volcanic origin, etc. Soils of these ecoregions are input responsive, and high returns can be expected even with medium level of input. Intensification of agriculture in ecoregions of high potential does not mean, however, that marginal rainfed areas of medium potential should be ignored. On the contrary, a majority of resource-poor farmers in SSA cultivate these regions. Development and adoption of improved agricultural technologies in these regions is crucial to enhancement of agricultural production.

Principal soil-related constraints to crop production and enhanced yields in medium-potential rainfall areas are low soil fertility and frequent drought stress. Semiarid and arid regions are more suited for production of food crops {maize (*Zea mays* L.), rice (*Oryza sativa* L.), millet [*Pennisetum americanum* (L.) Gaertner subsp.], sorghum [*Sorghum bicolor* (L.) Moench], cowpea [Vigna unguiculata (L.) Wolp. subsp. unguimlala], peanut (*Arachis hypogaea* L.), etc.} than are the humid regions. In addition to soils of low water-holding capacity, the rains are highly variable and erratic. Drought stress, high soil and air temperatures, high evaporative demand, and high risks of wind and water erosion are serious problems. Conservation of soil and water resources, therefore, is a matter of high priority.

Once water availability is assured, through in-situ conservation or supplementary irrigation, soil fertility becomes a major yield-limiting factor. Nitrogen deficiency is a widespread problem among principal plant nutrients.

The rate of commercial fertilizer application is extremely low. In addition to increasing the availability of inorganic fertilizers, it is equally important to adopt fertility enhancing practices such as legume-based crop rotations, frequent use of cover crops and green manures, agroforestry systems, and mixed farming practices based on controlled grazing and use of farmyard manure.

CONCLUSIONS

Results of on-station research with regard to intensive land use and improved systems of soil and crop management in SSA are available. However, these results have not been widely adopted because they have not been translated into technological packages, of improved soil and crop management systems. Agrarian stagnation, decreases in per capita agricultural productivity, perpetual food deficit, and widespread problems of soil and environmental degradation are direct consequences of the lack of adoption of science-based improved agricultural technology. One of the reasons for the lack or slow adoption is the noninvolvement of farmers in the development of these technologies. The client-oriented philosophy as a basis for the research is an important strategy to ensure farmer involvement at all stages of the decision-making process. Furthermore, development priorities should concentrate on intensification of agriculture in regions of high productive potential. Development and adoption of improved agricultural technologies in rainfed areas of medium potential also is crucial because a majority of resource-poor farmers of SSA cultivate these regions. Improved agricultural technologies should be directed to alleviating soil-related constraints of accelerated soil erosion, rapid fertility depletion, nutrient imbalance, and drought stress. Furthermore, essential inputs must be made available at affordable prices and on time. Beyond these farmers must be adequately rewarded for their produce and be assured of returns.

REFERENCES

Dagg, M., and J.C. McCartney. 1968. The agronomic efficiency of the N.I.A.E. mechanized tied ridge system of cultivation. Exp. Agric. 4:279–294.

DeVries, P, R.W.T. and C.T. De Wit. 1983. Identifying technological potentials for food production and accelerating progress in productivity. In Accelerating agricultural growth in sub-Saharan Africa, Int. Food Policy Rec. Inst. Conf., Zimbabwe. Conf. 29 August to 1 September. Washington, DC.

Food and Agriculture Organization. 1984. Land food and people. FAO, Rome.

Food and Agriculture Organization. 1986. African agriculture: The next 25 years. FAO, Rome.

Food and Agriculture Organization. 1990. Production year book. FAO, Rome.

Higgins, G. 1984. African resources of terrain, land and water for the population of future management for development. Advancing agriculture production in Africa. Commonwealth Agriculture Bureau, England.

Jurion, F., and J. Henry. 1969 Can primitive farming be modernized? Hors. Ser., Brussels, Belgium.

Kesseba, A.M. 1993. Strategies for developing a viable and sustainable agricultural sector in sub-Saharan Africa: Some issues and options. p. 211–243. *In* J. Ragland and R. Lal (ed.) Technologies for sustainable agriculture in the Tropics. ASA Spec. Publ. ASA, CSSA, and SSSA, Madison, WI.

Lal, R. 1988. Soil degradation and the future of agriculture in sub-Saharan Africa. J. Soil Water Conserv. 43:444–451.

Lal, R. 1990. Low resource agriculture alternatives in sub-Saharan Africa. J. Soil Water Conserv. 45:437–445.

Lal, R. 1992. Institutional and manpower constraints on soil research in Africa. p. 45–48. *In* K. Tato and H. Hurni (ed.) Soil conservation for survival. Soil Water Conserv. Soc., Ankeny, IA.

Lal, R. 1993. Technological options towards sustainable agriculture for different ecological regions of sub-Saharan Africa. p. 295–308. *In* J. Ragland and R. Lal (ed.) Technologies for sustainable agriculture in the Tropics. ASA Spec. Publ. ASA, CSSA, and SSSA, Madison, WI.

Lele, U. 1991. Aid to African agriculture: Lessons from two decades of donor experience. World Bank Report, The John Hopkins Univ. Press, Baltimore, MD.

Pereira, H.C., and V.R.S. Beckley. 1953. Grass establishment on eroded soil in a semi-arid African reserve. Emp. J. Exp. Agric. 21:1–14.

Prentice, A.N. 1946. The tied ridges. East African Agric. J. 12:101–108.

Vlek, P.L.G. 1993. Strategies for sustaining agriculture in sub-Saharan Africa: The fertilizer technology issue. p. 265–277. *In* J. Ragland and R. Lal (ed.) Technologies for sustainable agriculture in the Tropics. ASA Spec. Publ. ASA, CSSA, and SSSA, Madison, WI.

World Bank. 1984. 10th annual review of project performance audit results. World Bank Rep. 5248. Vol. 3. World Bank, Washington, DC.

World Bank. 1985. Annual review of project performance results. 11th Audit Rev. World Bank Rep. 5859. World Bank, Washington, DC.

World Bank. 1986. Annual review of project performance results 1985. 12th audit rev. World Bank, Washington, DC.

World Bank. 1985. Sustainability of projects: First review of experience. Rep. 5718. World Bank, Washington, DC.

17 Strategies for Sustaining Agriculture in sub-Saharan Africa: The Fertilizer Technology Issue

Paul L. G. Vlek

Institute of Agronomy in the Tropics
Georg-August University
Goettingen, Germany

This chapter addresses "Technology Issues" raised in the chapter by Kesseba (see Chapter 14 in this publication), and of land degradation and its consequences in terms of food production in sub-Saharan Africa (SSA). It also attempts to reconcile these data with what is manifest from the African production statistics. Finally, attempts also are made to elaborate on the two-prong strategy proposed in the chapter by Kesseba (1993) vis-a-vis the marginal and better endowed regions of SSA.

SOIL DEGRADATION

A recent study by the United Nations Food and Agriculture Organizations (FAO) classified the lands in the Tropics according to their production potential. Although SSA is by far the largest tropical region, covering 2.22 billion hectares (World Bank, 1989), only 30% of it is considered cultivable, of which again only 35% is actually in use (FAO, 1981). Although the land available per capita in SSA in 1975 was 1.2 times that for Latin America, by the year 2000 this ratio will have decreased to 0.87. Moreover, of the total harvested land of SSA, 64% is considered low potential or problem land (FAO, 1989a). The handicap of SSA in comparison to the other tropical regions is well illustrated by its almost fourfold higher fraction in low-potential (drought-prone) land over Latin America or two-fold over Asia, thus negating in part the advantages of a larger area. The fraction of problem lands (26%) in SSA is close to that of the Tropics as a whole (Fig. 17–1).

Irrigation can serve as an important modifier in the production capacities of land. Over 165 million hectares or 21% of the cropland is irrigated in the Tropics and sub-Tropics. This allows double or triple cropping and

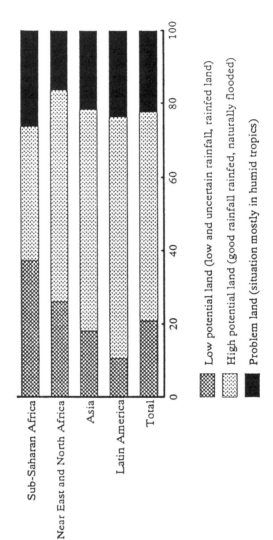

Fig. 17–1. Shares of total harvested land of different potentials, 1982 to 1984 (93 developing countries). Source is the FAO, World Agriculture: Toward 2000.

supplemental irrigation in a poor wet season allows a wider spectrum of crops to be cultivated. The available water resources are again widely different around the world with Asia claiming 88% of the irrigated land, as compared to SSA with merely 3%. For SSA this represents no more than 5.3 million hectares of land (OTA, 1988), or 4% of the land under cultivation. However, 70% of this is in only three countries: Madagascar, Nigeria, and Sudan (FAO, 1986). It is possible to extend the irrigated area in SSA to about 20 million hectares (World Bank, 1989), but this expansion is slow to be realized due to a variety of technical, financial and socioeconomic constraints. The limited water endowment of SSA is a serious restriction to increased agricultural output.

The soils of the Tropics are the key element in determining the productivity of the land. They differ widely in their characteristics as a result of different parent materials and pedogenesis. Most tropical regions have some representation of all soil types, but the dominant ones are different in the different ecological zones. The distribution of soils by climatic regions is given in Table 17–1 (Sanchez, 1976), and shows the predominance of the highly weathered soils (51%) and dry sands (17%) in the Tropics as a whole. Younger and productive soils are relatively more common in Asia, while Africa has a high proportion of its land in predominantly old, acid or poorly buffered soils. Although the land area of tropical Africa is three times that of Asia, 69% of it consists of Oxisols, Aridisols and Psamments, which are generally fragile or infertile, and representative of the low-potential and problem lands of SSA discussed above. The distribution of the soil orders in SSA is given in Fig. 17–2 (Okigbo, 1989).

Table 17–1. Distribution of soils in the Tropics by climatic regions.

Soil groups (soil taxonomy equivalents)	Rainy (9.5–12)†	Seasonal (4.5–9.5)†	Dry and desert (0–4.5)†	Total	Percentage of Tropics
	million ha				
Highly weathered, leached soils (Oxisols, Ultisols, Alfisols)	920	1540	51	2511	51
Dry sands and shallow soils (Psamments and lithic groups(	80	272	482	834	17
Light-colored, base-rich soils (Aridisols and aridic groups)	0	103	582	685	14
Alluvial soils (Aquepts, Fluvents, and others)	146	192	28	366	8
Dark-colored, base-rich soils (Vertisols, Mollisols)	24	174	93	291	6
Moderately weathered and leached soils (Andepts, Tropepts and others)	5	122	70	207	4
Total area	1175	2403	1316	4896	100
Percentage of Tropics	24	49	27	100	

† Numbers in parentheses refer to number of months with an average rainfall greater than 100 mm [Sanchez (1976)].

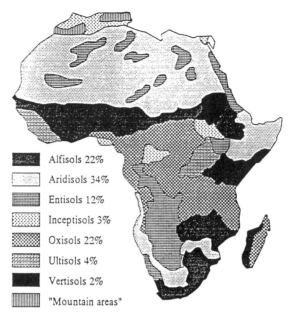

Fig. 17-2. Soil map of Africa-distribution of orders. Source is soil map of the world prepared by the Soil Geography Unit of the U.S. Conservation Services, Washington, DC.

With few exceptions, the classification of a soil does not necessarily provide an indication of the productivity of the land, but some soil types are known as resilient and have the capacity to hold up under long-term cultivation. Young alluvial soils often are rich in weatherable minerals and are regularly rejuvenated. Such soils constitute 8% of the soils in the Tropics. On the other hand, the older continental shields of Africa and South America have lost their weatherable minerals over eons of exposure to the elements, resulting in acidic soils that are difficult to manage under long-term cultivation because their fertility is largely stored in the fragile organic matter fraction.

Arable and permanently cropped land in SSA covers around 156 million hectares, or 6.7% of the total land. The crops grown in 1987 are listed in Table 17-2 in order of production, together with the estimated quantity of nutrients in the harvested product, using the nutrient levels estimated by Stoorvogel and Smaling (1990). Many crops, such as vegetables and most fruits are not included. However, based on this listing, it is concluded that as a minimum the equivalent of around 4 million Mg of fertilizer nutrients are being harvested on an annual basis. Roughly 8% of this, or around 300 000 Mg, is exported annually, largely to the West. Even though this represents only one-third of the actual fertilizer consumption of 1 million Mg (N, P_2O_5, K_2O) per year (Vlek, 1990), much of the harvested nutrients that are consumed in the city are not returned to the field as well. In fact, one-third of the SSA population is urban (World Bank, 1989).

Table 17–2. Production and export of key crops and nutrients in sub-Saharan Africa (1987).

Crop†	Production	Export	N	P_2O_5	K_2O	Harvested nutrients	Exported nutrients
				1000 Mg			
Cassava	58 300	15	240	53	295	588	0
Plantain	22 600	199	20	6	95	121	1
Yam	22 250		103	16	80	199	0
Maize	16 350	785	275	150	95	520	25
Sorghum	11 700		170	150	50	370	0
Millet	10 100		195	140	65	400	0
Rice	7 600	4	90	60	30	180	0
Sweet potato	5 850		27	4	20	51	0
Pulse	5 050	155	101	40	70	211	7
Groundnut	4 100	125	150	55	40	245	8
Potato	4 100	71	20	12	35	67	1
Sugar	4 000	1800	20	18	50	88	40
Taro	3 000		15	2	10	27	0
Citrus	2 200	590	4	1	6	11	0
Cotton	2 700	540	50	60	30	140	28
Coconut	1 700	50	100	30	20	150	4
Palm oil	1 600	720	5	3	8	16	7
Coffee	1 250	916	44	7	20	71	52
Cocoa	1 180	895	47	25	30	102	77
Pepper	1 180	26					0
Tobacco	300	175	17	6	25	48	28
Tea	255	218	9	2	4	15	13
Rubber	240	217	2	1	2	5	5
Total						3625	296

† Cassava = *Manihot esculenta* C., plantain = *Musa balbisiana* C., yam = *Dioscorea* spp., maize = *Zea mays* L., sorghum = *sorghum bicolor* L., millet = *Pennisetum americanum* L., rice = *Oryza sativa* L., sweet potato = *Ipomoea batatas* L., groundnut = *Arachis hypogea* L., potato = *Solanum tuberosum* L., taro = *Colocasia esculenta* L., citrus = *Citrus* spp., cotton = *Gossypium hirsutum* L., coconut = *Cocos nucifera* L., palm oil = *Elaeis guiniensis* Jacq., cocoa = *Theobroma cacao* L., pepper = *Capsicum anuum* L., tobacco = *Nicotiana rustica* L., tea = *Camellia sinensis* L.

At the current rate of fertilizer use, 8.5 kg nutrients ha^{-1}, the soils of Africa are effectively being mined for their nutrients, and have been for decades.

A detailed study was undertaken for FAO by the Winand Staring Center in the Netherlands, in which the nutrient depletion rates for the various countries and Land/Water classes found in SSA were estimated (Stoorvogel & Smaling, 1990). The results, summarized in Fig. 17–3, reinforce the compelling message arrived at above—soils in SSA are losing their fertility at exceedingly high rates. Ironically, the $2.2 billion earned in 1987 in coffee (*Coffea arabica* L.) sales alone would have been adequate to purchase approximately 8 million tons of fertilizer nutrients.

An important part of what constitutes soil productivity is determined by the fertility of the soil, which in turn is largely determined by the organic matter and stored nutrients of the soil. In many soils, a significant fraction of the nutrient store is in the soil organic matter (SOM), an extremely important but fragile soil constituent. On the basis of a rough estimate (Bohn, 1982) of the SOM resources worldwide, it appears that of the 22×10^{14} kg

Classes of nutrient loss rates (in kg/ha)

	Class	N	P_2O_5	K_2O
	Low	<10	<4	<10
	Moderate	10-20	4-7	10-20
	High	21-40	8-15	21-40
	Very high	>40	>15	>40

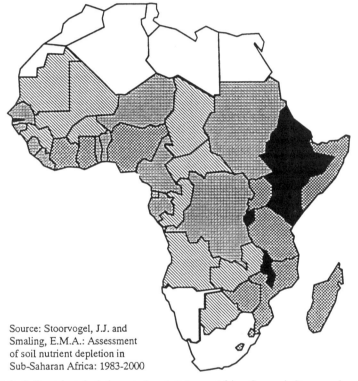

Source: Stoorvogel, J.J. and
Smaling, E.M.A.: Assessment
of soil nutrient depletion in
Sub-Saharan Africa: 1983-2000

Fig. 17-3. Soil nutrient depletion rate in sub-Saharan Africa. Source is Stoorvogel and Smaling: Assessment of soil nutrient depletion in sub-Saharan Africa: 1983 to 2000.

C, 69% is located in the temperate regions of Europe, North America, and North and Central Asia (Table 17-3).

Soil organic matter levels vary with environment and soil type, as well as management. Sanchez et al. (1982), based on a limited number of samples, demonstrated for some important soil orders in the Tropics (Ultisols, Alfisols and Mollisols) that the variability in SOM also is common in Africa. For instance, the median value for 570 topsoils in Eastern Africa was 4% (Birch & Friend, 1956). In comparison, a study of some 50 representative soils of West Africa gave values ranging from 0.1 to 3% with a median

Table 17–3. Mass of organic C in mineral and Histosol (organic, peat) soils.†

Continent	Mineral soils	Histosols	Total
	Organic C $\times$ 10^{14} kg		
Africa	2.6	0.02	2.6
America			
Central	0.3	0.04	0.3
North	4.8	1.6	6.4
South	2.2	0.09	2.3
Asia			
North and Central	4.4	1.6	6.0
South	0.6	0.004	0.6
Australasia	1.0	0.02	1.0
Europe	2.4	0.4	2.8
Total	18.3	3.77	22.0

† Bohn (1982).

of 0.5% (Bationo & Mokwunye, 1991) while a similar sample from the Sahelian zone ranged from 0.12 to 0.17% (Penning de Vries & Djieye, 1982).

The loss of productivity is often related to the loss of organic matter as a result of cultivation. Jenny and Raychaudhuri (1960) studied 522 soils in India and calculated "cultivation indices" reflecting losses of SOM between 0 and 80% as a result of cultivation. A number of more recent studies have described the process of SOM loss. An interesting example was given by Siband (1974). Sampling an unreal time series in Senegal of continuous cultivation, he noted a decline in topsoil SOM from 2.85% under forest to 0.84% after 90 yr. Pieri (1989) modeled this decline and established that even after 90 yr a true equilibrium had not been established (Fig. 17–4). It would be interesting to know if the East African soils in the area sampled by Birch

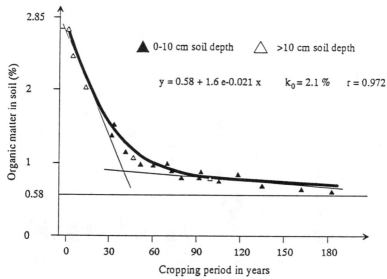

Fig. 17–4. Relation between soil organic matter and cropping period after land clearing in the topsoil, source is Siband (1974).

and Friend also have lost some of their SOM over the past 35 yr due to the increasing population pressures.

The decomposition of organic matter in the soil, as illustrated by Jenkinson and Ayanaba (1977), is a continuous process, leading to the loss of SOM when organic matter inputs are drastically reduced following cultivation. Added to this problem is the concomitant acceleration in the loss of topsoil (Dunne & Dietrich, 1982), and possible exposure to subsoil acidity. The principles involved here are responsible for farmers adopting shifting cultivation techniques to allow restoration of soil productivity (Sanchez, 1976).

AGRICULTURAL PRODUCTION

Given the role of organic matter as a source of energy for the soil heterotrophs and nutrients for the plant and soil fauna, as well as in promoting soil structure, one would expect declining yields when SSA land and soil resources are placed under cultivation. Some data from different tropical environments were summarized by Sanchez (1976). In fact, many studies have been carried out in different parts of SSA, demonstrating the decline in yield after a few years of cultivation (Pieri, 1989). One of our own studies from the Sahel (Fig. 17–5; Vlek, 1990) indicates the importance of organic matter in sustaining yields. But farmers have many uses for their produce and are rarely inclined to conserve organic matter in a region where land seems abundant and restoration of SOM is left for the fallow vegetation to accomplish.

However, the situation is changing. There can be no doubt that the population pressure is real. Per Caput land availability in SSA will change from 0.64 to 0.39 ha during the last quarter of this century. Per Caput food produciton in SSA has declined steadily since the 1970s despite 14 million additional hectares being taken into cultivation over that period. One would expect this pressure on the land to have its repercussions in yields. Indeed, some

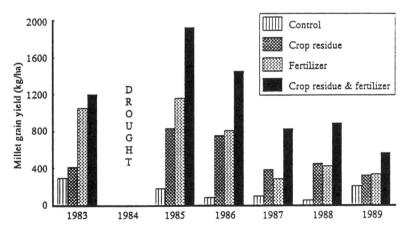

Fig. 17–5. Effect of restitution of crop residues and use of fertilizer N-P-K or both on millet grain yield in Sudorc, Niger, source is Vlek (1990).

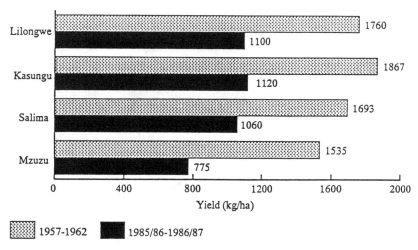

1957-1962   1985/86-1986/87

Fig. 17-6. Average yields of unfertilized local maize in Malawi, source is FAO.

regions in SSA report declining productivities (Fig. 17-6). Yet on the whole, yields in SSA have remained steady or actually improved (Table 17-4, FAO, 1974, 1989b).

Although the reliability of such statistics can always be questioned, the trends are clear—whatever soil degradation has taken place in SSA over the past decades, it has yet to have its resonance in the productivity of the land. Indeed, the FAO (1989a) concludes that, "All attempts to provide a clear assessment of the condition of the world's natural resources soon run up against the sparseness, inaccuracy and noncomparability of available data, and the weakness in our understanding of some of the processes."

There is hardly reason for optimism in SSA. Between 1973 and 1988, arable and permanently cropped land has increased by 14 million hectares or 10%. Simultaneously, forest and woodland declined by 40 million hectares or 6%, while pasture lands were kept stable. The difference is counted as "other land," desertified or otherwise laid to waste. Given these land-use statistics, the likely reason for the apparent yield stability lies in the mobility of the SSA farmer. The extent of soil degradation is temporarily hidden when land is destroyed and subsequently abandoned, as long as "new land" is avail-

Table 17-4. Average crop yield changes between 1961/1965 and 1987/1989 for sub-Saharan Africa.

Crop	1961–1965	1987–1989
	kg/ha	
Maize	934	1208
Sorghum	757	937
Millet	606	750
Cassava	7000	7250
Yam	8975	9530
Taro	4880	3970
Seed Cotton	350	800

able to take its place. With three times as much land taken out of forest as was taken into cultivation, SSA is buying time before the destruction is obvious from declining yields. There are indeed regions where such pressures have created production declines and environmental deterioration so severe (e.g., Mossi plateau in Burkina Faso), that it has led to massive migration out and urbanization. The average annual growth rate in urban populations (6.9%) is more than double that of SSA's entire population (3.1%).

The process leading to destructive land use may be gradual. As a first step, the unnecessary regeneration years for rebuilding soil fertility are being eliminated followed with increasing pressure by further shortening the necessary regeneration years. The negative consequences of these actions will be noticeable only after some time, dependent on the initial fallow cycle length. But once started on this downward path, it is difficult to reverse the process. Eventually, people abandon degraded land entirely and open up new lands in an entirely new region. Unfortunately, the abandoned lands often are problem lands. As fallow periods are shortened or eliminated, as has been the case more and more over the past decade, the true extent of soil abuse will eventually become apparent in the yield statistics.

SOIL FERTILITY MAINTENANCE

As Kesseba's (1993) chapter (Chapter 14 in this publication) shows, there is little hope that SSA will feed itself without some major adjustments in its development approach. Projections from International Food Policy Research Institute (IFPRI) for the year 2000 sketch net deficits in food production of 50 million metric tons, or 25% of the food needs. The bulk of the deficits are not in the poorest of the poor countries, but in those with per capita GNP's of over $500. Assuming there will be a donor willingness to provide it, the logistics of delivering the necessary quantities of food using the current infrastructure are mind boggling. An enormous expenditure could be avoided if a greater part of the food were produced in situ and in a sustainable manner.

Obviously, the low potential area cannot be ignored. Many improvements are indeed possible. However, the chances that these areas will produce the food surplus necessary to permit the development of a substantial nonagrarian sector are bleak in the short run. And for SSA to develop, economic activity cannot be restricted to agriculture, rather, agriculture must serve as the motor for broad-based economic development. Since this motor has to be both reliable and efficient, it will need to be based on the favorable ecological zones of Africa, e.g., the Savanna region.

There is no reason why a "green revolution" could not take place in Africa. However, what holds true for Asia applies as well to SSA—miracle crops do not grow better than the landraces under "low-input" conditions. Given the limited scope for irrigation and the dismal record of the fertilizer sector in SSA, the failure of the green revolution is not surprising. In fact, at current consumption rates, fertilizer use in SSA does not even come close to compensating for the harvested nutrients (Table 17-2). The development of Africa will require a drastic integration of external inputs to supplement and protect the internal inputs that are available in order to boost production in a sustainable manner.

Farmers cite various reasons for nonadoption of fertilizers such as risk, lack of credit, and untimely supply (Vlek, 1990). The first constraint is negligible on the more favorable land, assuming farmers have access to a market. The latter two constraints require investments that should, in the first phase, be targeted to favorable regions in order to prime the fertilizer sector. The policy of nationwide, or worse, continentwide fertilizer availability based on equity principles is untenable given the infrastructure of SSA and would involve enormous subsidies. The two-phased approach proposed by Vlek (1990) foresees concentrated development efforts and investments in a favorable environment where intensive food production is feasible and sustainable through integrated soil fertility management. Much of the technology is available, and adoption will be rapid if the government is able to provide the proper market access. Such nuclei of growth can eventually nourish the development of the less favorable regions.

Continued research investments should also be directed toward the low potential and problem areas of SSA in order to arrest soil degradation and promote efficient types of extensive farming. Government policies should aim at protection of these lands and link these zones with the growth nuclei through off-farm employment and, in a second phase, with improved infrastructure. The stakes are high, as many flourishing civilizations have experienced sudden decline when their agricultural base disappeared due to land degradation (Dale & Carter, 1954).

Why was SSA not in tune with the policies of Asia and Latin America and seemingly unable to recognize the need to establish a fertilizer sector? There are many reasons suggested including: the food situation in Africa lacked an urgency relative to Asia, the lack of intrastructure made delivery systems unmanageable, a greatly increased food production would not have found an off-farm market, most countries lacked the resources and represented too small a market to attract investments, the meddling of government bureaucracies, and the fertilizer raw material resources were poorly known. The situation may be evolving in to a more favorable situation in SSA, and the time may be ripe for the sector to come into its own. Developments in Nigeria over the past 10 yr are showing that, provided farmers do not have to compete with subsidized grains from the West, they are willing and able to produce for the market and purchase the fertilizer to do so. The Government fertilizer policies that would promote such developments were recently discussed by Makken (1991).

The suggestion that fertilizer use should be stimulated in SSA is, of course, anathema for many environmentalists. The dangers of indiscriminate use of chemicals in agriculture are known. However, failing to enhance fertilizer use in SSA might actually lead to an environmental disaster, as it will cause stagnation in economic development with millions of farmers trading their exhausted and irreversibly degraded lands for the still remaining problem lands, leaving behind a denuded landscape. Although hard evidence of land degradation and declining productivity is difficult to find, the threat is real. The challenge for agronomists concerned with or working in SSA is not to block this necessary evil, but to apply their science and the experiences gained from the excesses in the West to finding ecologically acceptable ways in which this modern input can be integrated in the multitude of soil conservation measures that guarantee sustainable, profitable farming and the basis for economic development.

REFERENCES

Bationo, A., and A.U. Mokwunye, 1991. Role of manures and crop residue in alleviating soil fertility constraints to crop production: with special reference to the Sahelian and Sudanian zones of West Africa. Fert. Res. 29:117–125.

Birch, H.F., and M.T. Friend. 1956. The organic matter and nitrogen status of East African soils. J. Soil Sci. 7:156–167.

Bohn, H.L, 1982. Estimate of organic carbon in world soils: II. Soil Sci. Soc. Am. J. 46:1118–1119.

Dale, T., and V.G. Carter. 1954. Topsoils and civilization. Univ. Oklahoma Press, Norman, OK.

Dunne, T., and W. Dietrich, 1982. Sediment sources in tropical drainage basins. p. 41–55. In W. Kussow et al. (ed.) Soil erosion and conservation in the Tropics. ASA Spec. Publ. 43. ASA and SSSA, Madison, WI.

Food and Agriculture Organization. 1974. FAO production yearbook. Vol. 28. FAO, Rome.

Food and Agriculture Organization. 1981. Agriculture: Toward 2000. FAO, Rome.

Food and Agriculture Organization. 1986. African agriculture: The next 25 years. ARC/86/3.

Food and Agriculture Organization. 1989a. The state of food and agriculture. FAO Agric. Ser. 22. FAO, Rome.

Food and Agriculture Organization. 1989b. FAO production yearbook. Vol. 43. FAO, Rome.

Jenkinson, D.S., and A. Ayanaba, 1977. Decomposition of carbon-14 labeled plant material under tropical conditions. Soil Sci. Soc. Am. J. 41:912–915.

Jenny, H., and S.P. Raychaudhuri, 1960. Effect of climate and cultivation on nitrogen and organic matter reserves in Indian soils. Indian Council of Agric. Res., New Delhi.

Kesseba, A.M. 1993. Strategies for developing a viable and sustainable agricultural sector in sub-Saharan Africa: Some issues and options. 1993. p. 211–243. In J. Ragland and R. Lal (ed.) Technologies for sustainable agriculture in the Tropics. ASA, CSSA, and SSSA, Madison, WI.

Makken, F. 1991. Alleviating fertilizer policy constraints to increased fertilizer use and increased food production in West Africa. Fert. Res. 29:1–7.

Okigbo, B.N. 1989. Development of sustainable agricultural production systems in Africa. Distinguished African Scientist Lecture Ser. 1. Int. Inst. Trop. Agric., Ibadan, Nigeria.

Office of Technology Assessment. 1988. Enhancing agriculture in Africa: A role for U.S. development assistance. U.S. Congress, Office of Technol. Assessment. OTA-F-356. U.S. Gov. Print. Office, Washington, DC.

Penning de Vries, F., and M. Djitèye. 1982. La productivité des pâturages Sahéliens. Pudoc, Wageningen, the Netherlands.

Pieri, C. 1989. Fertilité des terres de savannes. CF/CIRAD, Documentation Francaise. Paris Cedex, France.

Sanchez, P.A. 1976. Properties and management of soils in the tropics. John Wiley and Sons, New York.

Sanchez, P.A., M.P. Gichuru, and L.B. Katz. 1982. Organic matter in major soils of the tropical and temperate regions. p. 99–114. In Trans. 12th Int. Congr. of Soil Sci. New Delhi. Symp. Pap. I. 8–16 February. Indian Soc. Soil Sci. and Indian Agric. Res. Inst., New Delhi.

Siband, P. 1974. Evolution des charactéres et de la fertilité d'un sole rouge de Casamance. L'Agron. Trop. 29:1228–1248.

Smith, R.M., G. Samuels, and C.F. Cernuda. 1951. Organic matter and nitrogen built-ups in some Puerto Rican soils. Soil Sci. 72:409–427.

Stoorvogel, J. and E. Smaling. 1990. Assessment of soil nutrient depletion in sub-Saharan Africa. Vol. 1 to 4. Rep. 28. The Winand Staring Centre, Wageningen, the Netherlands.

Vlek, P.L.G. 1990. The role of fertilizers in sustaining agriculture in sub-Saharan Africa. Fert. Res. 26:327–339.

World Bank. 1989. Sub-Saharan Africa, from crisis to sustainable growth. World Bank, Washington, DC.

18 Short Season Fallow Management for Sustainable Production in Africa

V. Balasubramanian

International Institute of Tropical Agriculture
Yaounde, Cameroon

Nguimgo K. A. Blaise

Institut de Recherche Agronomique
Yaounde, Cameroon

In the densely populated areas of Africa, fallow periods between the 2- to 3-yr cultivation phases have been sharply reduced (to 1 yr or less in highlands, and less than 6 yr in lowlands) (Balasubramanian & Egli, 1986; Kang et al., 1990; Mutsaers et al., 1981; Prinz, 1987; Rocheleau et al., 1988). Short duration natural regrowth is not effective in restoring soil fertility and suppressing weeds (Agboola, 1980; Ruthenberg, 1971). In addition, heavy grazing of the sparse vegetation in natural fallow increases the risk of soil erosion. The African continent is thought to suffer the worst soil erosion problems in the world (Beets, 1978). Soil fertility and crop yields, therefore, decline with time (FAO, 1984; Nye & Greenland, 1960). In the absence of corrective measures, damage to the environment can occur quickly, thereby undermining the basis of livelihood. Restoration is usually a slow and costly process (Anonymous, 1989a).

The need for a long fallow period to maintain soil productivity can be reduced by the addition of fertilizers (Young & Wright, 1980). However, some reports indicate that soil productivity in the Tropics declines even with supplementary fertilizer use (Allan, 1965). Others show the maintenance of soil fertility under continuous cultivation with the judicious use of fertilizers alone (Heathcote & Stockinger, 1970; Singh & Balasubramanian, 1979). In either case, resource-poor farmers cannot depend entirely on costly inputs such as fertilizers to keep their fields productive at least not until farm income is high enough to afford inputs. Other suggested measures to maintain soil fertility under reduced fallow include legume species—food crop rotation or intercropping, alley cropping/farming, crop rotation, contour farming, etc. (Balasubramanian & Sekayange, 1986; Francis et al., 1986; Greenland, 1985;

Kang et al., 1990; Nhadi & Haque, 1986; Prinz, 1986). A number of researchers have shown how improved fallow management technologies can help speed up soil restoration and intensify crop production (Bowen et al., 1988; Prinz, 1986; Rocheleau et al., 1988). Woody perennials can be included if fallow periods are longer than 1 yr (Hulugalle & Lal, 1986; Rocheleau et al., 1988). Properly managed, planted fallow adds substantial amounts of fixed N and organic matter to the soil, recycles nutrients from the subsoil, provides effective ground cover against erosion, suppresses weeds and pests, and improves soil physical conditions (Balasubramanian & Sekayange, 1992; Francis, 1989; Hulugalle, 1988; IITA, 1989; Blaise & Balasubramanian, 1992; Prinze, 1986). A wide variety of species can be chosen for rotation fallow without much consideration for interspecies competition (Rocheleau et al., 1988).

In areas of acute land shortage, farmers cannot afford to leave their land fallow. In such areas, fallow legumes and food crops are often intercropped, either simultaneously or in a relay pattern. Intercropping grain legumes with nonlegumes is common in Africa (Okigbo & Greenland, 1977) and other developing countries (Francis et al., 1986). This system can be extended to fallow legumes or cover crops by interplanting the fallow cover with food crops to improve the productivity of land, without losing a season of food crop cultivation. Similar to rotation fallow, the beneficial effects of relay fallow depend on the N-fixing capacity, plant N content and biomass yield of the intercropped legumes (Nhadi & Haque, 1988). Species selected for the fallow legume + food crop intercropping should minimize competition with food crops.

It should, however, be emphasized that legumes add only N to the system. Other nutrients are only recycled from within the soil. Therefore, legume-based systems cannot assure long-term sustainability, unless nutrients extracted from the soil because of increased harvests in intensive systems are replenished by external additions.

The purpose of this chapter is to bring together and critically evaluate experimental data and farmer experiences with improved fallow management systems in selected agroecosystems of Cameroon and Rwanda in Africa, and to assess (i) the biomass production, N cycling, and crop response in fallow systems, (ii) the degree and duration of ground cover by fallow species, and (iii) farmers' reactions to and farm-level adoption problems of the planted fallow technologies. Both fallow rotation and intercropping systems of legumes with food crops are discussed.

ROTATION FALLOW SYSTEMS

Biomass Production, Nitrogen Addition, and Food Crop Response

In fallow rotation systems, fallow legumes occupy the land for one season only in short-term fallow and for more than one season in long-term fallow. The length of growing period in large part determines the quality and quantity of biomass produced. For certain species such as Mucuna [*Mucuna pruriens* (L.) DC], leaving the plant on the ground for more than 18 wk resulted in the shedding of leaves and the reduction of legume N contribution to soil in the semiarid highlands of Rwanda. However, at 16 wk, Mucuna produced the largest amount of biomass (5.3 Mg ha^{-1}) and N (153 kg ha^{-1}). For species such as sunnhemp (*Crotalaria ochroleuca* G. Don.) and pigeonpea [*Cajanus cajan* (L.) Millsp.], the amount of biomass produced and N in top growth increased substantially from the 16th to 24th wk after planting (Balasubramanian & Sekayange, 1992). These authors observed that in one season (14–18 wk) fallow in Rwanda, Mucuna produced the greatest amount of biomass, and leaf N and Sesbania [*Sesbania sesban* (L.) Merr.] the least, with sunnhemp a close second to Mucuna at two sites. Similar production potential and N yield by Mucuna (Milton, 1989) and sunnhemp (Rocheleau et al., 1988) have been reported from other parts of the Tropics. Grain yield of maize (*Zea mays* L.) in the following season was higher by 89% with Mucuna and 52% with sunnhemp green manures compared to maize gain yield (827 kg ha^{-1}) after natural fallow at Dihiro. Mean increase over 2 yr of sorghum (*Sorghum vulgare* Pers.) grain production due to planted fallow ranged between 22 and 61% compared to the effect of natural fallow at Kagasa, sorghum gain yield after natural fallow was 763 kg ha^{-1}. In this study, the effect of Mucuna was the most pronounced, followed closely by pigeonpea and sunnhemp. Increases in food crop yields due to rotation of short season fallow with food crops were reported from the subhumid zones of Nigeria (Agboola, 1980; Lal, 1983; Wilson et al., 1982) and Tanzania (Rupper, 1987). Farmers in the Ruvuma region of Tanzania started using the sunnhemp fallow as an alternative or a supplement to fertilizers in the cultivation of food crops.

In observation plots at Kagasa (1985–1987), a 1-yr Tephrosia (*Tephrosia vogelii* Hook. f.) fallow produced dry weight equivalents of 4.8 Mg ha^{-1} of leaf litter, 2.6 Mg ha^{-1} of foliage, and 9.5 Mg ha^{-1} of woody stems at harvest. The leaf litter plus fresh foliage returned 238 kg N ha^{-1} yr^{-1} to soil in contrast to 38 kg N ha^{-1} yr^{-1} of natural regrowth. Maize grain yield increased by 72% over natural fallow control (yield of 755 kg ha^{-1}) in the second season after the incorporation of fallow vegetation (Balasubramanian & Sekayange, 1992). Soil N improvement potential of Tephrosia fallow has been noted in the Nyabisindu Project area located in another part of Rwanda (Prinz, 1987) and in the Western Highlands of Cameroon (Prinz & Rauch, 1987). In addition to improvement in soil N status, the Tephrosia fallow produced 9.5 Mg ha^{-1} of woody stems which can be used as firewood. Rocheleau et al. (1988) reported that a 4-yr old sole crop of Sesbania could produce firewood and high-quality fodder, as well as recycle soil nutrients.

When the fallow vegetation is cut and removed from the site for fodder or other purposes, a significant amount of plant N that would have otherwise been returned to the soil is exported from the field. Such removal reduces the extent to which fertility is restored and hence the benefit to subsequent food crops (Table 18-1) (IITA, 1989). The N contained in the roots of fal-

Table 18-1. Effect of fallow species and biomass utilization on crop yield in subsequent seasons at Kagasa, Rwanda, 1985 to 1987.†

| Fallow treatment | February 1985 | | October 1985 | | February 1986 |
	Dry biomass	N content	Maize	Soybean	Sorghum
	Mg ha^{-1}			kg ha^{-1}	
Fallow species					
	--	--	598	680	1883
Sweet potato	2.07	34.5	517	608	1555
Natural fallow	3.00	80.0	536	696	1756
C. ochroleuca G. Don	3.56	103.6	571	652	1848
(fodder sunnhemp)					
M. prurieas	1.64	43.9	372	543	1289
C. cajan	4.42	120.8	477	703	1787
Mixture (Cr + Mu + Ca)					
SE	0.23	5.2	66	61	190
P <‡	0.001	0.001	NS	NS	NS
Biomass Utilization					
Exported as fodder	2.77	--	457	562	1224
Incorporated into soil	3.10	79.8	566	733	2148
SE	0.13	--	35	36	65
P <	NS	--	0.05	0.01	0.001

† The results were similar in 1986 to 1987 (not shown here).
‡ P = probability or significance level in this and subsequent tables.

low legumes appears to be insufficient to effect a significant increase in grain yields of the test crops. Nhadi and Balasubramanian (1978) reported that the N content and the rate of N mineralization of legume roots varied considerably not only among different species but also among cultivars of the same species. Root N contribution of legumes to food production systems needs further study. The effect of above-ground biomass removal on crop yield in the succeeding season was similar to that of continuous cropping as shown by Balasubramanian & Sekayange (1992). In this trial, incorporation of the mixed fallow vegetation increased the maize yield by 22% compared to production (3145 kg ha^{-1} of maize grain) after natural fallow, which in turn gave higher yield than continuously cropped plots (mean grain yield = 1915 kg ha^{-1}).

FALLOW LEGUME PLUS FOOD CROP
INTERCROPPING SYSTEMS

Biomass and Nitrogen Yield of Intercropped Legumes and Their Effect on Food Crops

When fallow legumes were simultaneously interplanted with sweet potato (*Ipomoea batatas* Poir) in Rwanda, there was a significant reduction (*P* < 0.01) in the tuber yield of the associated crop (Table 18-2). Crop response in the following season was not significantly different between treatments.

Table 18-2. Planted fallow systems (20 wk) and their effect on grain yield of subsequent test crops at Kagasa, 1983 to 1985.

Fallow treatment	1983 October season			1984 October season	
	SP tuber yield	Leaf biomass	N content	Maize	Soybean
	— Mg ha^{-1} —			— kg ha^{-1} —	
Natural fallow	--	ND†	--	920	405
Sunnhemp (SH) sole	--	1.84	47.8	1619	768
Mucuna (MU) sole	--	6.08	176.9	1683	756
Sweet potato (SP) sole	7.5	--	--	1236	750
SP + SH‡	3.5	1.45	37.7	1747	950
SP + MU‡	2.7	3.61	105.0	1640	836
SE	0.7	--	--	267	62
$P <$	0.01	--	--	NS	0.05

† ND = not determined.
‡ simultaneously interplanted (Source is IITA, 1989).

However, it is noteworthy that plots previously planted to sweet potato gave relatively high yields of maize and soybean [*Glycine max* (L.) Merr.]. Farmers in Rwanda report that they get better yields of grain crops after sweet potato than after one season natural fallow or other crops. The observation that a sweet potato crop may confer a beneficial effect on subsequent crops requires further examination.

In the sweet potato + fallow legume relay intercropping trials conducted in Rwanda during 1985 to 1987 (Balasubramanian & Sekayange, 1992), dry matter and N yields of sunnhemp and Mucuna decreased with delay in planting time, and their effect on subsequent maize crop was positive but nonsignificant as compared to maize yield after sweet potato. The sole crop Mucuna shed its leaves and produced grain at harvesting 24 wk after planting and therefore its biomass production and effect on subsequent maize yield also were low. When the planting time of legumes was delayed to 5 or 6 wk after sweet potato planting, sunnhemp did not survive and Mucuna produced very low amounts (26–166 kg ha^{-1}) of dry matter, due to the severe competition from the fast growing sweet potato (Balasubramanian & Sekayange, 1992).

In sorghum + fallow legumes intercropping trials, it was observed that the reduction in sorghum grain yield was much higher when Mucuna was planted at the same time as sorghum than when they were planted 2 to 4 wk after sorghum (Fig. 18-1). Relay planted sunnhemp behavior was similar to Mucuna. In these trials, dry matter and N contribution of fallow species decreased with delay in planting time and thus planted fallow effect on subsequent crops became less marked (Balasubramanian & Sekayange, 1992). When planted 2 to 3 wk after sowing of sorghum, Mucuna produced more biomass than sunnhemp and pigeonpea (Table 18-3), but Mucuna competed severely with associated sorghum and significantly reduced its grain yield. These examples show that because of the limited growing period available to the relay-planted fallow species in semiarid zones of Rwanda, biomass

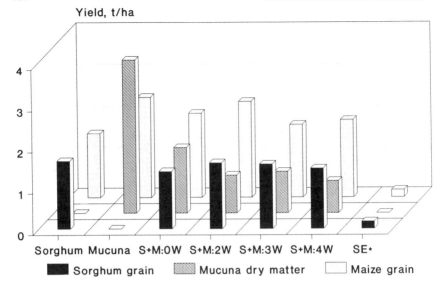

Fig. 18-1. Effect of simultaneously and relay intercropped Mucuna on the grain yield of associated sorghum, fallow biomass production, and the grain yield response of maize in the following season at Kagasa, Rwanda 1986 to 1987.

production in relay systems was generally low. As a result, it will take longer to improve soil fertility using this method. There also is a risk of competition for moisture between intercropped species during dry spells.

However, in the humid lowland forest zones with bimodal rainfall distribution, it is possible to intercrop a slow-growing legume (e.g., Sesbania)

Table 18-3. Dry matter production by relay (2 wk after sorghum) planted fallow species and effect on grain yields of associated sorghum and subsequent maize at Kagasa, Rwanda, 1987 to 1988 (Source is IITA, 1989).

Treatments	1987 February season			October 1987
	Sorghum grain	Dry biomass†	N content	Maize grain‡
	kg ha^{-1}	Mg ha^{-1}	—— kg ha^{-1} ——	
Sorghum (S)	3989	--	--	1256
Sunnhemp (SH)	--	4.49	116.7	639
Mucuna (MU)	--	10.06	292.9	1181
Pigeonpea (PP)	--	1.40§	37.6	875
S + SH	3295	0.22	5.6	1201
S + MU	2746	1.53	44.4	1122
S + PP	3501	0.17	8.5	1336
SE	251	1.10	--	182
$P <$	0.05	0.000	--	NS

†SH and MU were cut down 16 wk after planting and PP 27 wk after planting.
‡ Maize test crop planted in 1987 October suffered a drought in December 1987 to February 1988 and hence there was no response to fallow treatments.
§ Pigeonpea yielded 2.65 and 0.21 Mg ha^{-1} of dry stems and 375 and 35 kg ha^{-1} of dry grain in addition to leaf biomass for sole and relay crop, respectively.

with food crop (e.g., maize) in the first season and allow the full growth of the fallow species in the second season to be incorporated as green manure in the first season of the following year. In Sesbania + food crops intercropping trials conducted in the forest zone of Cameroon (1989–1990), maize yield in the first season was not affected, but the groundnut (*Arachis hypogaea* L.) planted in association with the well-established Sesbania in the second season suffered significant yield losses due to interspecific competition (Blaise & Balasubramanian, 1992). Thus, it is better to allow the interplanted fallow species to grow freely in the second season without any association of food crops. In the following year, grain yields of test crop maize were significantly higher in the previously intercropped Sesbania plots (Table 18–4). The Chromolaena broadleaf weed [*Chromolaena odorata* (L.) R.M. King and Robinson] contributed almost similar quantities of N to the system and produced similar test crop response as the Sesbania intercrop treat-

Table 18–4. Effect of type of fallow and residue management on grain yield of subsequent maize at two sites in southern Cameroon, 1990 (source is Blaise and Balasubramanian, 1992).

1989 treatments		1990 March			1990 March		
March	September	Residue management	Biomass	Plant N	Grain		
			Mg ha^{-1}		kg ha^{-1}		
			Minkomeyos				
MZ†	NF	B‡	4.93	56.8	4931	6084	5508
MZ	NF	I	4.55	59.9	4864	5632	5248
MZ	NF	M	4.90	65.4	4297	5458	4878
MZ	SS	I	4.07	43.3	3645	5467	4556
MZ	SS	M	3.53	40.9	3041	5770	4406
MZ + SS	GN + SS	I	1.56	56.7	4766	5449	5107
MZ + SS	GN + SS	M	1.88	68.8	3803	4872	4337
MZ	GN	I	0.22	2.9	4006	4725	4366
Mean			--	--	4169	5432	--
SE			0.42	10.7	154		460
P <			0.000	0.01	0.000		NS
Interaction: NS							
			Yoke				
MZ	NF	B	5.19	39.1	2232	3982	3107
MZ	NF	I	4.20	32.1	1580	3303	2441
MZ	NF	M	5.38	40.9	1092	2757	1924
MZ	SS	I	2.98	22.4	2417	3879	3148
MZ	SS	M	2.84	27.5	2180	3957	3069
MZ + SS	GN + SS	I	1.63	53.8	5312	5657	5484
MZ + SS	GN + SS	M	1.85	61.4	4103	4547	4325
MZ	GN	I	2.13	29.8	3414	4208	3811
Mean			--	--	2791	4036	--
SE			0.64	6.4	129		336
P <			0.001	0.01	0.000		0.000
Interaction: NS							

† MZ = maize, GN = groundnut (peanut), NF = natural fallow, SS = Sesbania.
‡ B = Burn, I = incorporate, M = mulch on soil surface.

Table 18-5. Effect of intercroped Sesbania/natural fallow and residue management on dry weight (kg ha^{-1}) of weeds in the following season at three sites in southern Cameroon, 1989 to 1990 (Blaise & Balasubramanian, 1992).

1989 treatments		1990 March	1990 March			
			Dry weight of weeds			
		Residue management	At planting		1st weeding at 4 wk	2nd weeding at 9 wk
March	September		Chromo§/ Imper¶	Other		
			— kg ha^{-1} —			
Minkomeyos (mostly *Chromolaena odorata*)						
MZ†	NF	B‡	3020	1910	67	1745
MZ	NF	I	3730	820	379	1556
MZ	NF	M	4060	920	647	1248
MZ	SS#	I	3410	660	355	971
MZ	SS#	M	2190	1350	757	669
MZ + SS	GN + SS	I	0	0	45	365
MZ + SS	GN + SS	M	0	0	243	574
MZ	GN	I	0	0	159	614
SE		--			120	203
P <		--			0.01	0.000
Yoke (mostly *Imperata Cylindrica*)						
MZ	NF	B	4260	940	601	--
MZ	NF	I	3670	540	621	--
MZ	NF	M	4700	680	576	--
MZ	SS#	I	2640	350	695	--
MZ	SS#	M	3110	480	502	--
MZ + SS	GN + SS	I	0	0	131	--
MZ + SS	GN + SS	M	0	0	199	--
MZ	GN	I	0	0	267	--
SE		--			100	--
P <		--			0.01	--

† MZ = maize, GN = groundnut (peanut), NF = natural fallow, SS = Sesbania.
‡ B = burn, I = incorporate, M = mulch.
§ Chromo = *Chromolaena odorata*.
# Imper = *Imperata cylindrica*.
# Sesbania planted after maize harvest without any land preparation was smothered by weeds and hence no effect on weeds in the following season.

ment at one site (Minkameyos), whereas Imperata grass weed [*Imperata cylindrica* (Anderss.) C.E. Hubbard] was poorer in N yield and test crop response than Sesbania at another site (Yoke) (Table 18-5). Other species under study for this system are pigeonpea (*C. cajan* L.), sunnhemp (*C. juncea* L.), Desmodium [*Desmosium distortum* (Aubl) Macbride], Mimosa [*Mimosa invisa* Mart. (v. inermis)], and Tephrosia.

In the humid and subhumid western highlands of Cameroon with monomodal rainfall distribution (March–November), simultaneous intercropping of Tephrosia with maize reduced the food crop yield due to fast growth of and serious competition from the fallow species (J. Kikafunda, 1990, personal communication). In this case, some kind of relay planting and appropriate proportional density and arrangement of fallow species to food crops have to be worked out to minimize the interspecies competition. Farmers in western Cameroon protect the voluntary crop of *Sesbania* spp. (about 500–1000 plants ha^{-1}) scattered in maize fields in the Bui highlands and a similar population of Tephrosia in the Bamenda plain. During land preparation in December, they cut down the fallow plants and bury them along with weeds and crop residues underneath the ridges to maintain soil fertility. The fallow plants would have scattered some seeds before they are cut down and buried, and this sustains the regeneration of fallow plants in the succeeding year. This system is being studied intensively with a view to make further possible improvements in biomass production and N yield by naturally growing legumes in this region.

DEVELOPMENT OF GROUND COVER

Rapid development of ground cover is necessary to protect the soil from erosion (Derpsch et al., 1986). In the semiarid highlands of Rwanda, sunnhemp was rated as the best species for early ground cover, followed by Desmodium, Mucuna, pigeonpea, Sesbania and Tephrosia, percentage cover at 50 d after planting ranged between 14 and 100 (Fig. 18-2). Maintenance of vegetation cover during the dry season and at the beginning of rains in October is essential to reduce soil erosion in the semiarid eastern Rwanda (IITA, 1989). Pigeonpea was found to be the most suitable species for this purpose because it is more drought resistant than others; at Kagasa, sunnhemp dried

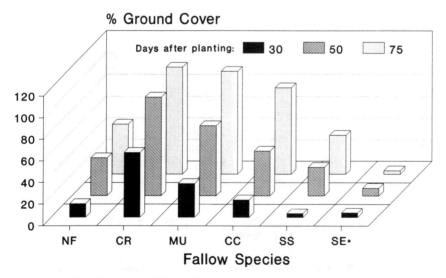

Key: NF=Natural fallow; CR=Crotalaria ochroleuca; MU=Mucuna pruriens; CC=Cajanus cajan; SS=Sesbania sesban; SE=Error

Fig. 18-2. Percentage ground cover by planted fallow at different times after establishment at Kagasa: Rwanda, 1986 to 1987.

up after 6 wk without rains, while Mucuna and Desmodium gradually shed their leaves during the dry season.

WEED SUPPRESSION

In the fallow management trials conducted in the lowland forest zone of Cameroon, it was observed that the weeds were effectively suppressed in plots planted to Sesbania + maize/groundnut or maize/groundnut in the previous year. One-year growth of Sesbania could effectively shade out problem weeds such as *C. odorata* and *I. cylindrica*, at least for one cropping season (Nguimgo & Balasubramanian, 1992). Weed density in plots previously planted to maize/groundnut was only slightly higher than that of Sesbania + maize/groundnut plots, but far less than that of natural fallow (Table 18–5). Sesbania planted after maize harvest in the first season was completely smothered by natural regrowth and therefore had negligible effect on weed infestation in the following season. Similar results on weed suppression were obtained elsewhere with Mucuna and Pueraria [*Pueraria phaseoloides* (Roxb.) Barth.] species (Anonymous, 1989b; IITA, 1985; NAS, 1979). Weed suppressing effect of other species such as *M. invisa* cv. Inermis, Desmodium, and pigeonpea may be equally good and worth studying.

LEGUME FALLOWS AND PRODUCTION SUSTAINABILITY

It should be noted that intensification of cropping systems, with or without improved fallow, will export greater amounts of nutrients (P, K, S, Ca, Mg, and trace elements) from soil, assuming enough N is fixed into the system by legumes. Under these conditions, it is imperative to add nutrients from external sources to maintain soil fertility and to sustain crop yields over long term. It is, therefore, necessary that legume fallows should be combined with other methods of soil fertility management to improve long-term sustainability. A cost-effective, integrated nutrient management package will involve (i) recycling of all crop residues, (ii) appropriate use of grain, green manure and fodder legumes, (iii) addition of manure, compost and other wastes, (iv) utilization of indigenous nutrient sources such as rock phosphate, liming materials, etc., and finally (v) skillful employment of costly fertilizers to supplement the above sources.

ON-FARM FALLOW MANAGEMENT TRIALS
AND FARMER REACTIONS

Biomass Production and Crop Yield Response

Mean biomass production in the planted fallow plots was significantly higher than in the natural fallow in two out of three regions of Rwanda (Table 18–6). The yield of sorghum in the following season was better in the plant-

Table 18-6. Biomass production of fallow treatments and their effect on subsequent sorghum yield on farmers' fields, Rwanda 1986 to 1987 (source is IITA, 1989).

	Fresh biomass, October			Sorghum yield, February		
Farmers/area	Trt. 1	Trt. 2	Trt. 3	Trt. 1	Trt. 2	Trt. 3
	——— Mg ha^{-1} ———			——— kg ha^{-1} ———		
	Bugesera Region, 1986–1987					
1. Mangano	--	3.6	5.2	2436	504	3248
2. Safari	--	2.8	6.8	1627	980	1764
3. Butoyi	--	6.4	3.2	745	784	352
4. Mupenzi	--	14.0	14.0	3304	3024	4704
5. Rwabukambiza	--	4.0	10.0	1251	598	1741
6. Shikama	--	10.4	15.0	995	820	1050
7. Myirabutu	--	8.0	10.0	625	675	760
8. Mutabazi	--	10.0	25.0	1560	1600	2200
9. Munyampeta	--	30.0	43.0	2600	2550	2650
Mean		9.8	14.6	1683	1282	2052
SE		1.4			174	
$P <$		0.05			NS	
	Rusumo commune, 1987 to 1988, 9 farmers					
Mean	--	14.7	18.0	1522	2211	2321
SE		1.4			316	
$P <$		0.05			NS	
	Nasho Basin area, 1987 to 1978, 11 farmers					
Mean	--	6.5	8.7	116	215	231
SE		0.8			44	
$P <$		NS			NS	

† Trt. 1 = continuous cultivation of bean to sorghum, Trt. 2 = natural fallow to sorghum, Trt. 3 = planted fallow to sorghum.
‡ Low yield due to drought in the Nasho basin area.

ed fallow plots, but the differences between treatments were not significant. This implies that it may take longer than one season to significantly improve soil fertility through planted fallow on farmers' fields where soils are generally infertile and/or degraded.

Farmers' Reactions to Planted Fallow Technology

During the regular farm visits by researchers, and farmers' visits to on-station research sites on Farmers' Day, Rwandan farmers were encouraged to react to the planted fallow technology. According to them, rotation fallow was less attractive due to: (i) the need for additional labor to plant and manage fallow species which do not yield edible/economic products, (ii) loss of additional food crop seasons in rotation fallow plots, and (iii) very slow build-up of soil fertility and crop yields in planted fallow systems. They also felt that the relay planting of fallow species in the broadcast-sown mixed cropping systems would be difficult and the food crops and fallow plants could compete for moisture during dry weather. Milton (1989) observed that green manure technology has not been commonly accepted by small farmers in tropical countries, probably because they are not willing to sacrifice cropland to grow a soil amendment. Rocheleau et al (1988) point out that if labor used in fallow species establishment and maintenance is to be compensated for by direct economic benefits, planted fallows should yield diverse products, such as leafy vegetables, animal fodder, bee forage, fuelwood, building poles, etc., in addition to soil fertility improvement. Some soil-enriching legumes are excellent food sources, e.g., pigeonpea.

Implications for the Cost-Effective Establishment of Planted Fallow

It is challenging to design appropriate planted fallow methods that will be easily incorporated into the local cropping systems without using much labor and/or inputs and that are attractive to farmers.

Fallow species planted at the end of the cropping cycle or the beginning of the fallow period are usually smothered by aggressive weeds and natural regrowth, particularly in humid forest zones (as discussed earlier). Some kind of land preparation before planting fallow species and at least one early weeding are required to ensure good establishment of fallow legumes in the humid Tropics. Sanchez (1982) noted that legume–cereal intercropping is extremely site specific and hence the different management aspects of the fallow–food crop systems should be validated locally in each agroclimatic zone. The following methods are useful.

1. Use of self-generating legumes (Mucuna, Mimosa, Sesbania) to eliminate the need to replant fallow legumes every year—in these systems, some fallow plants will be allowed to produce seeds before cutting, and selective weeding can control the population density of fallow species in relation to food crops, minimize the interspecies competition, and assure no tangible reduction in food crop yields.

2. Interseeding fallow plants at the time of first weeding of a food crop which does not involve much additional labor and allowing the full growth of fallow species in the second season in humid zones with bimodal rainfall distribution.

3. Overseeding fallow species just before or immediately after the harvest of one of the crops in a mixed cropping system. For example, fallow species can be planted with minimum labor at the time of groundnut harvest in the groundnut + cassava (*Manihot esculenta* Crantz) + maize + cocoyam (*Xanthosa sagittifolium* Schotti) system common in the lowland forest zone and at the time of potato (*Solanum tuberosum* L.) harvest in the maize + bean + potato system in the humid highlands.

CONCLUSIONS

The rate of improvement in soil fertility and crop yields due to planted fallow depends on the amount of biomass produced and recycled N in natural

Table 18-7. Suggested fallow legume species for different agroclimatic zones of Africa.

Legume species	Ecoregions		Remarks
Aeschynomene indica L. *A. uniflora* *A. sensitiva*	SHL,† HL, SHH, HH		Slow growth, good for rice (*Oryza sativa* L. or *O. glaberrima* Steud.) paddies
Astragallus sinicus	SHL, HL, SHH, HH		Well drained soil; poor competition with weeds; cold tolerance; food, GM‡, fodder, bee forage
Cajanus cajan (L.) Millsp.	SAL, SHL, HL, SAH, SHH		Short & long season types; GM, fodder & food; drought res.
Centrosema pubescens Benth.	SAL, SHH		Perennial; GM, fodder.
	SAH, SHH		Good ground cover
Crotalaria juncea L. *C. ochroleuca* G. Don. *C. anagroides*	SAL, SHL, HL, SAH, SHH		Early ground cover; GM, fodder; control nematodes
Desmodium distortum (Aubl) Macbride	SAL, SHL, SAH, SHH		Good ground cover; GM, fodder
D. tortuosum (Sw.) DC. *D. gyroides* (Roxb. ex Link) DC.	SHL, HL SHH, HH		GM, fodder
Indigofera spicata Forsk. *I. hirsuta* L.	SHL, HL, SHH		Good ground cover
Lablab purpureus (L.) Sweet	SAL, SHL, SAH, SHH		Adapted to poor soils

† SAL = semiarid lowland, SHL = subhumid lowland, HL = humid lowland, SAH = semiarid highland, SHH = subhumid highland, HH = humid highland.
‡ GM = green manure.

fallow or N fixed with fallow legumes. Among the species tested, sunnhemp, Mucuna, Desmodium, and early maturing pigeonpea are suitable for the short season (14–24 wk) fallow, while Sesbania, Tephrosia and late-maturing pigeonpea are good for the long duration (1 yr or more) fallow. Suggested fallow legume species for different ecoregions are listed in Table 18-7.

Applied research on fallow management should explore in detail the soil processes that govern fertility regeneration in cropping systems that rely on planted or natural fallow. This knowledge is required to help exploit the potential benefits of planted fallow. Equally important attention should be paid to the poor adoption rate of the fallow management technologies at the farm level. Improved fallow systems, in combination with other methods of soil fertility management, will help productive and more sustainable cropping systems.

REFERENCES

Agboola, A.A. 1980. Effect of different cropping systems on crop yield and soil fertility management in the humid tropics. p. 87–105. *In* Organic Recycling in Africa. FAO Soils Bull. Vol. 43. FAO, Rome, Italy.
Allan, W. 1965. The African husbandman. Oliver and Boyd, London.

Anonymous. 1989a. Promoting environmentally sustainable development. World Farmers' Time 7:10–12.

Anonymous. 1989b. Experiences of world neighbors in Central America. ILEIA Newslett. 5(2):9–10.

Balasubramanian, V., and A. Egli. 1986. The role of agroforestry in the farming systems in Rwanda with particular reference to the BGM region. Agrofor. Systems 4:271–290.

Balasubramanian, V., and L. Sekayange. 1986. Biological soil fertility management in African highlands with examples from Rwanda. p. 675–676. in Trans. Meet. Congr. 13, Vol. 3. Int. Soil Sci. Soc. 13–20 August. ISSS, Hamburg, Germany.

Balasubramanian, V., and L. Sekayange. 1992. Five years of research on improved fallow in the semi-arid highlands of Rwanda. p. 405–422. In K. Mulongoy et al. (ed.) Biological Nitrogen Fixation Conf., Ibadan, Nigerria. 24–28 Sept. 1990. John Wiley & Sons, London.

Beets, W.C. 1978. The agricultural environment of eastern and southern Africa and its use. Agric. Environ. 4:5–24.

Blaise, N.K.A., and V. Balasubramanian. 1992. Effect of fallow and residue management on biomass weed suppression and soil productivity. p. 463–473. In K. Mulongoy et al. (ed.) Biological nitrogen fixation and sustainability of tropical agriculture, Ibadan, Nigeria. 24–28 Sept. 1990. John Wiley & Sons, London.

Bowen, W.T., J.O. Quintana, J. Pereira, D.R. Bouldin, W.S. Reid, and D.J. Rathwell. 1988. Screening legume manure as nitrogen sources to succeeding non-legume crop: 1. The fallow method. Plant Soil 111:75–80.

Derpsch, R., N. Sidiras, and C.H. Roth. 1986. Results of studies made from 1977 to 1984 to control erosion by cover crops and no-tillage techniques in Parana, Brazil. Soil Tillage Res. 8:253–263.

Food and Agricultural Organization. 1984. Improved product systems as an alternative to shifting cultivation. FAO Soils Bull. Vol. 52. FAO, Rome, Italy.

Francis, C.A. 1989. Biological efficiencies in multiple cropping systems. Adv. Agron. 42:1–42.

Francis, C.A., R.R. Harwood, and J.F. Parr. 1986. The potential of regenerative agriculture in the developing world. Am. J. Alter. Agric. 1:65–73.

Greenland, D.J. 1985. Nitrogen and food production in the tropics: Contributions from fertilizer nitrogen and biological nitrogen fixation. p. 9–38. In B.T. Kang and J. van der Heide (ed.) Nitrogen management in farming systems in humid and subhumid tropics. IITA, Ibadan, Nigeria.

Heathcote, R.G., and K.R. Stockinger. 1970. Soil fertility under continuous cultivation in northern Nigeria. II. Response to fertilizers in the absence of organic manure. Exp. Agric. 6:345–350.

Hulugalle, N.R. 1988. Effect of cover crop on soil physical and chemical properties of an Alfisol in the Sudan Savannah of Burkina Faso. Arid Soil Res. Rehab. 2:251–267.

Hulúgalle, N.R., and R. Lal. 1986. Root growth of maize in a compacted gravelly tropical Alfisol as affected by rotation with a woody perennial. Field Crops Res. 13:33–44.

International Institute of Tropical Agriculture. 1985. Control of Imperata cylindrica. p. 174. In IITA annual report for 1984. IITA, Ibadan, Nigeria.

International Institute of Tropical Agriculture. 1989. Sustainable crop and livestock production for Rwanda. Final report of the first phase (1983–88) of the farming systems research project of the Institut des Sciences Agronomiques du Rwanda (ISAR), Rubona, Rwanda. IITA, Ibadan, Nigeria.

Kang, B.T., L. Reynalds, and A.N. Atta-Krah. 1990. Alley farming. Adv. Agron. 43:315–352.

Lal, R. 1983. No-till farming—Soil and water conservation and management in the sub-humid tropics. IITA, Ibadan, Nigeria.

Milton, F. 1989. Velvet beans: An alternative to improve small farmers' agriculture. ILEIA Newslett. 5:8–9.

Mundy, H.G. 1953. Some problems of arable farming in southern Rhodesia. Emp. J. Agric. 21:123–130.

Mutsaers, H.J.W., P. Moouemboue, and M. Bomoyo. 1981. Traditional food crop growing in Yaounde area (Cameroon): 1. Synopsis of the system. Agroecosystems 6:273–287.

National Academy of Sciences. 1979. Tropical legumes: Resources for the future. Natl. Acad. Press, Washington, DC.

Nhadi, L.A., and V. Balasubramanian. 1978. Root nitrogen content and transformations in selected grain legumes. Trop. Agric. (Trinidad) 55:23–32.

Nhadi, L.A., and I. Haque. 1988. Forage legumes in African crop-livestock production systems. p. 10–19. Int. Livestock Center for Africa. ILCA Bull. 30. ILCA, Adis Ababa, Ethiopia.

Nye, P.H., and D.J. Greenland. 1960. The soil under shifting cultivation. Tech. Comm. 51. Common Bureau of Soils, Harpenden, England.

Okigbo, B.N., and D.J. Greenland. 1977. Intercropping systems in tropical Africa. p. 63–101. *In* R.I. Papendick et al. (ed.) Multiple cropping. ASA Spec. Publ. 27. ASA, CSSA, and SSSA, Madison, WI.

Prinz, D. 1986. Increasing the productivity of smallholder farming systems by introduction of planted fallows. Plant Res. Dev. 24:31–56.

Prinz, D. 1987. Improved fallow. ILEIA Newslett. 3(1):4–7.

Prinz, D., and F. Rauch. 1987. The Bamenda model: Development of a sustainable land-use system in the highlands of West Cameroon. Agrofor. Syst. 5:463–474.

Rocheleau, D., F. Webber, and A. Field-Juma. 1988. Improved fallows. p. 114–122. *In* D. Rocheleau et al. (ed.) Agroforestry in dryland Africa: Science and practice of agroforestry 3. ICRAF, Nairobi, Kenya.

Rupper, G. 1987. Sunnhemp. ILEIA Newslett. 3:11–12.

Ruthenberg, H. 1971. Farming systems in the tropics. Oxford Univ. Press, London.

Sanchez, P.A. 1982. A legume-based pasture production strategy of acid infertile soils of tropical America. p. 97–120. *In* W. Kussow et al. (ed.) Soil erosion and conservation in the tropics. ASA Spec. Publ. 43. ASA and SSSA, Madison, WI.

Singh, L., and V. Balasubramanian. 1979. Effects of continuous fertilizers use on a ferruginous soil (Haplustalf) in Nigeria. Exp. Agric. 15:257–265.

Wilson, G.F., R. Lal, and B.N. Okigbo. 1982. Effects of cover crops on soil structure and on yield of subsequent arable crops grown under strip tillage on an eroded Alfisol. Soil Till. Res. 2:223–250.

Young, A., and A.C.S. Wright. 1980. Rest period requirements of tropical and subtropical soils under annual crops. p. 197–268. *In* Land resources for populations of future. FAO, Rome, Italy.

19 Technological Options Towards Sustainable Agriculture for Different Ecological Regions of sub-Saharan Africa

Rattan Lal

Department of Agronomy
The Ohio State University
Columbus, Ohio

The enthusiastic response to "sustainable agriculture" by scientific community and policy makers is due to severe problems of soil and environmental degradation, pollution of water and environment, and overdependence on nonrenewable sources of energy. However, sustainability is often perceived as a moral or an ethical issue which has taken on an emotional air. Consequently, the topic of sustainability has become a political issue rather than a rational approach; an emotional rhetoric rather than a practical science, a religious myth rather than a generalizable concept, and an interesting theme to discuss and debate rather than a measurable system to evaluate and quantify.

It has been the high-energy influx of agricultural ecosystems that broke the yield barriers and resulted in an unprecedented boom in agricultural output in the post-World War II era in North America and Western Europe. These technologies also have been successfully applied to the prime agricultural land in South Asia. The concern is whether this technology can be applied to impoverished soils of the humid, subhumid and semiarid Tropics. In addition to impoverished soils, these regions also are characterized by nonavailability of inputs at affordable prices.

In view of perpetual food deficit and severe problems of soil and environmental degradation in sub-Saharan Africa, sustainable agriculture is not necessarily synonymous with low-input, organic or regenerative agriculture in this region. Scientifically speaking, ecosystems utilized by human societies are only sustainable in the long-term if the outputs of the components produced balance the input into the system. Because demand for output from agricultural ecosystems is greater now than ever before, and it is rapidly increasing due to high demographic pressure, no-input or even low-input agriculture is a nonsolution. The issue to be addressed, however, is how to balance the inputs required so as to maximize efficiency and cost effectiveness of inputs, reduce risks of soil and environmental degradation, maximize the per capita productivity, and maintain/sustain an increasing trend in productivity.

SUB-SAHARAN AFRICA

The region of sub-Saharan Africa (SSA) consists of 39 countries extending from approximately 25 ° N to 25 ° S, and from 17 ° W to 51 ° E. The population of the region has been estimated at about 382 million in 1982, 433 million in 1986, and 490 million in 1990. At an annual rate of increase of 3.2% per year, the population is expected to approximately triple from 433 million in 1986 to 1263 million by the year 2020. The population may eventually stabilize at 10 times its present number (Table 19-1).

The region is characterized by a huge diversity of climate, soils, geology, hydrology, topography, ethnic groups and cultural heritage. Considering Africa as a whole, the Mediterranean region occupies 2.6% of the total area, the northern desert 33%, Horn of Africa 5.2%, semiarid regions 7.2%, tropical wet/dry regions 17.5%, equatorial wet climate 6.8%, eastern Africa 6.4%, Ethiopian Highlands 2.3%, south savanna plateau 5.7%, Mozambique 1.9%, Malagasy 2.2%, and southern Africa 9.2% (Griffith, 1972; FAO, 1978). Using Thornthwaite's classification, about 37% of Africa is arid, 13% is semiarid, 23% is subhumid, and another 13% is humid. Arid and semiarid regions are characterized by low, erratic and highly variable rainfall. Depending on these ecological regions, the climax vegetation varies widely depending on the amount and distribution of rainfall. Tropical rainforest prevails where rainfall exceeds 1500 mm in 240 to 300 d. Tropical xerophytic woodland occurs in regions with rainfall between 200 to 500 mm received in 90 to 150 d. Tree-savanna vegetation occurs in regions with rainfall of 500 to 1500 mm with grass-savanna predominating in regions with rainfall of 500 to 800 mm.

Equally diverse are soils of Africa. The predominant soils of the humid, subhumid and semiarid regions are Alfisols, Ultisols, Oxisols and Entisols-Inceptisols. The highly weathered Oxisols, Ultisols and Alfisols occur in the humid regions, and have low nutrient and water reserves available to plants.

Table 19-1. Projected population of sub-Saharan Africa.†

Region	1986	1990	2000	2025	2050
			millions		
Western Africa	189.7	215.2	294.9	648.1	1357.0
Eastern Africa	133.5	151.4	207.5	456.0	954.8
Central Africa	48.1	54.6	74.8	164.4	344.2
Southern Africa	61.3	69.5	95.2	209.2	438.0
Sub-Saharan Africa	432.6	490.7	672.4	1477.7	3094.2

† Rate of increase: (i) 1986 to 2000, 3.2% yr; (ii) 2000 to 2025, 3.2% yr; and (iii) 2025 to 2050, 3.0% yr.

Alfisols are relatively fertile compared with Oxisols and Ultisols, but are often characterized by poorer soil physical properties. Alfisols have weak structure and are prone to crusting, compaction and accelerated erosion. Entisols and Inceptisols, being of recent origin, are found mostly along flood plains, but also on steeper slopes and eroded areas. These soils have high fertility and favorable moisture regime. Vertisols, the black cotton soils, are heavy textured. They have high swell–shrink capacity, poor physical properties, and are difficult to manage. Vast areas of these soils occur in semiarid regions including the Nile Valley in Sudan, the Lake Chad basin, the Accra plains, and some isolated regions in eastern Africa. Soils of volcanic origin are found in the vicinity of Mount Cameroon and in the highlands of eastern Africa. Some arid regions of West Africa are characterized by the widespread occurrence of hardened laterite or plinthite at shallow depth, i.e., 5 to 30 cm. Although a large proportion of the soils of SSA are highly weathered and of low inherent fertility, these soils respond to ameliorative/restorative input of fertilizer nutrients, water input, and to new cultivars and farming/cropping systems.

ARABLE LAND AREA, WATER RESOURCES, AND AGRONOMIC POTENTIAL OF SUB-SAHARAN AFRICA

Total arable land area of SSA is estimated at 131 million hectares (Table 19-2). The average per capita land area of 0.27 ha is only slightly lower than the world average of 0.33 ha. However, for the expected population of 1478 million by the year 2025, the per capita land area may be as low as 0.09 ha with no additional land brought under cultivation (Table 19-2), and 0.18 ha if new land is cleared at the rate of 0.6% per year of the existing rainforest. The data in Table 19-3 also are based on the assumption that there is no loss of arable and due to soil degradation.

The regions with greatest yield potential for grain crops lie within the rainfall regime of 800 to 1500 mm (Table 19-4). Tropical root crops and

Table 19-2. Arable land resources of tropical Africa assuming no further deforestation (calculated from FAO, 1986).

Region	Total arable land in 1990	Per capita arable land†		
		1990	2000	2025
	10^6 ha	ha		
Western Africa	62.0	0.29	0.21	0.10
Eastern Africa	37.7	0.25	0.18	0.08
Central Africa	10.8	0.20	0.14	0.07
Southern Africa	20.5	0.29	0.22	0.10
Sub-Saharan Africa	131.0	0.27	0.19	0.09

† Assuming no additional land is brought under cultivation, and that population continues to increase at 3.2% yr^{-1}.

Table 19-3. Arable land resources of tropical Africa assuming deforestation at 0.6% yr^{-1}.†

Region	1990 Total	1990 Per capita	2000 Total	2000 Per capita	2025 Total	2025 Per capita
	10^6 ha	ha	10^6 ha	ha	10^6 ha	ha
Western Africa	62.0	0.29	72.5	0.25	88.6	0.14
Eastern Africa	37.7	0.25	48.7	0.23	65.5	0.14
Central Africa	10.8	0.20	31.4	0.42	64.1	0.39
Southern Africa	20.5	0.29	32.0	0.34	49.6	0.24
Sub-Saharan Africa	131.0	0.27	184.6	0.27	269.8	0.18

† Assuming population continues to increase at 3.2% yr^{-1}.

Table 19-4. Grain yield potential of various rainfall zones (FAO, 1978; Wright, 1986).†

Rainfall	Yield potential	Crop failure rate
	Mg ha^{-1}	
300–500	1.3	1 in 3 or more
500–800	3.0	1 in 5
800–1000	5.0	1 in 7
1000–1500	8.0 (12.0)‡	1 in 10
1500–2500	3.0 (20.0)‡	1 in 12

† Assuming medium level of fertilizer input and pest control measures.
‡ Root crops.

perennial plantations are most suited for the humid regions with rainfall exceeding 1500 mm per year. Potential yield of most crops can be increased two to four times by judicious use of off-farm input, e.g., chemical fertilizers, appropriate farm tools, improved varieties, etc. (FAO, 1978) (Table 19-5). With traditional systems of resource-based agriculture, agronomic yields of most crops are low. An important reason for low yields is the widespread system of no-input, resource-based, subsistence farming. For example, the average fertilizer use in SSA, although more than doubled over the decade ending in 1987, is merely 8 kg ha^{-1} of major nutrients (Table 19-6). There is a potential for irrigation to mitigate the drought. However, currently only 5 million hectares of land is being irrigated (Tables 19-7, 19-8, 19-9, 19-10). Furthermore, use of improved cultivars and of high production systems is currently limited to merely 5 to 6% of the arable land.

Table 19-5. Yield potential of crops.†

Land capability	Input	Millet	Sorghum‡	Maize	Soybean	Bean	Sweet potato	Cassava
					Mg ha^{-1}			
Very suitable	Low	0.9	1.1	1.6	0.7	0.7	2.2	2.0
	High	3.5	4.6	6.4	3.0	3.0	9.1	12.2
Suitable	Low	0.6	0.8	1.0	0.4	0.4	1.5	1.0
	High	1.8	3.0	4.2	2.0	2.0	6.0	8.1
Marginal	Low	0.3	0.4	0.5	0.2	0.2	0.7	0.3
	High	1.2	1.5	2.1	1.0	1.0	3.0	4.0

† From FAO (1978).
‡ *Pennisetum americanum.*

Table 19-6. Total fertilizer use in West Africa (arable land area = 54 298 000 ha).†

Year	Total consumption	kg ha^{-1}
	1000 Mg	
1976–1977	202	3.7
1979–1980	254	4.7
1983–1984	417	7.7
1986–1987	447	8.2

† From IFDC (1987).

Table 19-7. Water resources of Africa (km^3).†

Surface water	
Rivers	195 km^3
Lakes	30 262 km^3
Reservoirs	1 240 km^3
Groundwater	
Total continent	2 500 000
Saharan Basins	60 000 km^3
Total runoff	4 225 km^3 yr^{-1}

† From Wright (1986).

Table 19-8. Irrigated crops in Africa.†

Crops	Area	%
	1000 ha	
Food		
Rice	360	16
Maize, sorghum, millet, wheat	700	31
Groundnut (*Arachis hypogaea* Linn.), vegetables, fruit, oils	700	31
Cash		
Cotton (*Gossypium hirsutum* Linn.)	350	16
Sugarcane (*Saccharum officinarum* Linn.)	120	5
	2 230	100
Fallow	309	
Total	2 539	

† From Wright (1986).

Table 19-9. Irrigaton in sub-Saharan Africa.†

Country	Area
	1 000 ha
Senegal	198
Gambia	27
Mali	95
Cape Verde Island	2
Mauritania	10
Burkina Faso	5
Niger	34
Chad	2
Sudan	1 700
Ethiopia	96
Somalia	78
Kenya	33
Tanzania	157
Mozambique	68
Angola	28
Botswana	6
Total	2 539

† Source: From Stern (1986).

Table 19-10. Physiographic characteristics of the humid region and sustainable management options.

Characteristics	
Rainfall	>1500 mm yr^{-1}
Growing period	>270 d
Crop failure rate	1 in 15
Potential yield	3 Mg ha^{-1}, 15 to 20 Mg ha^{-1} (root crops)
Preferred crops	Perennials
Soils	Oxisols, Ultisols, Inceptisols
Constraints	Soil acidity, erosion/mass movement, high humidity, crop pests, crop storage and drying, nutrient imbalance
Ecoregion	Equatorial wet climate of central and western Africa including humid regions of Zaire, Congo, Cameroon, Nigeria, Ghana, Ivory Coast and Senegal
Sustainable management options	
Land clearing	Manual/chain saw
Biomass disposal	In-situ burning
Crops	Oil palm, plantain (*Masa spp.*), rubber, cassava, yam, colocassia (*Colocassia esculenta*), pineapple [*Ananas camosus* (L.) Merrill]
Fertility maintenance	Cover crops, residue mulching, supplemental dose of N and P
Erosion control	No-till, cover crops, vegetative hedges

MYTHS AND FACTS OF AGRICULTURE
IN SUB-SAHARAN AFRICA

The literature is replete with pessimism and negative rhetoric about SSA. However, some of the widespread beliefs are mere myths not supported by facts. On the basis of resources available and the encouraging trend observed in grain production even in the central region of the Sahel, it is justified to conclude that drought, overgrazing and emphasis on export-oriented crops have not drastically affected all regions of Africa. Even in the worst affected regions of the West African Sahel, there is a potential for bringing about major improvements in agricultural production. Cereal production has steadily increased in many regions including the central and eastern Sahel. In addition to increase in area under production, real gains also have been made in crop yields and in labor efficiency. Both 1988 and 1989 produced bumper harvests even in the Sahel. The severe cause of famine in the Sahel and elsewhere in the Horn of Africa is perhaps as much due to adverse climate as due to man-made calamities including poverty, poor and inefficient system of food distribution among those who need it, and war and other regional and ethnic conflicts.

Economic and agrarian stagnation of Africa is not necessarily due to lack of foreign aid either. The West African Sahel, the region characterized by perpetual food crises, received an estimated $90 billion as aid for the decade ending in 1988 (World Bank, 1988). Some of this unsolicited foreign aid may have been counterproductive through hampering creativity, distorting national priorities, and creating dependency. Several of the national research institutes (NRI's) in SSA are not sustainable. This is not necessarily due to lack of trained personnel. Many have been trained in Western Europe and the USA (Lal, 1990), a large proportion of which are unemployed or underemployed, who feel frustrated because of the lack of prioritization on part of the governments of issues relevant to investments in agricultural research. Low output of NRI's also is not mainly due to a dearth of modern equipment and laboratory facilities. It is a conducive environment and congenial atmosphere that generates commitment, dedication and loyalty that have been long in coming. The agriculture profession itself, right from the farmer to the research scientist, must be given the long awaited respect and incentive it deserves.

The issue of drought needs a special mention. For the quarter of a century ending in 1985 (1960–1985), average rainfall in several regions of SSA and especially the Sahel has been significantly lower than in the previous 60 yr (1900–1960) by as much as 50 to 60% in the arid zone and 30 to 35% in the semiarid zone. Some climatologists believe that the recent drought may be due to a natural cyclic occurrence whose adverse effect has been aggravated by anthropogenic perturbations. The frequency of cyclic occurrence is believed to be about 80 yr, with a quasi 20-yr cycle of wet and dry decades.

GREENING OF AFRICA

The decade of the 1980s has witnessed an enthusiastic response to the concept of sustainability. The literature both in agriculture and other subjects, is replete with a whole array of definitions. Some of these ideas are vague, contradictory, debatable and confusing. It is a fact, however, that there is no single blueprint of sustainability that can be applied even within a seemingly uniform ecosystem. Nonetheless, it is important to realize that concepts of sustainability are likely to be very different in temperate vs. tropical environments, subsistence vs. market-oriented commercial agriculture, and in management-intensive vs. capital-intensive agriculture. Sustainability, therefore, is a term that is still in the mind of the user. For the purpose of this article, the term sustainability refers to any technology that works, is based on an efficient use of inputs, leads to an increasing trend in per capita productivity, and does not cause degradation of soil and water resources.

It is widely believed that SSA has sufficient land and water resources, and an adequate number of trained manpower to bring about the much needed Green Revolution in Africa. The most critical question is how do you transform peasant, resource-based, subsistence farming into market-oriented commercial agriculture? In fact, Africa has a history of at least 50 yr of some excellent and useful agronomic data from on-station research experiments supporting the argument that technical solutions to bring about even a quantum jump in food production are available. However, a major limitation has been the validation of this technology under realistic on-farm conditions and with full participation of the farming community. It is the on-farm testing of on-station technology that will prove whether or not the proposed technological solutions are feasible. Some cutting edge technology may have to be created by involvement of farmer input from the planning stage. This may necessitate investment in long-term research with focused priorities and sharply defined achievable objectives.

On the basis of available literature (Lal, 1986), several of the popular technological options developed and tested by on-station experimentation are listed in Table 19–10 through 19–14 for different ecological regions. Sustainable management options listed in these tables are synthesis made by the author on basis of personal experience and the literature review. In addition to the 25 yr of research information from International Institute of Tropical Agriculture (IITA), the synthesis also reflects extensive literature generated by the research conducted in the Sahel and semiarid regions by several regional, national and international organizations.

The basic approach adopted in arriving at the conclusions listed in these tables is that of addressing the production and environmental constraints of soil and water resources for each ecoregion through technology that has proven successful in on-station research. Needless to say that soil and site-specific adaptation of these technologies would require fine tuning through

Table 19-11. Physiographic characteristics of the subhumid regions and sustainable management options.

Characteristics	
Rainfall	1000–1500 mm yr^{-1}
Growing period	200–270 d
Potential yield	8.0 Mg ha^{-1} (grains), 10–15 Mg ha^{-1} (root crops)
Crop failure rate	1 in 10
Preferred crops	Maize, cowpea, upland rice, cassava, yam
Soils	Ultisols, Alfisols, Inceptisols
Constraints	Soil erosion, compaction, drought, nutrient imbalance
Ecoregion	Tropical wet/dry regions with biomodal or monomodal rainfall but with a distinct dry season of 90–150 d, e.g., parts of Nigeria, Cameroon, Ghana, etc.
Sustainable management options	
Land clearing	Manual, chain saw, shear blade
Biomass disposal	In-situ burning
Tillage	No-till, sod seeding
Rotations	Maize–cowpea, maize + cassava
Fertility maintenance	Mulch farming, cover crops [Mucuna (*Mucuna utilis*)], agroforestry, N and P
Erosion control	Conservation tillage, mulch, vegetative hedges, agroforestry
Imperata control	Allelopathy

adaptive on-farm research. Involving farmers in validation and adaptation is crucial to successful and widespread adoption of the technology.

Inventory of the physical resources of SSA indicate that the need for soil and water conservation, and restoration of degraded lands is rather universal. Biological techniques of erosion-preventive measures are effective and should be used in major ecological regions of SSA. Vegetation hedges, either of woody perennials or of grasses such as Vetiver [*Vetiveria zizanioides* (L.) Nash], planted on the contour can curtail runoff and erosion on arable

Table 19-12. Physiographic characteristics of the semihumid region and sustainable management options.

Characteristics	
Rainfall	800–1000 mm yr^{-1}
Growing period	150–200 d
Potential yield	5.0 Mg ha^{-1}
Crop failure rate	1 in 7
Preferred crops	Sorghum, cowpea, rice in fadamas, pastures, yam, groundnut
Soils	Alfisols, Psamments, Vertisols, Plinthitic subgroups
Constraints	Supraoptimal soil temperatures, drought, crusting, hard-setting, low soil fertility, soil erosion
Ecoregion	Semiarid regions with one growth season, e.g., Nigeria, Ghana, Kenya, Tanzania, Malawi, Zambia, south savanna plateau, etc.
Sustainable management options	
Seed bed	Contour ridges, tied ridges
Planting	Early sowing with onset of rains
Soil and water conservation	Mulch farming, vegetative hedges, early sowing
Fertility maintenance	Kralling, farmyard manure, N and P fertilizer
Imperata control	Soil inversion in dry season

Table 19-13. Physiographic characteristics and sustainable management options for the semiarid regions.

Characteristics	
Rainfall	500-800 mm yr^{-1}
Growing period	100-150 d
Potential yield	3.0 Mg ha^{-1}
Crop failure rate	1 in 5
Preferred crops	Sorghum, millet, groundnut, cowpea
Soils	Alfisols, Vertisols, Plinthitic subgroups, Psamment, Aridisols
Constraints	Wind and water erosion, sand blasting, dust storms, drought, supra-optimal soil temperatures, low soil fertility
Ecoregion	The West African Sahel, e.g., Niger, Burkina Faso, Mali, Senegal, Chad, etc.
Sustainable management options	
Seed bed	Rough plowing at the end of rains, tied ridges
Planting	Early planting, high seed rate, deep sowing
Soil and water conservation	Tied ridges, diggets/stone lines, water harvesting, vegetative hedges, supplemental irrigation
Cropping systems	Millet + cowpea, sorghum + groundnut, rice-cowpea (*Acacia albida*)
Fertility maintenance	Farmyard manure, kralling, N and P fertilizers

lands with undulating terrain (see Chapter 9 in this publication). Conservation tillage and mulch farming techniques, where applicable depending on soil type and the rainfall regime, prevent the raindrop impact, minimize crusting and formation of surface seal, enhance activity of soil fauna, maintain

Table 19-14. Physiographic characteristics and sustainable management options for the arid regions.

Characteristics	
Rainfall	300-500 mm yr^{-1}
Growing period	<100 d
Potential yield	1.0 Mg ha^{-1}
Crop failure rate	1 in 3
Preferred crops	Millet, cowpea, pasture
Soils	Alfisols, Aridisols, Plinthitic subgroups
Constraints	Wind erosion, drought, high soil temperature, sand dunes, low soil fertility
Ecoregion	Dry regions with short growing season, e.g., northern Sahel, Horn of Africa, dry regions of southern Africa
Sustainable management options	
Planting	Early
Water conservation	Microcatchments, water harvesting, ridge–furrow system
Farming system	Mixed farming with low stocking rate, mixed cropping
Erosion control	Vegetative hedges, shelter belts, ground cover
Soil fertility management	Organic amendments, chemical fertilizers, banding or fertilizer placement
Crops	New crops and cultivars with drought-escape mechanisms

high infiltration rate, conserve soil water and reduce diurnal fluctuations in soil temperature. Water harvesting, tied ridges, and stone-lines or diggets have proven effective in conserving water in arid and semiarid regions. Fallowing, weed- and crop-free period to conserve water in the root zone for the following crop, is another useful technique to improve productivity from drought-prone ecosystems. In Botswana, Whiteman (1975) reported drastic improvements in soybean (*Glycine max*) yield from the fallowed land.

The question of fertilizer use has been a debatable issue. The concept of low-input, alternative or regenerative systems recommended enthusiastically for North American and West European agriculture should be given a careful appraisal in the context of agriculture in SSA. The average rate of fertilizer use in Africa is presently about 8 kg of major nutrients (N, P, K) per hectare (Table 19-6). The use of total fertilizer is about 19 kg ha^{-1} in SSA compared with 73 kg ha^{-1} in Asia (IFDC, 1987). Africa's annual fertilizer consumption represents about 2.5% of world consumption. This amount is hardly enough to obtain grain yields of 1 to 1.5 Mg ha^{-1} even with conservation-effective measures. Once soil–water is not a limiting factor, the most severe constraint to high yield is the universal deficiency of N. Although most soils have low P and K requirement, application of 10 to 20 kg ha^{-1} of P and 20 to 30 kg ha^{-1} of K is essential to ensure high fertilizer use efficiency of N. Grain crops usually require 50 to 60 kg ha^{-1} of N for average yields, and 80 to 100 kg ha^{-1} for high yields of grain crops such as maize (*Zea mays*). Leguminous crops such as cowpea (*Vigna unguiculata*) and soybean require none or a low rate of N fertilizers. Root crops, such as cassava (*Manihoc esculenta*) need frequent use of potash. Because these soils are highly weathered and impoverished, there is no substitute for fertilizers. This is not to say that need for chemical fertilizers cannot be reduced by addition of compost, farmyard manures, live mulches, and agroforestry. These techniques are useful and can provide 20 to 30 kg ha^{-1} of N. In fact, effectiveness of chemical fertilizers is greatly enhanced when they are used in conjunction with organic amendments and other improved systems of soil and crop management.

Specific agricultural technologies for sustaining agricultural production and efficient utilization of soil and water resources are available for different ecological regions of SSA (Lal, 1986). Nonetheless, available technologies are mere guidelines and need to be validated and adapted under local-specific conditions through on-farm testing. The latter is necessary to involve African farmers in the decision-making process. Technical data available supports some basic concepts and principles rather than suggest local-specific technological packages. The latter are a soil-specific set of cultural practices to be developed from these principles for the region.

STRATEGIES FOR COMBATING DROUGHT

About 40 countries in SSA have an estimated 5 million hectares of irrigation (see Chapter 16 in this publication, Table 16-2). About one-half of the irrigated land is in Sudan and Madagascar (Wright, 1986; Foster, 1986). This much irrigated land represents about 15% of the irrigable potential and only 2% of the available land area (FAO, 1986). Expansion of irrigated land area using large-scale irrigation schemes is an expensive undertaking for several reasons. First, the domestic irrigation industry is in its infancy, lacks experience, and depends on the imported material. Second, surface water and renewable groundwater resources in SSA are finite. Furthermore, vast regions, including west and central Africa, are characterized by flat topography with few good dam sites. Consequently, long off-take canals are necessary resulting in costly and low-efficiency systems. Third, most soils have low irrigation potential and irrigable soils occur in small scattered patches. There are few large irrigable areas such as those in Sudan. Fourth, surface water resources are not large. Those that are available for irrigation are mismatched with irrigable soil resources. The average runoff coefficient for Africa is only 20%. Above all, groundwater reserves also are localized. There are no large rechargable aquifers similar to those available in Asia. Only a small fraction of groundwater is renewable annually (Stern, 1986).

Rather than invest in large-scale grandiose irrigation schemes, there is a tremendous scope for expansion of irrigated agriculture on the basis of small-scale irrigation utilizing surface waters in small valleys for gravity irrigation of 2 to 10 ha, and utilization of shallow groundwater of which large amounts are available by collector wells or bore holes. In addition, favorable water regimes and perched water tables exist in low-lying valleys called "fadamas" or "mares." "Mares" are small, shallow, natural depressions in which water collects during periods of intense and frequent rains. There are several hundred of these mares in the West African Sahel. There is a potential to develop small-scale irrigation using wind pumps or diesel pumps.

Agricultural potential in arid and semiarid regions also can be enhanced by water harvesting. It involves deliberate enhancement of surface runoff from marginal lands and storing this water for irrigation for favorable soils. The most favorable ratio of contributing to cultivated areas varies among soils, ecological regions and crops/cropping system. The ratio may range from 1:1 to 25:1 with size of contributing area ranging from 0.5 to 1000 m^2. The most desirable ratio for SSA is 5:1. It is important to recognize, however, that such schemes are successful only when other inputs (fertilizers) also are available (Jones, 1986).

INVESTING IN AGRICULTURE

It is time that agriculture in Africa goes beyond the survival syndrome. The scope of agriculture in SSA should be expanded to produce raw materials for agrobased industries in addition to the basic necessities of food, feed, fiber, and fuel. Agriculture, to be sustainable, must support other industries, and must become an indispensable link for holistic development of the society. To do so is to plan long-term, sustainable development of the natural resource base.

Rainfall is a given, and little can be done to alter its amount or distribution. However, there is much that can be accomplished to increase the effective utilization of rain. Meteorological services should be strengthened to improve regional forecasting especially regarding the onset of rains. It is the onset of rains that determines the time of planting and duration of the growing season. Early planting is critical to the success of crops. On the other hand, planting before rains are certain, can lead to wasting of seed and other inputs and crop failure.

Understanding potential productivity and constraints of soils and defining their land-use capabilities is the key to rational development of agricultural systems. The widespread degradation of soils of SSA is attributed to neglect and misuse of natural resources over a long period of time. Although several analytical laboratories have been established in SSA, these facilities are used primarily for soil classification rather than for providing guidelines on soil and fertility management.

Africa is endowed with a wide diversity of climate, vegetation, geography, terrain and soils. Yet the range of species grown is rather narrow. Introduction of new species could spread risks and increase options. There is no justification for ignoring cash crops. An appropriate mix of food and cash crops is essential to enhance the income of the impoverished farming community. Several crops traditionally grown in south Asia have been successfully introduced in Australia. Similarly, chickpea (*Cicer arietinum* Linn.), pigeon pea (*Cajanus cajan*), mung bean (*Phaseolus mungo*) and other minor legumes, have a tremendous potential in SSA. In addition, there is a scope to develop new cultivars that respond to input (Jones, 1986; Dogget, 1986). Presently, only 6 to 8% of the cultivatable area is planted to high-yielding and input-responsive improved cultivars (Brady, 1985). Total area of modern wheat (*Triticum aestivum*) and rice (*Oryza sativa* Linn.) varieties in SSA in 1983 was only 800 000 ha (Dalrymple, 1986a,b).

Reviving agriculture in SSA would require African governments to assign the top priority to investment in agriculture. There are several options, and the appropriate choice depends on ecological environment. Humid and subhumid regions are suitable for perennial cash crops [rubber, oil palm (*Elaeis guineensis* Jacq.), cocoa (*Theobroma cacao*), timber, fruit trees, etc.] and root crops [cassava, yam (*Dioscorea rotundata* Poir.), sweet potato (*Ipomoea batatas*), etc.]. While developing land resources for these crops, it is necessary that supporting industry is developed to absorb the produce. Cassava can be used to manufacture starch or cattle feed for export. There are several high-value pharmaceutical products that can be developed from species adapted to the humid ecosystems.

Semiarid regions (rainfall 800–1200 mm yr^{-1}) are the most suitable for production of grain crops. The potential of grain production, given the adequate input, is highest in this region. Seed industry is practically nonexistent, and is urgently needed. Arid regions (with rainfall of 500–800 mm yr^{-1}) are suitable for livestock production. There is a need, however, to introduce improved breeds of beef or dairy cattle. The improved breeds must be locally bred with desirable indigenous traits incorporated. There should also be a strong emphasis on breeding cattle for export out of the region. The successful development of cattle industry also depends on the improvement of rangelands and pasture. There is a tremendous scope for development of suitable grasses, legumes and browse plants.

It is vital that planning and decision-makers recognize that agriculture in SSA is a no-prestige profession. It is important to raise the prestige of this noble profession through better policies, freer marketing, providing incentive for excellence, and rewarding output with appropriate remuneration commensurate with the expected returns.

REFERENCES

Brady, N.C. 1985. Chemistry and world food supply. Science (Washington, DC) 218:847.

Dalrymple, D. 1986a. Development and spread of high-yielding wheat varieties in developing countries. USAID, Washington, DC.

Dalrymple, D. 1986b. Development and spread of high-yielding rice varieties in developing countries. USAID, Washington, DC.

Doggett, H. 1986. Crops, plant breeding and seed production. p. 63–72. In L.J. Foster (ed.) Agricultural development in drought-prone Africa. Overseas Development Inst., London.

Food and Agriculture Organization. 1978. Report on the agro-ecological zones project. Vol. 1. Methodology and results for Africa. FAO, Rome.

Food and Agricultural Organization. 1986. Irrigation in Africa south of the Sahara. FAO Investment Centre Tech. Pap. 5. FAO, Rome.

Foster, L.J. 1986. Agricultural development in drought-prone Africa. Overseas Development Inst., Tropical Agric. Assoc. Regent's College, London.

Griffith, J.F. (ed.) 1972. Climates of Africa. Elsevier, Amsterdam.

International Fertilizer Development Center. 1987. Africa fertilizer situation. IFCC, Muscle Shoals, AL.

Jones, M.J. 1986. Conservation systems and crop production. p. 47–59. In L.J. Foster (ed.) Agricultural development in drought-prone Africa. Overseas Development Inst., London.

Lal, R. 1986. Soil surface management in the tropics for intensive land use and high and sustained production. Adv. Soil Sci. 5:1–110.

Lal, R. 1990. Thematic evaluation of soil institution building projects in Africa. U.N. Dev. Program, New York.

Lal, R. 1993. Technological base for agricultural sustainability in sub-Saharan Africa. p. 257–263. In R. Lal and J. Ragland (ed.) Technologies for sustainable agriculture in the Tropics. ASA Spec. Publ. ASA, CSSA, and SSSA, Madison, WI.

Smyle, J.W., and W.B. Magrath. 1993. Vetiver grass—A hedge against erosion. p. 109–122. In J. Ragland and R. Lal (ed.) Technologies for sustainable agriculture in the Tropics. ASA, Spec. Publ. ASA, CSSA, and SSSA, Madison, WI.

Stern, P.H. 1986. Note on irrigation potential. p. 38–42. In L.J. Foster (ed.) Agricultural development in drought-prone Africa. Overseas Development Inst., London.

Whiteman, P.T.S. 1975. Moisture conservation by fallowing in Botswana. Exp. Agric. 11:305–314.

World Bank. 1988. Strengthening agricultural research in sub-Saharan Africa: A proposed strategy. World Bank, Washington, DC.

World Bank. 1989. Sub-Saharan Africa, from crisis to sustainable growth: A long-term perspective study. The World Bank, Washington, DC.

Wright, E.P. 1986. Water resources for humans, livestock and irrigation. p. 25–37. In L.J. Foster (ed.) Agricultural development in drought-prone Africa. Overseas Development Inst., London.

20 Towards Sustaining Agricultural Production in the Tropics: Research and Development Priorities

Rattan Lal

Department of Agronomy
The Ohio State University
Columbus, Ohio

John Ragland

Agronomy Department
University of Kentucky
Lexington, Kentucky

The information presented in this compendium and literature survey was collected over the 25 yr ending in 1992 and indicates the enormity of technical data available for addressing problems of agricultural sustainability in the Tropics. The wealth of research information includes genetic, chemical, physical and social approaches needed for achieving sustainability.

Plant scientists and geneticists have engineered improved varieties capable of doubling and even quadrupling yields. These varieties are reportedly tolerant and, in some cases even resistant to insects, pathogens, and nutrient imbalances, e.g., Al toxicity and P deficiency. Plant architecture and life cycles have been altered to increase plant population and complete the growth cycle within a short growing season. Advances in biotechnology permit the direct incorporation of desirable genes, thus shortening the period required by conventional breeding techniques. Soil scientists have developed improved systems of soil and water conservation, soil fertility maintenance, and providing balanced plant nutrients for high yields. Soil, previously considered unproductive, can be used for extensive and profitable agriculture by a judicious combination of off-farm inputs with managerial skills. Agronomists have developed improved cropping/farming systems to optimize the resource use, create the much sought after biodiversity, provide ground cover during periods of harsh climatic conditions, and maximize biomass production and agronomic returns. Economic analyses of improved technology are promising with a highly favorable output–input ratio. Powerful computer software exists for reducing the cast and telescoping the time required to find site-specific solutions. Sociologists and anthropologists have found ways to make improved agricultural technologies meet the social, cultural and political criteria of acceptability. If the claims of increase in yield and profitability of station-proven technology are correct, agronomic output in most tropical ecoregions can be increased substantially and immediately.

Despite the availability of research information, the problem of low productivity, resource degradation, and nonsustainability of agriculture remains a serious threat to human survival. The rate of adoption of new and improved technology is regrettably low. Farmers are either not aware of the improved technology or are not willing to take the risk of adopting it. In fact, the success stories are few. Even the Green Revolution of the 1960s and 1970s that prevented mass starvation in South Asia is reportedly slowing down. The "Greening of Africa" is yet to happen. Why?

Why is new technology not being adopted? Why have conservation tillage, live mulch, use of leguminous cover crops, and alley cropping not been adopted in spite of their technical successes on research stations? Why is the area under improved crops and cultivars in Africa still low? Why have African farmers not abandoned "the back breaking hoe" and adopted improved tools developed by national and international research centers? Why has the African and Asian homemaker not accepted the modern solar cookers powered by biogas or solar heat instead of continuing with the unhygienic and enviornmentally disastrous wood fuel?

These are important issues with which the world community has not come to grips. Is there a need to continue generating new data and technology when the precious and expensive information already available is being so poorly utilized?

TECHNOLOGY TRANSFER

A high priority should be given to identification of causes of low or non-adoption of improved technology by the farming community. An important reason might well be the noninvolvement of farmer in the planning process, e.g., the top–down approach. The researcher-perceived problem may, in fact, not be a problem from the farmer's perspective. That being the case, merely conducting on-farm trials will hardly bring about adoption of new technology. Identifying the causes of nonadoption may, however, be easier said than done. The problem is indeed complex, and requires a multi- and interdisciplinary approach. Problem-solving through a multidisciplinary approach with "farmer-back-to-farmer" strategies were the cornerstone of the farming system movement of the 1970s. The movement has almost become a history especially in the CGIAR system, with only a modest success.

There is an urgent need, however, to develop innovative and effective methods of technology transfer. The scientific community must identify technological innovations worthy of the name. Credibility of the scientific community must be restored, and farmer's trust won through genuine efforts in helping farmers alleviate their basic constraints.

IDENTIFICATION OF RESEARCH AND DEVELOPMENT PRIORITIES

Developing strategic plans, at least once every 5 yr, is a favorable pastime. It is often a political issue, because every new administration is likely to begin with a new strategic plan, often with an utter disregard for what was planned by the previous administration. There must be a systematic approach in prioritization and development of a strategic plan. The principal objective is to solve the practical problems faced by the farmer, and alleviate production related technical, social, cultural and logistical constraints.

A possible strategy for prioritization is outlined by the circuit/wiring diagram depicted in Fig. 20-1. The diagram comprises three modules and is described below.

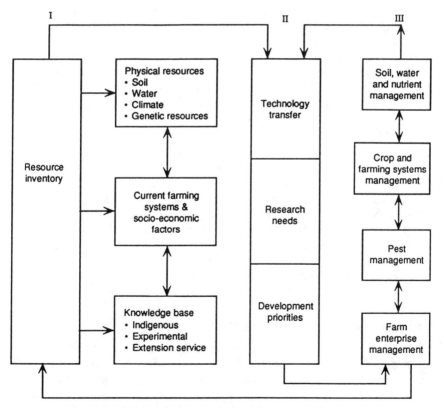

Fig. 20-1. A circuit diagram depicting interlinkages between current status and future research and development needs.

Resource Inventory

The first step is through resource inventory, or taking stock of what is known. This may involve a thorough literature survey, and development of an easily accessible information source. In addition to collecting information, the knowledge should be classified and interpreted for potential use in specific cases. In addition to scientific information, indigenous knowledge also must be collated, and interpreted in terms of specific uses.

Technology interpretation is greatly facilitated by an understanding of the physical resources, e.g., soil, water, climate, vegetation, genetic resources, and biodiversity. Potential and constraints of the physical resource base must be carefully assessed. Equally important is the knowledge and understanding of the current farming/cropping systems, their potential and constraints and sustainability, and specificity in relation to biophysical environments. Knowledge of future trends and changes in farming/cropping system also is helpful in defining priorities.

Technology Transfer and Development Priorities

Resource inventories will provide the much-needed information on soil, climate or ecoregion-specific technologies for addressing specific production-related constraints. These technologies, with farmer participation and support, are ready for transfer following on-farm validation and adaptation. The resource inventory also identifies knowledge gaps for research needs on farming systems, and biophysical or socioeconomic and cultural environments. Development priorities, in relation to institutional support and marketing or logistic support, also are identified on the basis of resource inventory.

Component Research

There is a need to develop long-term research program on individual components or subsystems. This would bring about an increase in productivity or alleviate specific constraints. There is nothing wrong in conducting even a narrow-based disciplinary research, provided that it is done with sustainability perspective and addresses practical problems of resource use and management. There are four specific components and subsystems that need to be addressed:

Soil, Water and Nutrient Management

The research objective is to enhance the nutrient capital and flow to maximize photosynthesis. Water conservation and management are crucial to efficient utilization of nutrients, and are especially critical in arid and semi-arid regions. Specific research objectives are to: (i) develop alternative sources of nutrients, (ii) develop methods for efficient techniques of nutrient utilization, (iii) improve retention and recycling of nutrients, (iv) evaluate effects of organic matter content on soil fertility and physical properties, (v) improve soil structure to effectively control erosion and reduce soil degradation, and (vi) restore productivity of degraded soils.

Crop and Farming Systems Management

There is a need to develop new and innovative cropping/farming systems. In spite of three decades of intensive research, there are few viable alternatives to shifting cultivation. Improved farming systems must be highly productive and fit within the socioeconomic, political and cultural framework of the farming community.

Pest Management

There is a need to develop integrated pest management systems, specific to local conditions. Pest management, encompassing weed and disease control plus reduced damage by insects, should be an integral part of cropping/farming system management. It involves the integration of all cultural, biological and chemical inputs to minimize damage by insects, diseases, and weeds.

Farm Enterprise Management

This involves a basic study of the farm household, including nonfarm activities. The economics of production and flow of goods and materials are especially important.

QUANTIFYING THE SUSTAINABILITY CONCEPT

Sustainability is presently a vaguely defined and a qualitative concept. There are numerous definitions and approaches, and the specific use depends on the user's understanding, needs, and resources. The concept has a subjective value. Consequently, there is a need to develop a quantitative and objective definition of sustainability so that the concept is clear and free of ambiguities and vagueness.

A high priority should be given to development of a set of quantifiable parameters and indices of sustainability. These indices should provide a quantitative measure of sustainability at different levels, e.g., crop, farming system, farm, community, ecoregion, and nation.

4J

① SUSTAINABLE (WHY)

② POLITIES < IN BRAZIL / RIO > GAP - FACTS

③ INTRO TO PIAUI / REGION

↳ PROB

↓

ACCESS
TO
LAND

→ TRAINING

FARMING
SYSTEMS
AND
ENVIRONMENT

↳ MIGRATION.

prefeitura